Principles and Applications of Metal Rolling

This is a book written by a practitioner. It is somewhat different from a normal textbook involving machines and equipment in the field of Mechanical Engineering. Generally stress is given to the theories and principles involved and processes are explained in great detail. There is no doubt that these are essential for an engineer. But it is equally important for an engineer to know the basic design, working principle and operation of various machines/equipment used for conversion of raw materials into desired products. The introduction of laboratories, workshops and industrial tours help fill this void to some degree. Yet, engineering curriculum retains the deficiency.

Meant for undergraduate and graduate students, this book attempts to fill the void in the sub-field 'Rolling Process'. It also intends serving as a reference book for practicing engineers. It begins with comprehensive coverage of rolling processes and the mechanics of rolling, which is the theoretical content of the subject. This is followed by a chapter on 'rolling practices' that highlights plant level procedures and practices employed by the rollers for producing desired products. The book concludes with the description, operation and design principles of various equipment, mechanisms and systems used inside a rolling plant.

Siddhartha Ray is professor in the Mechanical Engineering Department and Dean (UG Programmes) at Heritage Institute of Technology, Kolkata. Earlier he taught, for a decade, at the National Institute of Technical Teachers' Training & Research, Kolkata, where he also served as the Director for about two and half years. Before joining the academia in 2001 Ray spent more than thirty years in various industries, designing and developing machines and systems for rolling mills and other kinds of material handling plants. He has published a number of technical and research papers, and he also holds two patents. Besides teaching and research, he enjoys writing popular science books and articles in Bengali.

Principles and Applications of
Metal Rolling

Siddhartha Ray

CAMBRIDGE
UNIVERSITY PRESS

CAMBRIDGE
UNIVERSITY PRESS

4843/24, 2nd Floor, Ansari Road, Daryaganj, Delhi - 110002, India

Cambridge University Press is part of the University of Cambridge.

It furthers the University's mission by disseminating knowledge in the pursuit of education, learning and research at the highest international levels of excellence.

www.cambridge.org
Information on this title: www.cambridge.org/9781107076099

First published 2015

Printed in India by Thomson Press India Ltd., New Delhi 110001

A catalogue record for this publication is available from the British Library

Library of Congress Cataloging-in-Publication Data
Ray, Siddhartha.
Principles and applications of metal rolling / Siddhartha Ray.
pages cm
Includes bibliographical references and index.
Summary: "Discusses the entire gamut of rolling process and the practices followed in rolling industry and operation and use of various rolling mill equipment and systems"-- Provided by publisher.
ISBN 978-1-107-07609-9 (hardback)
1. Rolling (Metal-work) I. Title.
TS340.R39 2015
671.3'2--dc23
2015004528

ISBN 978-1-107-07609-9 Hardback

This book is dedicated to the memory of my mother

Sadhana Roy

She remains the main source of inspiration in my life.

Contents

List of Figures

List of Tables

List of Symbols

x,y,z	three rectangular coordinates
F	force
F_R	repulsive force
F_A	attractive force
r_o	inter-atomic distance
τ_o	ultimate shear stress
G	modulus of rigidity
Hi	abbreviation of "High", signifying number of rolls in a mill stand
h	thickness of stock/workpiece
h_1	input thickness
h_2, h_0	output thickness after deformation through rolling
h_n	thickness at neutral axis
w,b	width of stock/workpiece
w_1,b_1	Input width
w_2,b_2	output width after deformation
w_m,b_m	mean width
A_1	cross sectional area of stock/material at input
A_2	cross sectional area of output after deformation
L_1	length of stock/workpiece at input
L_2	length of stock at output after deformation
α	contact angle or angle of bite/contact
Δh,(h_1-h_2), δ	draught or reduction in height/thickness
E'	elongation factor
Δw, Δb_m, (b_2-b_1)	spread
R,r	roll radius
D	roll diameter
l	projected length of arc of contact
μ	coefficient of friction
N	neutral point
α_n, γ	neutral point angle
v_1	stock velocity at entry
v_2	stock velocity at output

h_g	set roll gap
γ	poisson ratio
E	Young's modulus
Δf	mill stretch
R', r'	Deformed roll radius
h_m	mean thickness of stock/workpiece
v_n, v_r	roll surface velocity
v_x	roll surface velocity in the direction 'x'
ϕ	angle subtended by a point on roll surface within roll bite
ϕ_x	Angle subtended by a point on roll surface at a distance x from roll axis.
σ_x, σ_y, σ_z	normal stresses
k, τ_s	shear yield stress
τ_x	contact shear stress
σ_1, σ_2, σ_3	principal normal stresses
τ_{xy}, τ_{yz}, τ_{zx}	Shear stresses
τ_1, τ_2, τ_3	principal shear stresses
σ	yield stress
σ_a	actual resistance to deformation (tensile yield stress)
σ_t	ultimate strength
σ_A	tensile stress due to back tension
σ_B	tensile stress due to front tension
η	coefficient of viscosity
F	horizontal projection of contact area
σ_s	tensile yield stress of annealed metal
T	absolute temperature
s	specific heat of metal
t	rolling temperature, ^{0}c
Δt	time interval, seconds
A	energy required to deform metal
W	weight of metal in kg
u, $\varepsilon^{\cdot}$	strain rate, sec^{-1}
ε	Strain
p	contact pressure between roll and stock,
a	Length of lever arm
λ	lever arm coefficient
T	Torque
kw	Kilo-watt
HP	horse power
HAGC	Hydraulic automatic gauge control
EDC	Edge drop control

Preface

Organization of this book is somewhat different from a normal textbook involving machines and equipment, in the field of mechanical engineering. Generally, stress is given to the theories and principles involved and the processes are explained to a great detail. There is no doubt that these are essential for an engineer. But it is equally important for an engineer to know the basic design, working principles and operations of the various machines and equipment which are used in the practical field for conversion of the raw materials into desired products. Baring the subject of 'Machine Tools', in most other fields of manufacturing processes, available textbooks seem to be rather miserly in thorough discussions on the description, design, working principles of various machines and systems involved and practices followed in actual operation.

An attempt has been made to bridge this gap by introducing laboratory exercises and workshops along with industry visits, in the engineering curricula. More often than not these prove to be inadequate. It is next to impossible that an equipment or machine like a turbine, an extrusion press or a rolling mill can be installed in an academic institution. By observing the operation of a machine or system during a visit to an industrial plant, definitely a lot can be learned about the manufacturing process, but seldom can it give an idea about the working principles of the various mechanisms, their design details or about the intricacies of operational practices.

While teaching 'Manufacturing Technology' to MTech students at the National Institute of Technical Teachers' Training and Research (NITTTR), Kolkata, I observed this shortcoming in my students and cherished the desire to bridge the gap as soon as possible. With 30 years' experience in the industry in design, development and commissioning to operation plant and machinery (out of which more than half the period was in the field of Rolling Mills), I decided to write this book titled *Principles and Applications of Metal Rolling.*

This book is intended to cover undergraduate and postgraduate engineering curricula for 'Rolling Technology' in India and in other countries. It is also meant to be a reference book for practicing engineers working in the field of rolling mills. The first two chapters cover the rolling process and mechanics of rolling comprehensively, which is the theoretical content of the process of rolling. The third chapter on 'Rolling Practices' highlights plant level procedures and practices employed by the rollers for producing desired products. The whole of the fourth chapter on 'Rolling Equipment and Systems' is devoted to the description, operation and design principles of various equipment, mechanisms and systems used in a rolling plant.

In preparing the book, help has been taken from some specialized books written and edited by experts and from the literature of equipment manufacturers. Acknowledgement and references to such books and literature have been made at appropriate places.

Suggestions and comments on the organization and contents of the book are welcome and may kindly be sent to the publisher or the author.

Acknowledgments

While writing this book I have received help, suggestions and encouragement from many of my ex-colleagues, friends and well wishers, which I gratefully acknowledge.

At the very outset I would like to express my appreciation to my MTech students of Manufacturing Technology course offered by the National Institute of Technical Teachers' Training and Research (NITTTR), Kolkata. While teaching and discussing the subject of Rolling Technology with them, I got impetus to write the present book. I am indeed thankful to them.

During the writing of this book I have drawn upon the knowledge and experience I received while working with M/s Davy Ashmore India Ltd., in close contact with their collaborators in the UK – Davy Lowey Ltd. and Loewy Robertson Engineering Company of the UK, and later with M/s Tata Construction and Projects Limited. I would like to acknowledge my deep sense of gratitude to the following persons, who gave me the opportunity to work in the field of design and operation of Rolling Mills and also gave me all possible help and cooperation during the writing of this book: Late P. Sen, Ex.-M. D. of Davy Ashmore India Ltd. and my ex-colleagues P. K. Bera and A. K. Mitra. I am particularly indebted to S. Bhattacharya from the same organization, who advised me while writing the topic on Mill Electrics.

I am also grateful to friends from the Davy Group of companies in the UK, namely: T. Shiemeld, T. Smith, A. F. Uff, D. Fretwell and others who helped me with information and material on the subject as and when I needed them. I am thankful to S. Majumdar, Ex President of Hindalco and colleagues at NITTTR, Kolkata and Heritage Institute of Technology, Kolkata for advising and encouraging me during the preparation of the book.

My thanks are also due to T. K. Dutta, an ex-colleague of Davy Ashmore, who joined later M/s. SMS Demag, India, for giving me valuable suggestions on Roll Lubrication and Cooling System.

I am thankful to Kingshuk Ghosh, the DTP operator of my department at NITTTR, who untiringly finished the entire manuscript, working beyond his working hours. My thanks go to G. Patra for preparing some of the diagrams using CAD.

I am grateful to my wife Dipali Ray and other family members for their constant encouragement and sacrifices during the preparation of the book.

Finally, my thanks are to the Cambridge University Press for readily accepting to publish the book.

CHAPTER 1

Introduction to Rolling Process

1.1 Definition of Rolling Process

Rolling is the most important metal forming process. More than 95% of ferrous and non-ferrous metals and alloys are processed to their usable shapes by rolling. Usable shapes of rolled metals are plate, sheet, strip, foil, different sections like rail, beam, channel, angle, bar, rod, and seamless pipe, etc., as shown in Fig. 1.1 and Fig. 1.2.

In the rolling process, permanent deformation is achieved by subjecting the material to high compressive stress by allowing the material to pass through the gap between two rotating cylindrical rolls.

The rolls may be flat or grooved, and are kept at a fixed distance apart from each other. The rolls are rotated in opposite direction by means of electrical drive system (motor, gearbox, spindle and couplings).

Depending on the direction of rotation of the rolls, the input material enters the gap between the rolls from one end and comes out from the other end with a reduced cross-section, the roll gap area being kept less than the cross-sectional area of the input material (rolling stock). For obtaining the desired final shape of rolled material, it is generally necessary to pass the material through the rotating rolls several times. During each of the passes, the roll gap is adjusted by bringing the two rolls closer to each other, or by allowing the material to pass through different set of roll gaps with diminishing cross sectional area.

The entire assembly of the rolls mounted on bearings is held in bearing blocks (called chocks), which in turn are held between the gaps of two cast frames (called housings), complete with roll gap adjustment facilities and roll driving arrangement. The entire set up is called a *rolling mill stand.* One or more number of rolling stands in combination with other necessary and related equipment to obtain finished rolled products from one or similar group of input materials is called a *rolling mill* or *rolling plant.*

Rolling process can be classified based on various conditions/methods employed in rolling. These are:

(i) Temperature of the material- thus we can have *hot rolling* (temperature above the recrystallization temperature), *warm rolling* and *cold rolling.*

(ii) Shape of the rolled product- *flat, sections* or *hollow sections rolling.*
(iii) Direction of rolling- *lengthwise, transverse,* and *skew rolling.*
(iv) Mode of rolling mill operation- *continuous* (unidirectional), and *reverse rolling,* where direction of rotation of rolls are reversed.

When two rolls of equal diameter and with axis lying in same plane rotate in opposite direction with same rotational speed, and the material being rolled is homogeneous in its mechanical properties and is acted upon only by the forces from the rolls, the process is called *simple rolling.*

1.2 Hot and Cold Rolling Processes

From metallurgical point of view, rolling process can be classified under two broad categories, namely (i) hot rolling and (ii) cold rolling.

1.2.1 Hot rolling

In hot rolling the material is rolled at a temperature higher than its recrystallization temperature. The advantage of hot rolling is twofold. First, at elevated temperature the strength of any metal or alloy is reduced. Thus the compressive force required for deformation is comparatively less and therefore smaller capacity rolling stand can be used for rolling operation . The second advantage of rolling a material at a temperature higher than its recrystallization temperature is that a large amount of plastic deformation can be imparted without getting it *strain hardened* . With strain hardening, the deformation stress increases as more and more deformation takes place rendering the material hard and brittle. As a result, the material becomes more and more difficult to be deformed, and beyond limit, deformation leads to various faults or defects.

The ferrous raw material for rolling various shapes is the *ingot* which is cast out of molten metal. In case of low carbon steels the ingot is quite large. It is first rolled into *blooms.* The blooms are rolled into smaller sizes, called *billets.* Large structural sections such as rails, beams, girders, channels, angle sections, and plates are rolled out of blooms, while billets are rolled into smaller structural sections, bars, plates, and strips. Alloy steel and stainless steel, produced in mini steel plants, are generally cast into smaller sizes of ingots. Non-ferrous metals like aluminum are cast into wide slabs, from which plates, sheets, and strips are obtained.

The above practice still goes on in older plants. However, the present trend is to install continuous casting units to cast smaller sections directly from liquid metal and thus eliminate bloom rolling. In some continuous casting plants, billets of small cross sections may also be continuously cast, thus eliminating even the billet rolling mill. Installation of continuous casting results in substantial saving in capital cost of the rolling plant as well as rolling process cost. Non-ferrous alloys of aluminum, brass, nickel-silver, etc., are continuously cast in bar or strip form.

Rolling of ingots to blooms and blooms to billets, and further rolling of blooms and billets to different usable products like structural sections, bars, plates, and strips are all rolled through hot rolling. Some of the products produced through hot rolling process are illustrated in Fig. 1.1.

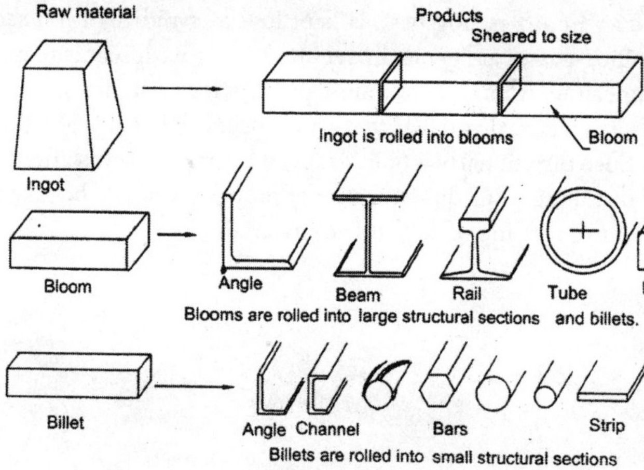

Fig. 1.1 Products produced by hot rolling

1.2.2 Cold rolling

When rolling of a material is done at room temperature or below the recrystallization temperature of the material, it is called cold rolling. Obviously, the advantages of hot rolling is absent in cold rolling. The resistance to deformation is more. Furthermore, during rolling, strain hardening takes place, i.e., the strength of the material progressively increases with increase in degree of deformation in the original material.

However, there are a few advantages also. The first one is about controlling the *grain size* and thereby achieving the desired mechanical properties of the finished rolled material. When the input material is cold rolled, the grains of the input material get elongated along the direction of rolling. Thus the effective grain size is reduced, as the surface area of each grain increases whereas their volume remains the same. With subsequent passes of rolling, the elongated grains break and the grain size becomes progressively smaller and the material gets harder and harder. After a certain percentage of volumetric deformation, the cold rolled material becomes too hard and brittle to be rolled further profitability. At this stage, the cold rolled material is *annealed*, which is nothing but heating the material in a neutral atmosphere (heating in presence of oxygen is avoided to prevent oxidation) above its recrystallization temperature. By adjusting the time for which the rolled material is kept at this higher temperature (soaking time), the size of the newly formed grains of the annealed material can be closely controlled.

Cold rolling is generally done to produce flat rolled products like sheet, plate, strip, and foil. When the length of the rolled product is too large, the material is wound and used in the coil form.

In cold rolling, since the degree of deformation, i.e., reduction in thickness of the flat product in any rolling pass, is kept low to avoid high roll separating force, several rolling passes are generally required along with requisite number of intermediate annealing. A number of rolling passes with grain deformation in the same direction gives a directional bias to the various mechanical properties of cold rolled products. Such directional bias (anisotropy) often remains even after annealing after final pass of rolling. This directionality of properties has to be taken care of during subsequent processing of cold-rolled material.

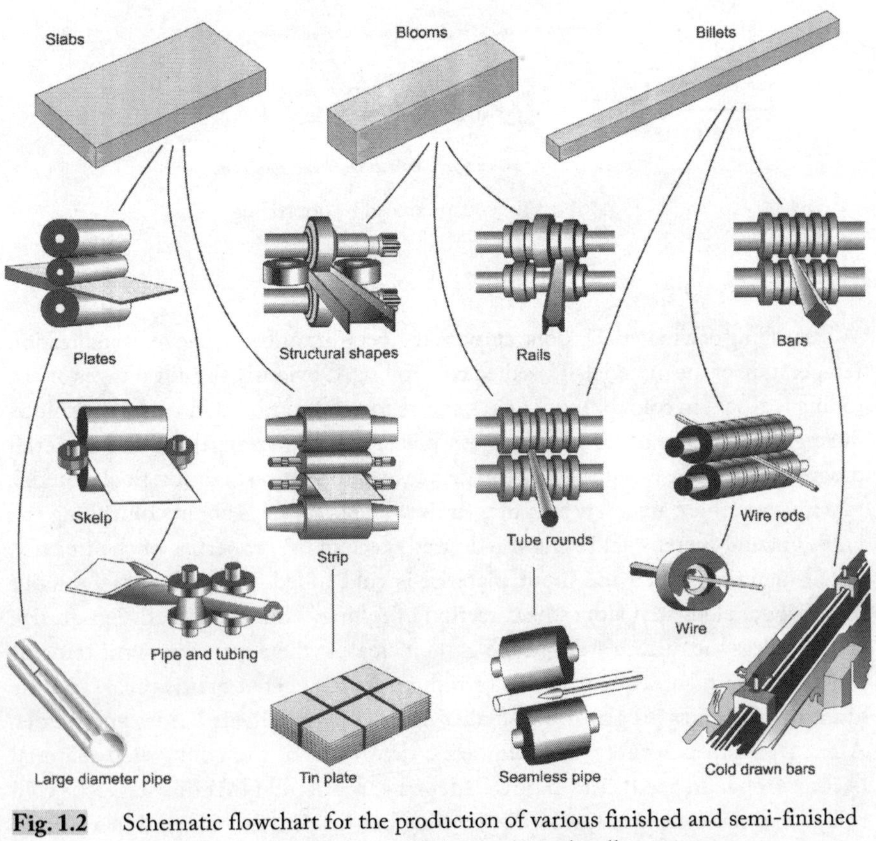

Fig. 1.2 Schematic flowchart for the production of various finished and semi-finished steel products which pass through rolling process

During cold rolling of flat products, the material is passed between two flat cylindrical rolls of the mill stand. Furthermore, various advanced techniques and systems are employed to keep the rolled material flat and the thickness of the finished product within close tolerance throughout the length and width of the product. These advanced techniques and systems have been discussed in Chapter 3 and 4 respectively.

1.3 Brief History of Rolling

1.3.1 The early history

It is a characteristic feature of engineering and technology that developments and innovations in different fields of engineering and technology are often not attributable to any single person, or a specific date, or even a single place, or country. Engineering and technological developments are induced by social needs, and are often result of many pioneers over a period of time and places. History of development of rolling technology is no different.

Rolling of soft metals like gold, silver and perhaps lead was first performed by goldsmiths for making jewelry. Hand driven rolls of about ½ inches (12.5 mm) in diameter were in use during the fourteenth century. However, the concept of a true rolling mill is first found as a sketch in the notebook of Leonardo da Vinci. There is no evidence that this was ever built.

Before the end of the sixteenth century, at least two mills incorporating basic ideas of rolling are known to have been in operation. In 1553, a Frenchman named Brulier, rolled sheets of gold and silver of uniform thickness for making coins. In 1578, a man called Bevis Bulmer, received a patent for operation of a two spindles slitting mill with series of disc cutters to slit flats into narrow strips.

As per record available, rolling of iron into thin flats was first done in Bristol, the UK in 1666. As per a pamphlet on 'British Iron Trade' published in 1725, all bars were made by hammering even at that date. By 1682, large rolling mills for hot rolling of ferrous materials were in operation near Newcastle, the UK. These mills were used for rolling bars into sheets, which were cut into square rods in slitting machines.

Hot rolling of ferrous material into thin sheets was practiced in many continental countries, but was monopolized by Germany from the beginning of the seventeenth century until Major Hanbury started production in Pontypool works in the UK sometime around 1720[1].

In 1728, a patent was issued to John Payne in England for producing different rolled shapes using grooved roll. However, there is no proof to support that such an idea was put to practice.

In 1766, John Purnell of England received a patent for grooved rolls and arrangement of driving both rolls in unison through use of coupling boxes and nut pinions. Until this time, rolls were driven individually by water mill.

During this period, the hot mills were evolving to the modern form. William Playfield's English patent filed in 1783[2] shows use for cast housings and one screw on each window for screw down.

By the middle of the eighteenth century, pack rolling came into practice for producing thin plates for tinning.

The eighteenth century also saw the advent of tandem mill in which metal is rolled in successive stands. Richard Ford's English patent in 1766 is the first record of a true hot tandem mill for producing wire rods. James Cockshutt and Richard Crawshay, about 1790, built a four-high tandem mill near Sheffield, the UK, with a capacity of 1 – 2 tons per day. Patents were obtained for tandem mill for rolling of plates and sheets.

The earliest record of a true continuous wide-strip rolling mill is that built in 1892 at Teplitz in Czechoslovakia[3] (then an Austrian territory) consisting of two three-high roughing stands and a finishing train consisting of five number two high continuous finishing stands. The two three-high train and the continuous five stand train was driven by 1000 H.P. steam engine each. Starting with 8″ (200 mm) thick steel slabs, the mill was reported to be producing sheets up to 50″ (1270 mm) wide strip of thickness between 0.08″ – 0.12″ (2 mm to 3 mm) and in length up to 60′ (18.29 m).

History of metalworking in the United States began with the arrival of colonists from Europe. The first American rolling mill was built in 1751 in Massachusetts. This was used for rolling 3″ (75 mm) wide and 3/4″(19 mm) thick hammered bar to 1/4″(6 mm) thickness suitable for slitting into nail rods, in 4 passes. In 1793 Isaac Pennoch established a rolling mill which by 1810 was rolling plates in mills with 16″ – 18″ diameter and 3 – 4 feet long rolls. In same plant around 1820, boiler plates were rolled by Dr. Charles Lukens (Pennoch's son-in-law).

The first rolling mill in America to have been powered by a 70 H.P. steam engine was built by Christopher Cowan in western Pennsylvania. In 1819, the first angle iron rolled in the US was produced at the Union Rolling Mill in Pittsburg. By 1825 five rolling mills were in operation in Pittsburg which eventually became focal point of metalworking industry. Rolling of corrugated plates was patented in 1850. After civil war ended in 1865, rapid expansion of railways gave a tremendous impetus to the American iron and steel making and rolling industry.

1.3.2 Modern steel rolling plants

Since the early twentieth century, rolling industry saw a continuous string of developments in terms of size and production capacity, new design of mill stands and ancillary equipment, increased working efficiency, greater instrumentation and automation for quality at higher mill speed.

The pioneering effort of Tredegar Iron and Coal Company of South Wales is worth mentioning. In 1905 Whitehead[4] decided to modernize the continuous hot mill operation. In one 8-hr shift, the mill could produce 103 tons of 8″ wide × 0.03″ (0.75 mm) thick material. The plant had continuous furnaces, electric cranes, flying shears to cut strip coming out from the mill at a speed of 3000 fpm (~900 mpm), specially designed coilers to coil at fast speed, and many other novel features.

Another milestone in continuous rolling of wide strip was achieved in the Butler plant established at the Columbia Steel Company of America, which went into production in 1926, designed and developed by Townsend and Naugle for over a period of 10 years. This plant[5] included, among others, annealing furnaces, long continuous pickling line, welding arrangement for joining coils for continuous operation, and other processing auxiliaries.

After successful operation of the Butler plant, a number of such integrated plants were erected in the UK and elsewhere. As per data on wide strip mills published by Davy and United Engineering[6] of the UK in 1960, the largest width capacity mill was a 108″ (2740 mm) mill in Voroshilov Works in Russia.

1.3.3 Modern non-ferrous rolling plants

Credit goes to the Revere Copper and Brass Company[7] of America for beginning to roll copper strip in a so-called continuous mill in 1926. In this plant 5″ thick × up to 36″ width copper ingot was hot rolled in a two-high reversing mill to about 3/16″ (4.75 mm) thick strip. This strip, after suitable pickling and cleaning was cold rolled in a 4 stand four-high continuous mill. The plant gradually introduced mechanized sheet conveying system. In Britain, Imperial Chemical Industries that has been rolling brass and copper even before the American company, put a highly mechanized plant in Birmingham in 1933. In 1949, the Scovill Manufacturing Company[8] in the USA erected one of the most modern brass-rolling plants in which continuous casting of brass ingots of 2.5″ (63.5 mm) thick and 29.5″(750 mm) wide and more than one ton weight were produced. The ingots were initially cold rolled to 0.415″ (10.54 mm) in eight passes with two inter-anneals in a massive two-high mill. The mill stand was equipped with quick roll changing facility and vacuum cup type stock lifting arrangement was introduced.

The first four-high mill[9] was installed and operated in the UK, in the nickel and nickel alloy rolling plant of Henry Wiggin and Co. by around 1932 for cold rolling of these strips.

With gradual increase in consumption of aluminum alloys from the third decade of twentieth century and specially during and after the World War II, a number of large and high speed hot and cold aluminum rolling mills were established.

Pioneer of aluminum rolling in Britain was the British Aluminium Company formed in 1894. By mid 1940s, they laid out a modern aluminum rolling plant at

Falkirk,[10] which included continuous casting units. Thickness of 8″(200 mm) and about one ton weight slabs were sawed, which were hot rolled, edge trimmed then coiled for annealing and further cold rolled in both two-high and four-high mills. The four-high stands using 20″ (508 mm) diameter work roll and 49″ (1245 mm) back-up rolls and driven by 1500 HP variable speed motor, and rolling 6′ (1829 mm) wide coils at 450 fpm (136 mpm) were built by Robertson of Bedford. This mill manufacturer, later known as Lowey Robertson's Engg. Company, played a significant role in the development of large and high speed aluminum strip and foil rolling mills with various modern features and quality control systems. The South Wales Rogerstone works of the Northen Aluminium Co. (Alcan Industries) installed one of the most modern plants operating in Europe in 1950. This plant included a three-stand, four-high cold tandem mill. In this mill, coil weighing 1.75 tons could be rolled to a maximum width of 54″ (1370 mm) and a reduction in thickness of 90% was achieved in one pass through the three stands. The speed of strip from the last stand was greater than 23 miles per hour (2024 fpm or 613 mpm). Similar mills with similar production methods and parameters were also established in America.

1.3.4 Modern cold rolling facilities

From 1930's, cold rolling of hot rolled steel strips became a process of prime importance. A comparative picture of width and thickness of strip that was cold rolled between 1923 and 1937 will make the point clear:

Year	*1923*		*1937*	
Strip width	22″ (560 mm)	10″ (254 mm)	90″ (2290 mm)	35″ (890 mm)
Minimum thickness	0.14″ (3.55 mm)	0.05″ (1.27 mm)	0.038″ (0.96 mm)	.01″(0.25 mm)

Typical strip mills built in the 1930's are three-stand 84″ wide tandem mill with about 20½″ diameter work roll, 56″ diameter back-up roll and driven by motors totaling 6850 HP rolling at speeds up to 540 fpm. Just before and after the World War II, four-stand tandem mills came into operation. However, since 1960's five-stand tandem cold strip mill for rolling ferrous alloys has become the industry norm.

Although most of the tandem mills employ conventional four-high stands, a unique tandem mill designed for rolling of stainless steel strips up to 50″ (1270 mm) wide commissioned in 1969 by Nisshin Steel Corporation at the Shunan Works[11] located in Nanyo, Japan, utilizes a train of 6 stands, whose first and last is two-high stand while the intermediate stands are multi-roll Sendzimir mills.

With limited market of stainless steel, silicon steel, and foil gauge copper alloys, single stand Sendzimir mills proved to be quite popular for rolling these materials.

In order to meet the demands of high strip quality along with high productivity (i.e., high rolling speed), substantial improvement has been achieved since 1960's in the mill instrumentation, controls and auxiliaries like introduction of Automatic Gauge Control, Flatness Control and improved Roll Coolant systems.

1.4 Deformation of Materials

1.4.1 Elastic and plastic deformation

Metals are crystalline solids in which atoms are arranged in a well-oriented pattern. In the absence of any external force in a metal crystal, let the distance between a pair of atoms be r_o. The equilibrium interatomic distance r_o is that distance at which the attractive and the repulsive forces are equal in magnitude. The slope of the repulsive force curve is always more than that of the attractive force curve at the point of intersection of the curve. Therefore the equilibrium is of stable nature. The net inter-atomic force varies with the atomic spacing in a manner shown in Fig. 1.3, the positive side signifying attractive force. Under the application of an external tensile force, the interatomic distance increases beyond r_o to maintain the equilibrium. If the external tensile force is of magnitude F, then the interatomic distance should be r_a so that the net interatomic force is an attractive force of magnitude F, to balance the external force. If r_a is not very much different (of the order of 5%) from r_o, when the external force is removed, the atoms attain their original positions. A similar behavior is also observed with an external compressive force (when $r_a < r_o$). This behavior is called the elastic behavior of a solid material, and the associated deformation is termed as *elastic deformation*.

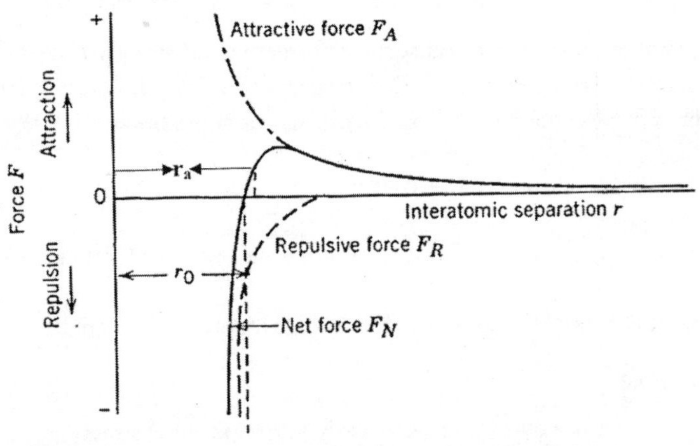

Fig. 1.3 Variation of net interatomic force with interatomic distance

This phenomenon described for a pair of atoms is also true for normal solids on a macroscopic scale. It may be noted from Fig. 1.3 that the tangent to the curve at the point r_o coincides with the curve over a small range on either side of point r_o. This indicates that the external force is proportional to the change in the interatomic distance. Hence, within this elastic range, most solids follow a linear force deformation rule, and are thus called linear elastic solids.

Now, let us consider a crystal lattice of a solid material with regularly spaced atoms as shown in Fig. 1.4a. Under the externally applied shear force, the upper layers of atoms will move to the right and the lower layers will move to the left.

When the applied shear stress reaches beyond a sufficiently high value called the shear yield stress of the material, the crystal lattice takes the shape as in Fig. 1.4b. Here, all the atoms are again in equilibrium and will remain so if the external stress is removed. Thus, a permanent deformation is produced in the crystal lattice. This permanent deformation is termed as *plastic deformation* and cannot be recovered by withdrawing of the external stress.

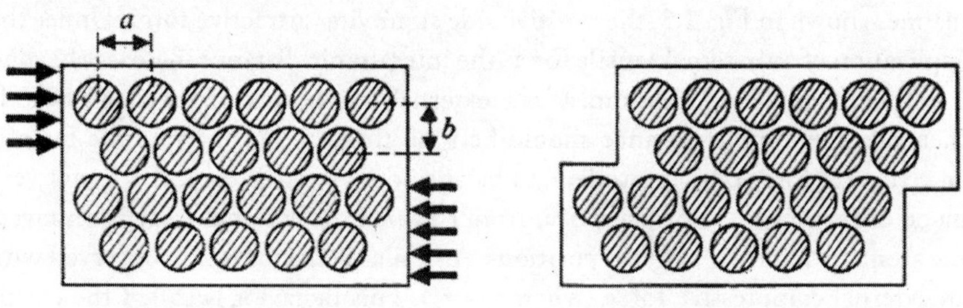

(a) Original position of atoms

(b) Position of atoms after slipping through distance 'a'

Fig. 1.4 Scheme of plastic deformation in perfect crystal

The amount of shear stress to cause the deformation between two layers of atoms in a perfect crystal can be estimated. Referring to Fig. 1.5(a), the shear stress, τ, and the amount of displacement of the top layer, x, may be approximated by the relation:

$$\tau = \tau_0 \sin\frac{2\pi x}{a} \qquad 1.1$$

This variation is shown in Fig. 1.5(b). For small values of x/a, $\sin\frac{2\pi x}{a}$ may be approximated as $\frac{2\pi x}{a}$. Hence, the foregoing relation can be rewritten approximately as

$$\tau \approx \tau_0 \frac{2\pi x}{a}, \text{ where } \tau_0 \text{ is the ultimate shear stress}$$

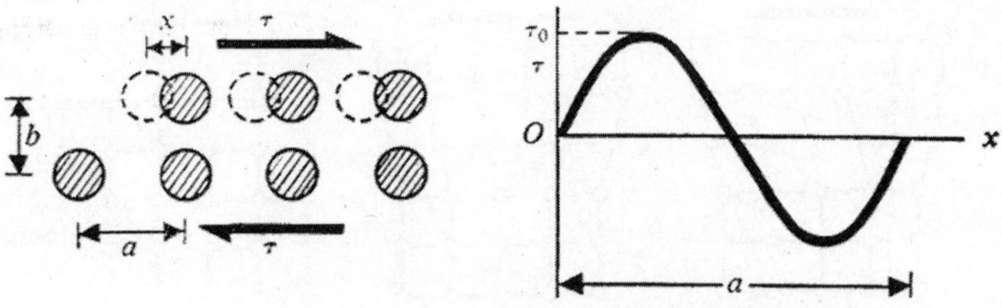

(a) Atomic arrangement for slip through x (b) Variation of τ with slip

Fig. 1.5 Mechanics of slip in perfect crystal

Now, by definition, shear strain produced is 'x/b'. Using linear stress–strain relationship, we can write

$$\tau = \tau_0 \frac{2\pi x}{a} = G\frac{x}{b} \quad 1.2$$

where the constant G is known as the shear modulus or modulus of rigidity. Thus, the ultimate shear stress necessary to cause the permanent shear deformation is,

$$\tau_0 = \frac{G}{2\pi}\frac{a}{b} \quad 1.3$$

In a metallic crystal structure, the values of a and b are of the same order, and as a rough approximation, we can take a ≈ b. Then

$$\tau_0 \cong \frac{G}{2\pi} \quad 1.4$$

However, the ultimate shear stresses for different solids and alloys, obtained through experimentation have always been found to be much less (in the order of 100 times) than the value of τ_0 as obtained above. This large difference can be explained only by considering the imperfections that generally always exist in an actual crystal structure (like point defects, dislocations, interfacial / grain boundaries, etc.).

Figure 1.6 explains how the movement of an edge dislocation rather than the movement of a whole plane of atoms can cause a slip by application of less shear stress (which also means by application of less energy). Here the material is deformed by breaking and reforming one row of bonds at a time as the dislocation moves along the plane of slip.

This phenomenon can be appreciated form the following analogy of progressive movement of a carpet wave.

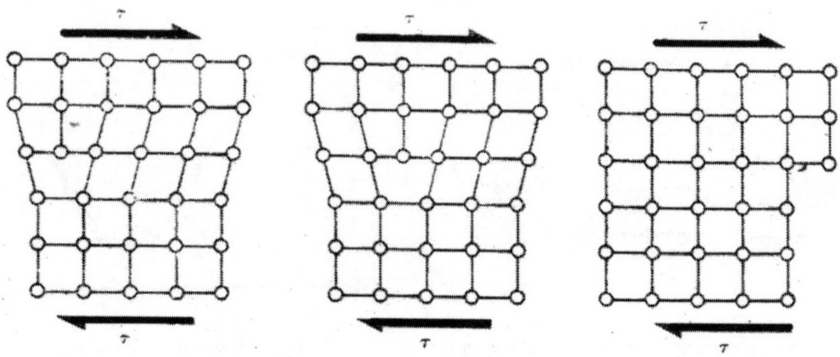

Fig. 1.6 Crystal deformation through movement of edge dislocation

Let us consider a thick, heavy carpet lying on a floor, which has to be moved through a distance δ (Fig. 1.7). It can immediately be visualized that a very large force will be needed if the whole carpet has to be bodily moved over the distance δ. The same effect, however, can be realized very easily if first a hump is made at one end of the carpet by moving that end by the distance δ and then this hump is moved the entire carpet length.

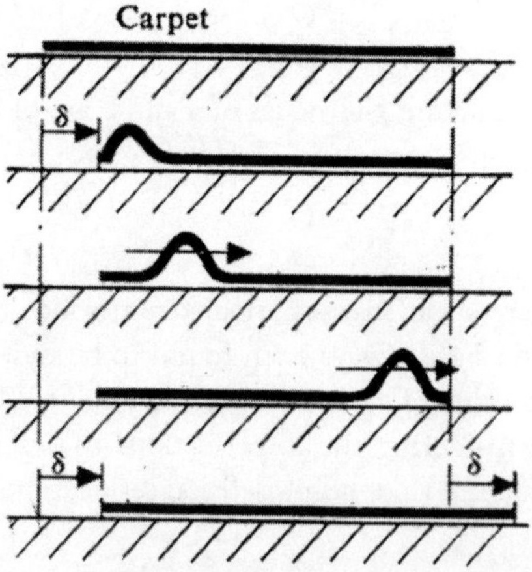

Fig. 1.7 Carpet analogy

1.4.2 Size effect and strain hardening

It has been seen that crystal imperfection like dislocation helps slip. Therefore, if a very small crystal of a solid that has got very little imperfection, is taken, it will be difficult to cause slip. Thus strength of a crystal decreases with its increase of size having more dislocations. This phenomenon is called *size effect.* However, when a material is deformed, more and more dislocations take place which interact with each other. These interactions have the tendency to block movement of various dislocations and as a result more and more force is required for further deformations, i.e., the strength of material increases. This gradual increase of strength with deformation of the material, caused due to interaction and immobility of dislocations is known as *strain hardening* or *work hardening*.

The effect of strain hardening can be understood clearly from a simple tension test curve, as shown in Fig. 1.8.

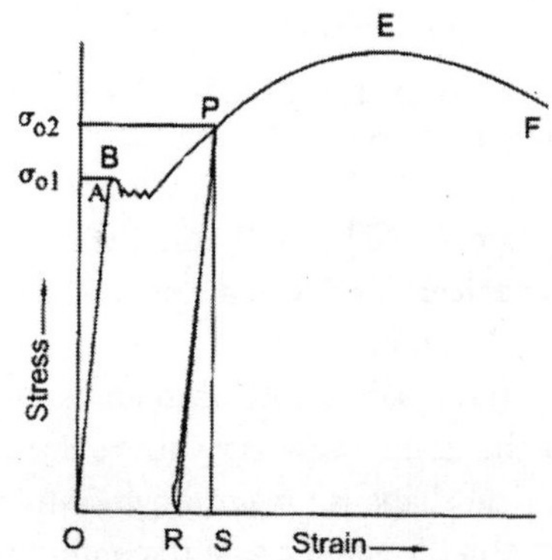

Fig. 1.8 Strain hardening effect

If the test piece is loaded beyond the yield point up to P, and then unloaded, the elastic deformation will recover via the unloading curve PR, which is more or less parallel to original loading line OA. Out of the total strain OS, the portion RS is the elastic strain which is recovered. OR is the plastic strain of the test piece which is not recovered.

If the test piece is again loaded, it will approximately follow the line RP, with a small deviation due to hysteresis. This indicates that the Young's modulus of the material does not change due to plastic deformation. It is interesting to note that with increased loading, the test piece will start yielding at point P, and will follow the same stress–strain curve, PEF, which would have been obtained without unloading, at point P.

Thus the yield point of the test piece after plastic deformation increases from σ_{01} to σ_{02} corresponding to point B and P, respectively, in the figure.

Metals and their alloys are polycrystalline solids. Small, continuous volumes of solid are known as grains of crystal, and the surfaces that separate the grains are known as grain boundaries. In a polycrystalline solid, the grain boundaries also interact with and impede movement of dislocation. In a definite volume of material, smaller is the size of grains, larger will be the boundary areas causing more impediments to dislocation movement, resulting in higher strength to deformation. Similarly, larger grain size with smaller boundary areas results in lower strength.

1.4.3 Mechanical working and recrystallization

Metals do not always posses all the properties required out of it. The material properties, mainly the mechanical properties, can, however, be changed and controlled to a large extent by various methods to make the material suitable for a given application. These methods include (i) alloying, (ii) heat treatment, and (iii) mechanical working and recrystallization.

As mentioned, the mechanical properties, e.g., strength and hardness of a polycrystalline material are also governed by the grain size of the material. Figure 1.9 shows the nature of variation of two mechanical properties with the average grain size. The grain size can be controlled by mechanical working and recrystallization. In the rolling process, mechanical working and recrystallization are involved. These processes are discussed briefly below.

When a near spherical grain is deformed into an elongated shape, its volume remains constant whereas its grain boundary surface area increases. Similarly, if a large grain of near spherical shape is broken down into a few pieces of smaller near spherical shapes, then also the volume to surface area ratio reduces. Both these cases are known as reduction of effective grain size. Different mechanical working processes deform and thereby reduce the grain size of parent material.

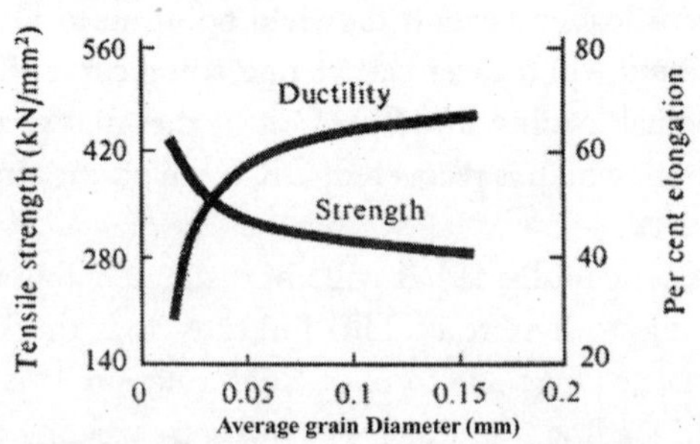

Fig. 1.9 Effect of grain size on strength and ductility of annealed brass

It is possible to restore the original grain geometry by heating the material above a temperature where the new fine grains are formed. This process is called *recrystallization* and the temperature needed to achieve this is known as the *recrystallization temperature*. If the material is kept for a considerable period of time at this temperature, the newly formed fine grains grow in size. If materials are worked below recrystallization temperature, which is known as *cold working*, the deformed grains remain in same shape and size. It should be noted that in the mechanical working processes, the grains are deformed, whereas in recrystallization, new grains are formed and, what is more, their size can be controlled.

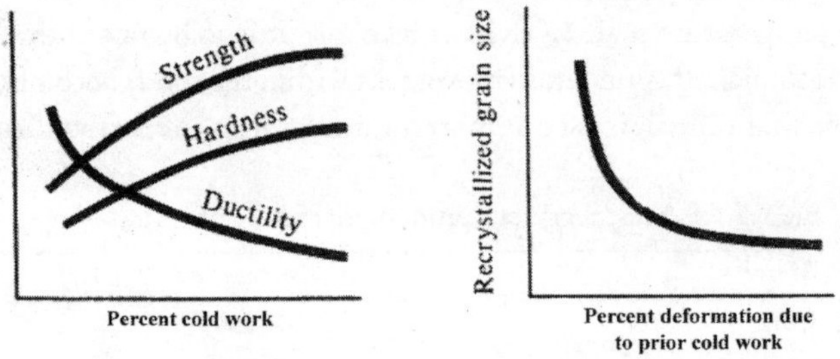

(a) Variations of mechanical properties with amount of cold working

(b) Dependence of recrystallized grain size on prior cold work.

Fig. 1.10 Effects of mechanical working

Figure 1.10(a) shows the variation of mechanical properties with the amount of cold working. Apart from changing the mechanical properties, the amount of cold work also governs the recrystallized grain size (see Fig. 1.10 (b)).

1.4.4 Hot and cold rolling and recrystallization

During plastic deformation, a portion of the deformation energy becomes stored within the material in the form of additional dislocations and increased grain boundary surface area*. If a deformed polycrystalline metal is subsequently heated to a high-enough temperature, the material will seek to lower its energy. New, spherical-shaped crystals will nucleate and grow out of the original structure. This process of reducing the internal energy through formation of new crystal is known as recrystallization. The temperature at which recrystallization takes place is different for each metal, and it also varies with the amount of prior deformation. Greater the amount of prior deformation, with more stored energy, lower is the recrystallization

* A sphere has the least amount of surface area of any shape to contain a given volume of material. When the shape becomes altered from that of a sphere, the surface area must increase. Consider a round balloon filled with air. If the balloon is stretched or flattened into another shape, the rubber surface is stretched further. When the applied load is removed, the balloon snaps back to its original shape, the one involving the least surface energy. Metals behave in an analogous manner. During deformation, the distortion of the crystals increases the energy of the material. Given the opportunity, the material will try to lower its energy by returning to spherical grains.

temperature. However, there is a lower limit below which recrystallization will not take place in a reasonable amount of time. Table 1.1 gives the lowest practical recrystallization temperatures for several metals. This temperature is approximately 0.4 times the melting point of the metal when the melting point is expressed in an absolute temperature scale (Kelvin or Rankin). It is to be noted that it is at this temperature atomic diffusion (atom movement within the solid) becomes noticeable, which means that diffusion is an important mechanism in recrystallization.

Table 1.1 Lowest recrystallization temperature of common metals

Serial No.	*Metal*	*Temperature [°F (°C)]*
1.	Iron	840 (450)
2.	Aluminum	300 (150)
3.	Copper	390 (200)
4.	Nickel	1100 (590)
5.	Silver	390 (200)
6.	Magnesium	300 (150)
7.	Gold	390 (200)
8.	Zinc	Room temperature
9.	Lead	Below room temperature
10.	Tin	Below room temperature

When metals are plastically deformed below their recrystallization temperatures, the process is called cold working. The strained metal hardens and the structure consists of distorted grains. If the deformation is continued, the metal may fracture. Therefore, it is a common practice to recrystallize material after certain amount of cold work. Ductility is restored, and the material is ready for further deformation. The heating process is known as *recrystallization anneal* and repeated anneals enable deformation to be carried out to great extent without the danger of fracture.

If metals are deformed at temperatures sufficiently above the recrystallization, the process is known as *hot working*. Deformation and recrystallization can take place simultaneously, no work hardening takes place and consequently large deformation is possible (see Fig. 1.11). As shown in Table 1.1, the minimum temperature for hot working depends on the material being worked.

Having already noted the correlation of fine grain-size with improved properties, it is easy to find recrystallization to be an attractive means of grain-size control. In metals that do not undergo allotropic changes, a coarse grain structure can be converted to a fine grain structure through deformation and recrystallization. The material must first be plastically deformed to store energy to provide the driving force for recrystallization. Control of the recrystallization process then establishes the final grain-size. Higher is prior deformation, lower is recrystallized grain-size.

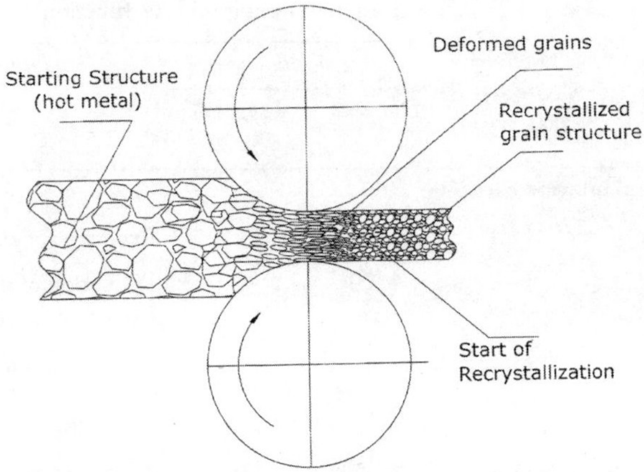

Fig. 1.11 Schematic representation of the hot-rolling process, showing the deformation and recrystallization of the grains of metal being rolled

The recrystallization process tends to produce uniform grains of comparatively small size. If the metal is held at or above the recrystallization temperature for long time, the grain size of the new structure will continue to increase. In effect, some of the grains will become large at the expense of their neighbors as the materials seek to further decrease the grain boundary surface area. Mechanical properties tend to diminish with increased grain size, so control of recrystallization time is of great importance. The material should be held at elevated temperature long enough to complete recrystallization. The temperature should then be decreased below recrystallization temperature to prevent further grain growth.

1.5 Classification of Rolling Mills and Rolling Stands

1.5.1 Classification

Rolling mills are classified according to their functions, i.e., the type of finished products they produce. It is to be noted that function of a mill does not always imply specific design of the mill stand in particular or the rolling mill (including all auxiliary equipment and their layout) in general. The design and layout of a rolling mill depends to a great extent on the mill output, both in quantity and quality.

Table 1.2 illustrates the classification of rolling mills by function. Since size of the rolling stand depends on the input stock to be rolled, the classification table often indicates the range of mill stand roll diameter and barrel length. For mills that roll heavy sections, roll diameter is important, while for flat rolling barrel length is more important. The table also indicates whether the rolling process involved is hot or cold.

Table 1.2 Classification of rolling mills by function

Mills	*Rolling Process (Hot /Cold)*	*Size of rolls, mm*		*Function*
		Diameter	*Length of barrel*	
1. Primary and intermediate mills				
(a) Blooming	Hot	800–1450	----	Reduction of ingots to blooms
(b) Slabbing	Hot	1100	----	Reduction of ingots to large flat semis called slabs.
(c) Billet	Hot	450–750	----	Rolling blooms to billets of cross-section 50×50 to 150×150mm
(d) Tube rounds	Hot	700–800	----	Rolling blooms to semis of round cross-section diam. 70 to 350mm
2. Section mills				
(a) Rail and Section	Hot	750–900	----	Rolling wide gauge rails, large beam 240–600 mm and higher, and other heavy sections
(b) Heavy Section	Hot	500–750	----	Rolling heavy sections: steel squares and round from 80 to 150 mm, beams and channels from 120 to 240 mm in height, and other sections
(c) Medium Section	Hot	350–500	----	Rolling medium sections: steel squares and rounds from 40 to 80 mm, beams and channels up to 120 mm in height, angle bars from 50×50 to 100×100mm, and other sections
(d) Light Section	Hot	250–350	----	Rolling small sections: steel squares and rounds from 8–40 mm, angles and bars from 20×20 to 50×50 mm, and others sections
(e) Strip (Skelp)	Hot	300–400	----	Rolling strip from 65 to 500 mm wide and 1.5–10 mm thick.

Mills	*Rolling Process (Hot/Cold)*	*Size of rolls, mm*		*Function*
		Diameter	*Length of barrel*	
(f) Rod	Hot	250–300	----	Rolling rod from 5 to 10 mm diam.
3. Flat mill				
(a) Plate	Hot	----	2000–5000	Rolling plates from 4 to 50 mm and more thick
(b) Hot sheet	Hot	----	700–2500	Rolling wide strips from 600 to 2300 mm
(c) Hot wide strip	Hot	----	700–1300	Rolling of thin sheets 0.2–4 mm thick and 500–1200 mm wide in coils
4. Tube rolling mills	Hot	----	----	Rolling seamless tube up to 650 mm diameter.
5. Cold rolling mills				
(a) Cold sheet	Cold	----	700–2800	Rolling sheets from 600 to 2500 mm wide
(b) Narrow strip	Cold	65–215	150–700	Rolling narrow (cold) strip 0.02–4 mm thick and 20–600 mm wide supplied in coils
(c) Foil	Cold	----	200–1500	Rolling foils from 0.08 to 0.006 mm thick
(d) Wire flattening	Cold	65–190	----	Rolling narrow strip 0.8–12 mm wide from wire
6. Special kind of rolling mills				
(a) Wheel	Hot	----	----	Rolling wheels for railway wagons and other purposes.
(b) Hoop and tyre	Hot	----	----	Rolling hoops for bearings and tyres for wheels
(c) Ball	Hot	----	----	Rolling balls
(d) For rolling variable sections	Hot	----	----	Rolling various kinds of variable sections
(e) Gear wheels	Hot	----	----	Rolling teeth of gear wheels

1.5.2 Rolling mill stands

A rolling stand usually consists of the following components: two rolls to do the job of rolling the material; the roll supporting bearings and bearing blocks, a structure called mill housing in which the rolls with its bearings and bearing blocks are housed; a drive system for rotating the rolls and a screw down mechanism for movement of one or both rolls for adjusting the gap between the rolls. These components of a rolling stand will be discussed in detail in the rolling mill equipment section.

Use of rolls for rolling of metals has had long history dating back from the sixteenth century and by 1682, large rolling mills for hot rolling of ferrous materials, powered by water, were in operation near Newcastle, England. Since then innumerable innovations in the design of mill stand have taken place and many types of mill stands are now successfully employed to perform different rolling operations.

The different types of rolling stands are categorized in various ways. Single stand mills are generally classified on the basis of their roll arrangements. Some special types of mill stands are known by their manufacturer's name or a generally accepted designation based on use or application.

1.5.3 Roll arrangement in mill stands

The arrangement of rolls in mill stands depends on the specific function of the mill and needs of the outputs. Based on roll arrangement, mill stands can be of following kinds: (i) with horizontal rolls; (ii) with vertical rolls; (iii) with horizontal and vertical rolls; (iv) with inclined rolls; and (v) other types of stands.

1.5.3.1 *Stands with horizontal rolls*

These are the most common mill stands. There are a number of types based on number of rolls employed. Table 1.3 shows some of the types of horizontal mill stands.

Table 1.3 Roll arrangement and use of stands with horizontal rolls

Arrangement of rolls	*Name of stand*	*Use*
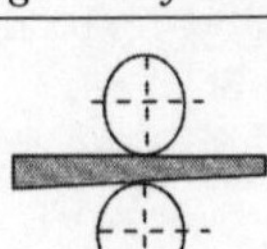 Fig. 1	Two-high stand	Reversing mills for rolling large shapes (blooms, beams, rails) and plates. High production mills for rolling semi products and sections. Strip mills (old method) Cold reduction mills for rolling wide and narrow strip (old method).

Arrangement of rolls	*Name of stand*	*Use*
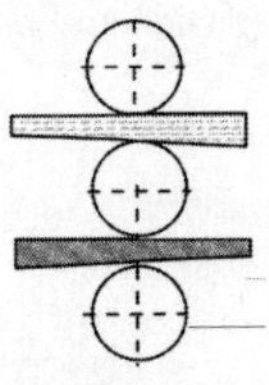 Fig. 2	Three-high stand	For rolling beams, rails, semi products, blooms and other large shapes starting with light weight ingot. For rolling sections in small tonnages.
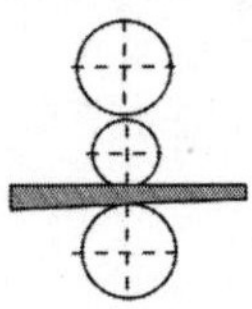 Fig. 3	Three-high stand with floating roller (Lauth mill)	For hot rolling medium and thick sheets and sometimes thin sheets.
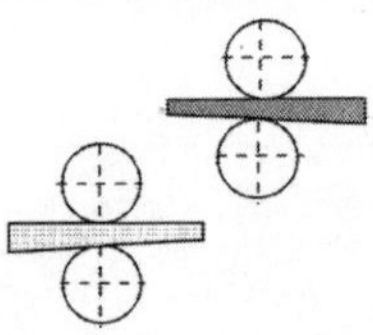 Fig. 4	Double-duo (or Dowlais) stand	For hot rolling light and medium sections in small tonnages
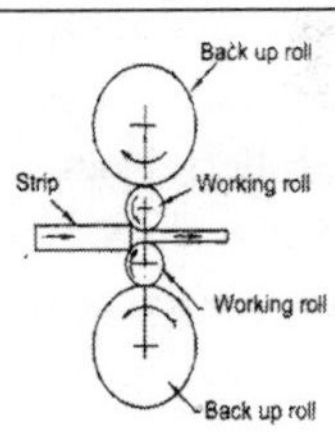 Fig. 5	Four-high stand	For hot and cold rolling sheets, wide strip and narrow strip
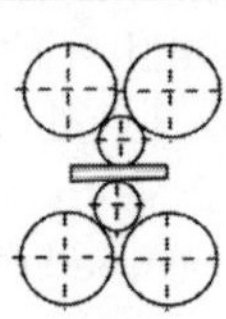 Fig. 6	Six-high stand	For cold rolling wide and narrow strip
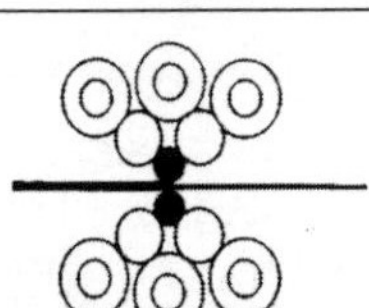 Fig. 7	Twelve-high stand	For cold rolling wide and narrows strip

Arrangement of rolls	*Name of stand*	*Use*
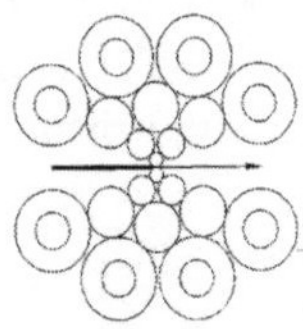 Fig. 8	Twenty-high stand	For cold rolling wide and narrow strip
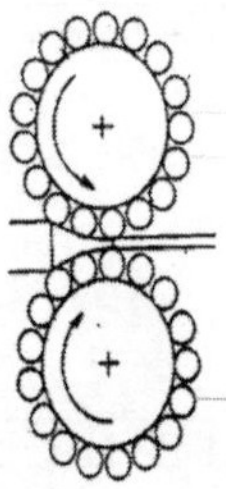 Fig. 9	Planetary mill	For hot rolling wide and narrow strip with high draught.

Two-high stand:

Two-high (designated as '2-Hi') stands have two horizontal axis rolls arranged one above other. Mills with 2-Hi stands are the most common. They are used as (a) *reversing mills* where rotation of the rolls is reversed after each pass. These mill stands are used for hot rolling of large cross section (blooms, billets, beams, and rails); (b) high production section and intermediate bar mills, which are composed of a number of stands in tandem, the stock having only one pass in each stand; (c) wide hot plate or strip mill; (d) stands for cold rolling of sheet and strip, etc.; and (e) skin pass or temper pass cold mill for annealed cold rolled strip.

A 2-Hi mill stand is specified by the original diameter and the barrel length of the rolls. Thus a '1150 mm × 2800 mm 2-Hi mill' means the mill has a 2-Hi stand having each roll of 1150 mm diameter and 2800 mm barrel length. It is to be noted that 1150 mm is the maximum diameter of the new roll. With progress of rolling, the rolls have to be machined on its diameter over and over, and the rolls are discarded when their diameter reaches a minimum value called discard diameter.

Three-high (3-Hi) stand:

In 3-Hi stands, there are three horizontal rolls arranged one above another. In these mills, stock is rolled in both directions without reversing the rolls. The stock is rolled in one direction between the bottom and middle roll, while rolled in the opposite direction when stock is introduced between middle and top roll. 3-Hi rolling stands are still employed (a) as three-high plate and sheet mill, called *lauth* mill, in which the center roll is not driven and is somewhat smaller diameter than the top and bottom driven rolls. After each pass, the middle roll is raised or lowered to exert pressure on top or bottom roll; (b) as rail or section mill where high production is not needed, where all the rolls are equal and driven, (c) as a blooming mill for reduction of small ingots weighing 1–1.5 tons.

For 3-Hi mill stands, arrangement of lifting or tilting tables are made on both sides of the mill stand for lifting and lowering of the stock between two levels of rolling.

Double-duo stand:

A double-duo (or Dowlais) stand uses two pairs of 2-Hi rolls housed in one stand, set in two different vertical planes. The bottom roll of one pair is higher than the top roll of the other pair, thus permitting passes in both directions with one unidirectional motor drive. Such mill stands are no longer built, and the existing ones are used particularly for rolling high quality steel sections with rolls of 300–400 mm diameter.

Four-high (4-Hi) Stand:

A 4-Hi stand is composed of four horizontal rolls all arranged in same or almost same vertical plane. Rolling takes place between the two middle rolls, called *work rolls*. Their diameter is considerably smaller than the two outer rolls placed one at the top and one at the bottom side. These larger rolls support and controls bending of the work rolls by the rolling load during rolling. For this reason the two outer rolls are called *back-up rolls*.

By using smaller diameter work rolls, the rolling load is kept low which in turn keeps the mill stretch small. At the same time, the tendency of bending of slender work rolls is prevented by supporting the rolls by large diameter back-up rolls. Because of this characteristics, 4-Hi mills are preferred for cold rolling of sheet, strip and particularly for thin gauge strips and foils. These stands are also used for hot rolling of plates and strips.

In most of the 4-Hi mills, the work rolls are driven and back-up rolls are idle and they rotate through frictional contact with work rolls. In some cases, when work rolls are small enough to take the required rolling torque, the back-up rolls are driven.

In the case of back-up driven 4-Hi stand, a very special design of the stand allows it to be run as a 2-Hi stand also. In the 2-Hi configuration, the two work rolls are removed, and the two driven back-up rolls are screwed down to operate as a 2-Hi mill. Such a mill is called a *two-high four-high combination mill,* and can be used both as a 4-Hi reduction mill and a 2-Hi skin-pass mill where quantum of production do not justify use of two separate mill stands.

A 4-Hi mill is specified, for example, as: '195 mm & 515 mm × 500 mm 4-Hi mill' which means a 4-Hi mill stand with the work roll and back-up roll diameters of 195 mm and 515 mm respectively, both rolls having 500 mm barrel length.

Six-high cluster mill stand:

In the six-high stand, also known as cluster mill stand, designed by Mathey, there are two work rolls and each supported by two back-up rolls. In these mills there is a practical limit to which the size of the work rolls can be reduced with respect to the backup rolls. As no significant advantage in rigidity over four-high mills was found, and with the limitation on lower side of work rolls, these stands were not very popular and were not built in big numbers after they were introduced in 1910.

Multi-roll stand:

These stands are used for production of thin strips.

In order to overcome the difficulty of reducing the work rolls without reducing the size of the backup rolls, W. Rohn of Germany, proposed a design, around 1925, of mills having 12 or 20 number of rolls. Figures 7 and 8 in Table 1.3 show these arrangements. A twelve-high mill of Rohn design with work rolls 0.4″ (10 mm) in diameter, intermediate rolls of 0.8″ (20 mm) diameter and peripheral (final back-up) rolls of 1.6″ (~ 40 mm) diameter was built and used for rolling of strip of nickel in coils from 0.02″ (0.5 mm) to 0.004″ (0.1 mm) in six passes.

The basic concepts of Rohn were successfully developed further by T. Sendzimir of the USA, whose mill stands known as 'Sendzimir cluster mills' having 8,12 or 20 rolls have become worldwide popular, particularly for producing thin gauge strip of hard materials like steel, alloy steel, stainless steel and super alloys (Ni, Cr, Tungsten based alloys).

Special feature of these multi-roll stands is use of very small work rolls coupled with high mill rigidly. This enables heavy draughts to be employed thus increase productivity, and high strip thickness accuracy. Constructional details of such mills are discussed in mill equipment section.

Planetary Mill Stand:

These mills have two large driven back-up rolls, and two systems of small diameter work rolls, which move around the back-up rolls. The bearing of the work rolls are housed in races which are connected with each other by gearing. The arrangement is similar to series of rollers which rotate about the inner race in a roller bearing.

These mills appeared about 1952 for hot rolling of wide strip with very large draught (up to 90–95%). These can be ideally coupled with continuous casting machine for producing weld free large diameter coils.

1.5.3.2 *Stands with vertical rolls*

Table 1.4 illustrates of different types of stands with vertical rolls.

Vertical rolls are used for reducing the rolling stock from the sides, when it is not desirable to twist the stock by 90^0.

Independent stands with only a pair of vertical rolls called vertical stands are used (a) for lateral reduction of stock in continuous intermediate and section mills and (b) for straightening edges during rolling of wide strip (Fig. 1 of Table 1.4)

Stands containing both horizontal and vertical rolls are called *universal stand.* These stands are 2-Hi or 3-Hi or 4-Hi stands with the addition of one or two pairs of vertical rolls for lateral reduction of the stock. Mills with such stands are used: (a) as slabbing mills – a 2-Hi mill with a pair of vertical rolls, which reduces ingots to slabs (Fig. 2 in Table 1.4); (b) as an universal mill for hot rolling wide strip (Fig. 3 in Table 1.4) ; (c) as a mill for rolling wide flange beams (Fig. 6 in Table 1.4). Universal 3-Hi mills are no more built (Figs. 4 and 5 in Table 1.4). In a continuous billet mill, seperate horizontal and vertical stands are placed alternately.

Table 1.4 Diagram of stands with vertical rolls and universal stands

Arrangement of rolls	*Name of stand*	*Use*
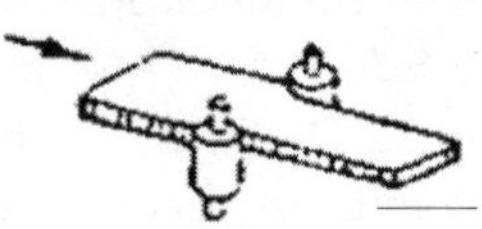 Fig.1	Vertical stand	For lateral reduction of metal being rolled. Example: billet mill.
 Fig. 2	Universal two-high (with one pair of vertical rolls)	For rolling slabs and wide flat products
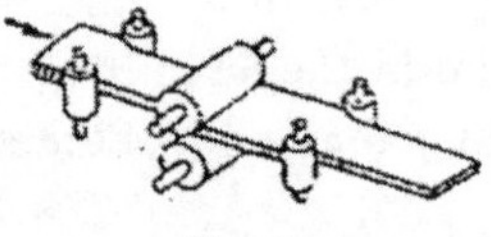 Fig. 3	Universal two-high (with two pair of vertical rolls)	For rolling wide flat products.

 Fig. 4	Universal three-high (with one pair of vertical rolls)	For rolling wide flat products.
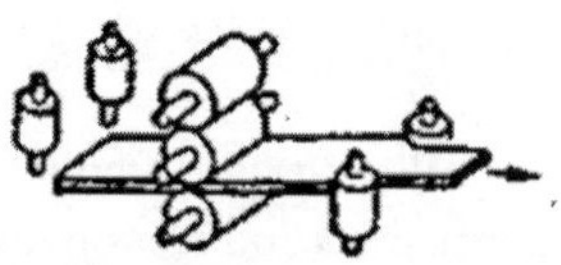 Fig. 5	Universal three-high (with two pairs of vertical rolls)	For rolling wide flats products.
 Fig. 6	Universal beam	For rolling wide flange beams from 300 to 1200 mm in height

1.5.3.3 ***Stands with Inclined Rolls***

The mills with inclined rolls are essentially hot mills in which the cylindrical stock being rolled rotates about its axis and simultaneously moves forward. This is called *helical rolling*, which is particularly employed for production of tubes. Some of the types of such mills with their use are tabulated in Table 1.5.

The first three types, as shown as Figs. 1, 2 and 3 in Table 1.5, are called *tube piercing mills*, which are used for piercing holes in solid round or ingots of round section. In piercing mills a mandrel on a rod is placed between the rolls. The solid piece of metal moves into it in course of its forward movement and is pierced. Helical rolling has a tendency to form a central cavity in the stock which greatly assists the process of piercing.

Plug rolling mills (Fig. 4 in Table 1.5) are used for elongating the pierced stock into a tube by reducing the thickness of the walls.

Three conical roll mill (Fig. 6 in Table 1.5) is used for making round articles with variable diameter along the length of round input stock. In this mill, as the stock is moved forward by an external drawing mechanism, the distance between the roll is altered which forms the variable diameter along the length of the stock.

Table 1.5 Mill stands with inclined rolls

Arrangement of rolls	*Mill*	*Use*
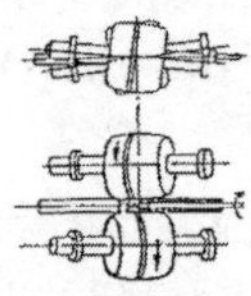 Fig. 1	Piercing mill, barrel shaped rolls	Piercing tubes with diameters 60–650 mm
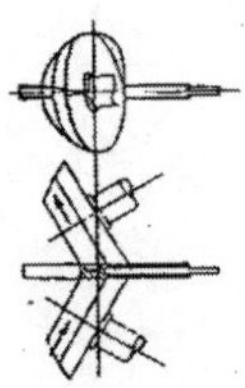 Fig. 2	Piercing mill, conical rolls	Piercing tubes with diameter 60–200 mm
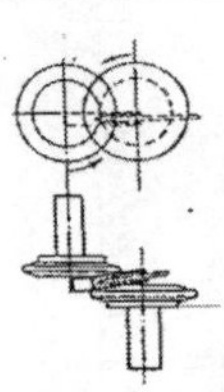 Fig. 3	Piercing mill , disc-shaped rolls	Piercing tubes with diameters 60–150 mm
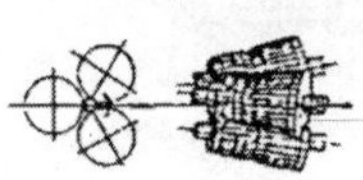 Fig. 4	Assel type plug mill	Elongation of tubes with reduction of wall thickness
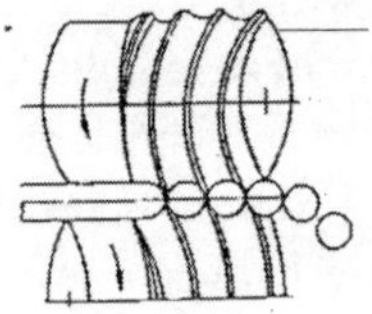 Fig. 5	Ball rolling mill (Grooved rolls with axes at an angle)	Rolling balls with diameter 18–125 mm
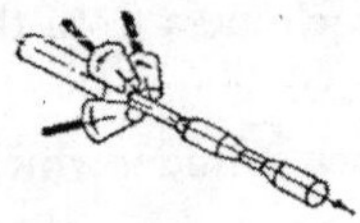 Fig. 6	Three roll variable section rolling mill	Rolling cylindrical pieces of variable cross section

1.5.3.4 *Other mill stands*

Parallel to the developments in constructional features of mills in arrangement and placement of rolls, another important trend of development took place mainly to achieve rigid but cheaper mill housing in place of housing of conventional 4-Hi mills. These are the pre-stressed mills.

Pre-stressed mill stand: Pre-stressing the specially designed bearing blocks by bolts and nuts, dispensing with conventional mill housing resulted in lighter and cheaper mill stand. Considerable reduction in size and weight of the mill is possible as shown in Fig. 1.12. This concept of pre-stressed mills has been widely exploited in case of 2-Hi hot mill stand.

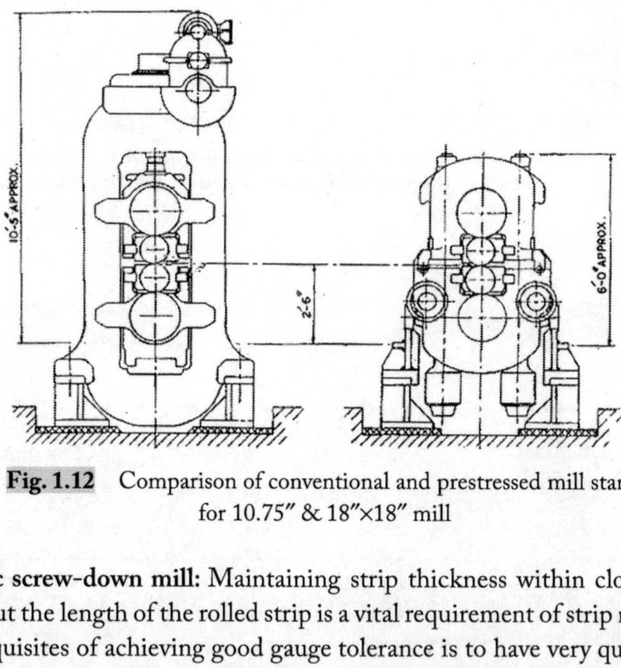

Fig. 1.12 Comparison of conventional and prestressed mill stand for 10.75″ & 18″×18″ mill

Hydraulic screw-down mill: Maintaining strip thickness within close tolerance throughout the length of the rolled strip is a vital requirement of strip mills. One of the prerequisites of achieving good gauge tolerance is to have very quick response of the screw-down adjustment. In order to achieve this, a hydraulic screw-down mill has been developed. In such mills the entire electro-mechanical screw down arrangement is replaced by two large sized position-controlled hydraulic cylinders, called *roll load cylinders,* which are fitted between the two housings and roll chocks, at the top or bottom side. The oil comes in or out of the cylinders from the oil power pack through a high response electro hydraulic proportionate valve (servo valve). LVDT type displacement transducers are fitted to each cylinder which can measure position of the cylinder ram up to one micron accuracy. The schematic arrangement of a hydraulic 4-Hi mill is shown in Fig. 1.13.

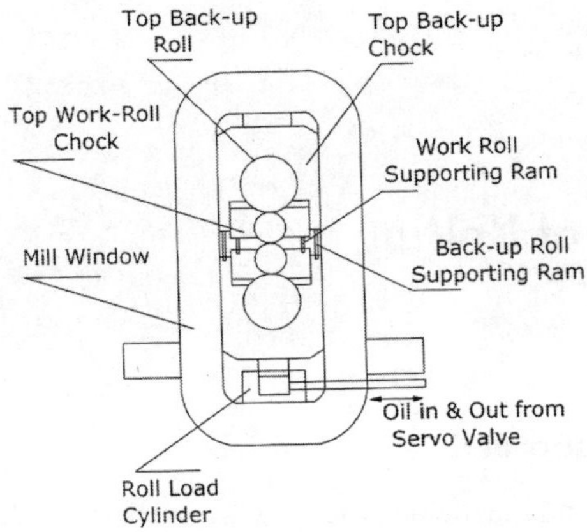

Fig. 1.13 4-Hi hydraulic stand with roll load cylinder

Automatic gauge control system is incorporated in a hydraulic mill which is known as *hydraulic automatic gauge control (HAGC)* system, which has been discussed later in details under section 4.3.6.

References

1. Fanning, F.H., "Wide strip Mills- Evolution or Revolution", Yearbook of the AISI, 1952.
2. "Rolling Mills, Rolls and Roll Making", Mackintosh-Hemphill Company, Pittsburgh, Pennsylvania, 1953.
3. Badlam, S. "The Evolution of the Wide Strip Mill", Yearbook of the AISI, 1927.
4. Whitehead, L.D., "The Continuous Rolling Mill", Proc. Staffs, Iron and Steel Inst., 1924-25.
5. "Continuous Rolling of Sheets at Butler, Pennsylvania", Iron Age, 1927.
6. Dowding, Sturdy and Thomas, "Wide Strip Mills", Davy united Engg. 1960.
7. Malborn. J, "The Development of the Continuous Rolling of Sheet and Tinplate", J.West. Scot. Iron and Steel Inst., 1937-38.
8. Hoben and Mulvey, "The New Continuous Brass Mill of the Scovill Manufacturing Co., Waterbury, Conn., USA." J.Inst. Met., 1950.
9. "100 Years of Progress (1835-1935)", Published by Henry Wiggin and Co. Ltd. 1935.
10. "Rolling Aluminium at British Aluminium Co., Falkirk", Met.Ind. (Lond.), 1947.
11. Ohama, Sasaki and Sendzimir, "World's First Sendzmimir Tandem Mill", Iron and Steel Engineer Year Book, 1973.

CHAPTER 2

Mechanics of Rolling

2.1 Characteristics of Rolling

In order to understand the rolling process, it is essential to understand some of the terms characterizing the rolling process, which are called characteristics of rolling. These are also known as *rolling process parameters*.

While explaining or defining these characteristics, for the sake of simplicity, the following rolling conditions are considered:

(i) The case of simple rolling
(ii) The rolls are cylindrical with plain surface
(iii) The material or stock is rectangular in cross-section

It is to be carefully noted that while discussing the characteristics of rolling or analyzing the mechanics of rolling (i.e., determination of rolling force, etc.) often the stock material will be considered to be rectangular in cross-section meaning the rolls will be in contact with two flat surfaces of the stock, which is known as *flat rolling*. However, in the field of rolling, different types of rolling operation takes place for rolling products of different cross-sections, and those are achieved by passing different sections of stock through various shaped grooves made by the pair of rolls, several times.

Understanding of rolling characteristics as well as analysis of mechanics of rolling becomes easier with rectangular cross-section material. That is why the above three conditions have been considered. However, at appropriate places, grooved rolling will be taken up and discussed.

Figure 2.1 illustrates the rolling process as considered above.

Initial height and width of the stock are 'h_1' and 'w_1', respectively, which enter the roll with a velocity of 'v_1'. The corresponding values of the outgoing rolled stock are 'h_2', 'w_2', and 'v_2'. Let lengths of ingoing and outgoing stock are 'l_1' and 'l_2', respectively.

The various rolling characteristics or terms are explained below, primarily keeping above simplified rolling condition in view. How these characteristics may get modified under different rolling conditions, have also been mentioned.

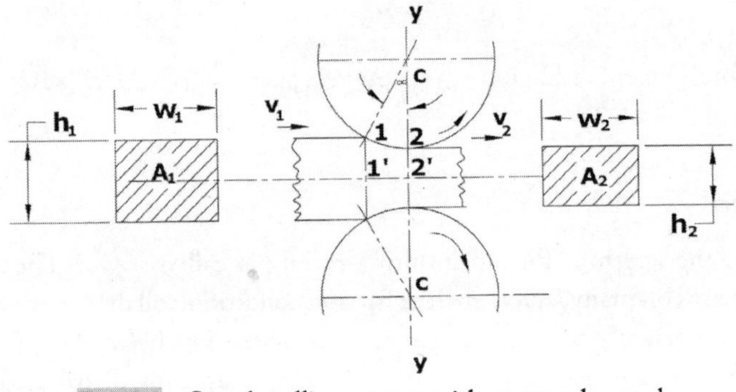

Fig. 2.1 Simple rolling process with rectangular stock

2.1.1 Draught and reduction

Draught means the reduction in height in a particular pass. Absolute draught is the difference of entry thickness and exit thickness, i.e., h_1-h_2. Relative draught is absolute draught expressed as a percentage of initial thickness, i.e., $[(h_1-h_2)/h_1] \times 100$ percent. While the height is reduced during rolling, so does the cross-sectional area of the stock. *Reduction* means the reduction in area from A_1 to A_2, in a particular pass. The percent reduction = $[(A_1-A_2)/A_1] \times 100$. If the stock is not rectangular, as in the case of rolling different shapes (section rolling), draught does not carry any meaning. In such a case reduction becomes the meaningful parameter of rolling.

Draught can be either direct or indirect. Direct draught, as defined above, is the reduction in height in the direction which lies in the plane containing the center lines of the two rolls. If the center lines lie in a vertical plane, the direct draught is the reduction in height in vertical direction. Indirect draught results when the rolls exert pressure on the stock which is not in the plane containing roll center lines. Indirect draught takes place in actual rolling in many instances when the rolling process is not a case of simple rolling or in many cases of section rolling.

2.1.2 Elongation

In rolling, the volume of material entering rolls remains equal to the one that is leaving*. Therefore, reduction in area essentially means elongation of the stock. *Elongation factor*, $E' = l_2 \div l_1$, and *percentage elongation* is defined as

$$[(l_2-l_1) / l_1] \times 100 \text{ percent} = (E'-1) \times 100 \text{ percent.}$$

* This is valid in most rolling processes, excepting in the early cogging passes of cast ingot where some reduction in volume takes place due to compacting effect of cast structure including piping and blowholes.

As $A_1 l_1 = A_2 l_2$ and $l_2/l_1 = A_1/A_2 = E'$

Therefore, percent reduction = $[(A_1-A_2)/A_1]\times100 = [(E'-1)/E']\times100 = [(l_2-l_1)/l_2]\times100$

2.1.3 Spread

Increase in the width of the stock during rolling is called *spread.* The degree of spread depends on many factors such as friction condition, roll diameter, pass shape, and most importantly on stock height to width ratio, i.e., h_1/w_1. When this ratio is high, the spread is not only large, it is also non-uniformly distributed over the height of the stock. Figure 2.2 illustrates this phenomenon. Spread of a rectangular stock is considered negligible when width is more than 12 times of the thickness i.e., $w_1>12h_1$.

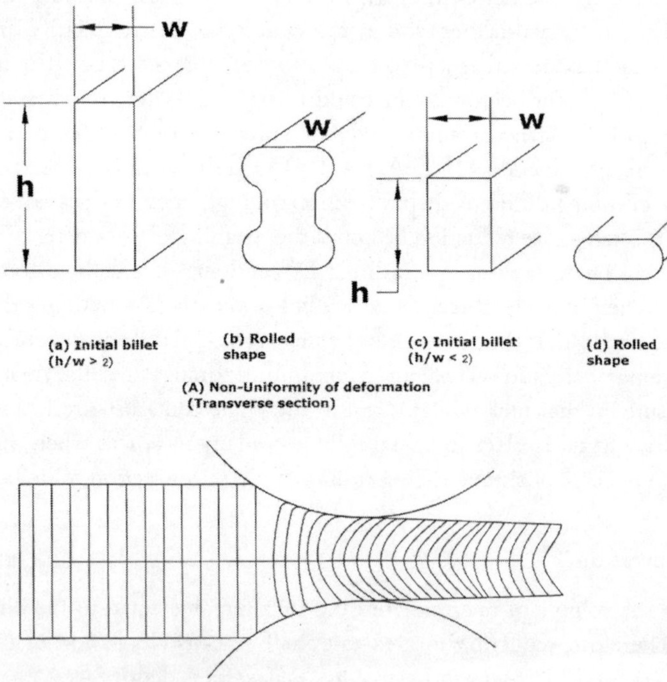

Fig. 2.2 Deformation of billet in rolling

Absolute spread is equal to the change in width = w_2-w_1. Spread is also expressed in terms of *spread factor* = w_2/w_1 and *percentage spread* = $[(w_2-w_1)/w_1]\times100$ percent.

Another expression which is commonly used is the *proportional spread* =

$$(w_2\text{-}w_1) / (h_1\text{-}h_2)$$

$$\text{or, } [(w_2\text{-}w_1) / (h_1\text{-}h_2)] \times 100 \text{ percent.}$$

In the case of rectangular stock passing through plain roll, it has *free* or *unrestricted* spread. But if the form of the pass keeps this spread within certain limits, then it is termed as *restricted* spread. In grooved pass rolling, the width of incoming stock must be kept smaller than that of the pass to avoid overfilling of the groove and excessive wear and side thrust.

2.1.4 Contact angle, angle of bite and area of contact

Referring to Fig. 2.1, it is seen that at the entry of the rolls, the stock contacts the roll at point '1'. If this point is joined with the center 'C' of the roll, then an angle 'α' is subtended by radius 1C and the axis of the rolls 'YY'. This angle is called the *contact angle.* If both rolls of the mill are equal, the contact angle at the bottom roll will also have the same value 'α'. This contact angle will increase if the height of ingoing stock increases or the diameter of the roll is decreased. There is a limiting maximum size of contact angle, beyond which the rolls will not be able to grip or bite the stock and feed it inside the rolls. This limiting angle is called *angle of bite,* which depends on the coefficient of friction between the stock and the rolls. The friction condition is a very important factor in rolling which determines the amount of maximum draught that can be taken in any particular pass.

From Fig. 2.1, it can be seen that deformation due to compression of the stock ends at the plane of exit YY, and the rolled stock leaves the roll at point '2'. The arc 1 to 2 is called the *arc of contact,* and its length is equal to αR where 'α' is in radian and R is the roll radius. The area over which the roll is in contact with the stock is called the *area of contact* or *contact area* that is equal to $\alpha R w_m$ where w_m is the mean width of the stock between points 1 and 2. The projection1′–2′ of arc of contact on the rolling plane (the plane perpendicular to axis YY) is called the *projected arc of contact.* This works out to be equal to

$$\sqrt{R^2 - \left(R - \frac{h_1 - h_2}{2}\right)^2} = \sqrt{R(h_1 - h_2) + \frac{1}{4}(h_1 - h_2)^2}$$

As R is much bigger than $(h_1 - h_2)$, neglecting higher orders of $(h_1 - h_2)$, this can be written approximately as $\sqrt{R(h_1 - h_2)}$.* The projected area of contact

* If R = 300 mm and $(h_1\text{-}h_2)$ = 10 mm, the error of approximation is 0.414%. This is small hence the approximate expression is quite acceptable.

becomes $w_m \times \sqrt{R(h_1 - h_2)}$, which is an important factor for calculating the force required for rolling, called *rolling load* or *roll separating force.*

2.1.5 Rolling friction

Figure 2.3 shows the forces acting on the stock as it contacts the rolls on entry side. The radial force of rolls pushes it away from rolls while the frictional forces try to pull it into the roll gap. Let 'α' be the angle of contact. The stock will enter the roll gap if there is net positive force in the direction of rolling.

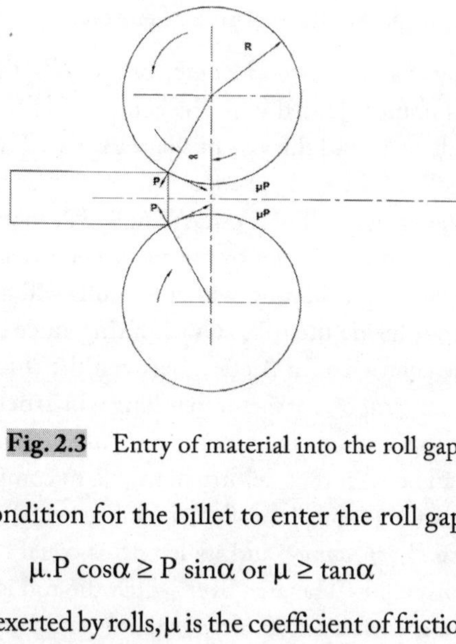

Fig. 2.3 Entry of material into the roll gap

Therefore, the condition for the billet to enter the roll gap is,

$$\mu.P \cos\alpha \geq P \sin\alpha \text{ or } \mu \geq \tan\alpha$$

where, P is the force exerted by rolls, μ is the coefficient of friction and α is the contact angle. In the limiting case for maximum value of α, i.e., α_{max}, the two sides of the above equation may be equated, which gives the following relation between α_{max} and μ.

$$\alpha_{max} = \tan^{-1} \mu$$

The maximum possible draught, $(h_1-h_2)_{max} = \Delta h_{max}$, in the stock is given by,

$$\Delta h_{max} = D\,(1- \cos\alpha_{max})$$

where D denotes the diameter of rolls. If it is desired to increase the draught (Δh_{max}) further we have to either (i) increase the coefficient of friction, or (ii) increase diameter of rolls, or (iii) increase both. In cold rolling the value of coefficient of friction is small. In hot rolling the coefficient of friction (μ) is generally high. For hot rolling of steel μ may be determined from the following empirical relation.

$$\mu = k_1.k_2.k_3.\,(1.05\text{-}0.0005t)$$

where t is the temperature of billet in ^{0}C; k_1, k_2 and k_3 are factors which take into account the following:

k_1 – a factor that takes into account the effect of material of rolls; its value is 1.0 for steel rolls and 0.8 for C.I. rolls.

k_2 – a factor for influence of rolling speed; it varies asymptotically from 1.0 to 0.4 for rolling speed of 2 m/s to 18 m/s and higher. This means coefficient of friction decreases with rolling speed. This is true for cold rolling also.

k_3 – a factor to take care of material being rolled; its value is 1 for mild steel and for alloy steel its value varies between 1–1.3.

From Fig. 2.1 it is seen,

$$\tan\alpha = \frac{\text{Projected arc of contact } 1' - 2'}{R - \dfrac{h_1 - h_2}{2}} = \frac{2\sqrt{R(h_1 - h_2)}}{2R - (h_1 - h_2)} = \frac{\sqrt{2D\Delta h}}{(D - \Delta h)}$$

where $\Delta h = (h_1\text{-}h_2)$ = draught of the pass

2.1.6 Neutral point and forward slip

When a stock is undergoing reduction under roll gap, the cross-sectional area is reduced. Therefore, the velocity of the stock progressively increases as it passes between the rolls. When the rolls are moving with uniform angular velocity, the surface velocity of the roll exceeds the stock speed at the entry section. As the speed of the stock is increasing as it progresses through the roll bite, there will be a point where the two speeds will be the same. This point where stock speed is equal to that of the roll surface is called the *neutral point.* At the neutral point there will be no frictional force, but beyond this point the direction of frictional force reverses and tries to obstruct the movement of the stock forward. This is shown in Fig. 2.4.

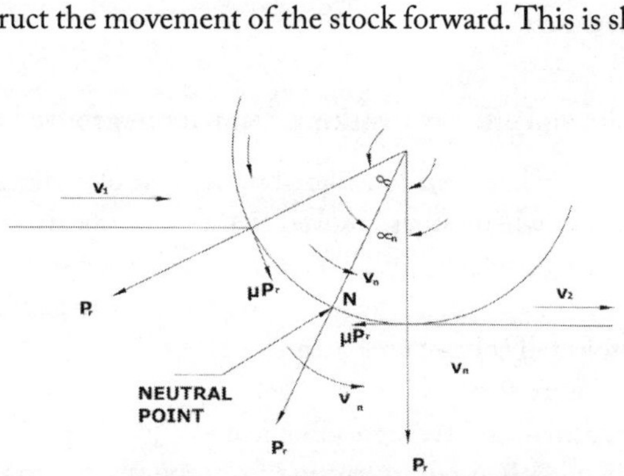

Fig. 2.4 Position of neutral point

As the speed of the stock continues to increase beyond the neutral point, the speed of the stock, as it comes out of the rolls at the exit side, will be more than surface speed of the rolls. This additional speed (v_2-v_n) is called the *forward slip* or the *speed gain* or the *extrusion effect* of rolling where 'v_n' is the surface speed of roll.

The position of neutral point is determined by the amount of forward slip, and is dependent on coefficient of friction and draught of the pass.

Assuming that 'μ' and the radial force 'P_r' remaining constant along the arc of contact, (which has been found to be not, in both theoretical analysis and in practice), and also assuming that a neutral point exists in contrast to a neutral zone, Ekelund calculated the angle 'α_n', that is subtended by the neutral point at the roll center as

$$\alpha_n = \sqrt{\frac{\Delta h}{2D}} - \frac{1}{\mu}\frac{\Delta h}{2D}, \text{ radian}$$

This equation shows that α_n is dependent on μ and Δh. As 'μ' increases, α_n increases while α_n decreases with increase of draught.

These derivations do not apply if external forces interfere with the rolling process as in the following cases.

(i) **Impeded Delivery:** If some force acts to impede the delivery of rolled stock, then this force must be overcome by additional force being set up in the roll gap. This is achieved through shifting of the neutral point towards exit plane and decrease in forward slip.

(ii) **Tension:** When tension is applied either in front or back of the stock, in both cases the rolling load and torque decreases. The *font tension* causes the neutral point to move towards point of entry (i.e., α_n increases) and hence forward slip increases. The *Back tension* causes the neutral point to move towards exit plane and forward slip decreases. In both cases, the tension is doing work on the stock favoring its reduction, hence rolling power decreases.

2.1.7 Forward slip and effective working diameter in grooved rolls

Previously, forward slip has been considered in the case of rectangular stock being rolled in flat rolls, but where other grooved shapes are concerned, the situation is very much changed. Consider an oval bar entering a circular grooved pass as shown in Fig. 2.5. The stock first contacts the roll at point 1 at roll dia. D_1, and as reduction takes place the projected contact area increases in width.

As the stock enters further into the roll gap, the contact area changes to something of the form illustrated, spreading from C′ to C″ so that at the exit plane the stock contacts the roll at points varying from roll dia. D_1 to roll dia. D_2. This is due partly to the draught and also partly to the spread of the bar. It is obvious therefore that as all points on the surface of the bar leave the rolls at the same speed, viz. the stock speed, the slip will differ from one point to another on the profile of the groove, due to the variation of peripheral speed with diameter. At the deepest point of the pass there will be most forward slip while near the roll gap or 'break' (i.e., at roll dia. D_2) the peripheral speed of the roll may exceed that of the stock giving 'backward slip'. At some point between the two, there will be no slip at all at the exit plane, or in effect the neutral point will be in the exit plane and the roll radius at this point on the pass profile may be referred to as the 'effective radius', the 'radius of grip', or the 'ideal radius'. The corresponding diameter is often termed the *effective working diameter* of the pass and it will be affected by the reduction taken in the pass.

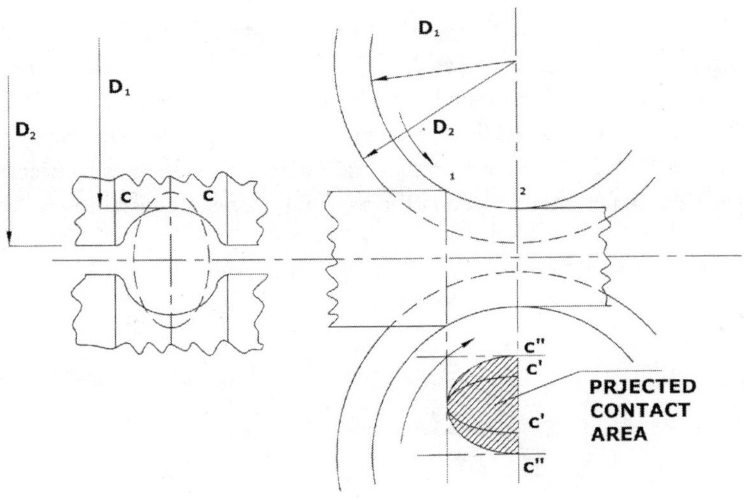

Fig. 2.5 Initial and final contact conditions-oval-round

It will be appreciated that if a bar is to be delivered from the rolls without any bending up or down, the effective working diameters of the two rolls for the particular groove should be equal. For reasons explained later, they are deliberately made slightly different in some cases so as to cause the bar to turn up or down on delivery as required. Unfortunately, the nature of the rolling process introduces some uncertainty into the calculation of the *effective working diameter* so that 'rules of the thumb' have been developed on the basis of experience. Obviously, tests may be carried out for certain standard shapes and the working diameter ascertained as follows: If N is the roll r.p.m., v_2 the measured delivery speed mpm, and D the working diameter required in mm, then

$$\pi DN = v_2 \times 1000$$

$$\text{Or, } D = \frac{1000 v_2}{\pi N} \text{ mm}$$

Furthermore, in a continuous mill the roll designer will wish to set the effective diameter so that the stand speeds may be balanced. This is particularly important where the speed of the stands cannot be adjusted individually, for in that case the roller has to effect the final balancing by adjusting the reductions on the various stands and of course this adjustment is limited, so the diameters must be fairly closely determined if a good quality stock is to result. Two main techniques have been suggested for determining the working diameter of a pass. Considering an oval groove in a roll, as shown in Fig. 2.6, one method stipulates that the required diameter D_1 is given by the horizontal line through the center of area of the groove, i.e., Area A_1= Area A_2. Another method says it is given by the line which bisects the profile of the groove, i.e., the arc length ab + cd = bc. The diameter given by the latter is smaller and has been found suitable in the case of light reductions while the former has been found better for heavy reductions. The effective working diameters may be used for balancing a continuous train with grooved passes by keeping the exit stock area × peripheral speed at the working diameter constant for all stands.

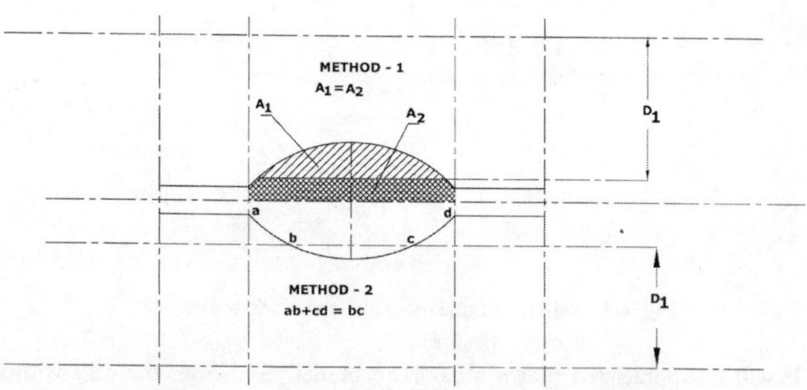

Fig. 2.6 Position of effective roll dia. for an oval groove

2.1.8 Pitch line and rolling line

In many grooved passes the section may be unequally distributed in the top and bottom rolls where the working diameters of two rolls will be different and accurate estimation of them are difficult. In such cases *pitch line* and *rolling line* concepts are often needed. Pitch line is a line parallel to the axis of the two rolls and passes through midway of the two axes. Rolling lines is a line which is also parallel to the pitch line but passes through the center of area of the pass.

The relative position of these two lines controls the direction of delivery of the stock. If the rolling line lies above the pitch line, it means the bottom roll has a larger effective diameter and this situation is termed *underdraught.* The peripheral speed of the bottom roll is greater, averaged over the profile cut on this roll, and as a result the stock will tend to turn upwards on delivery. When rolling line lies below the pitch line, the opposite is true, and it is called *overdraught,* when the stock will try to turn downwards. This phenomenon is shown in Fig. 2.7.

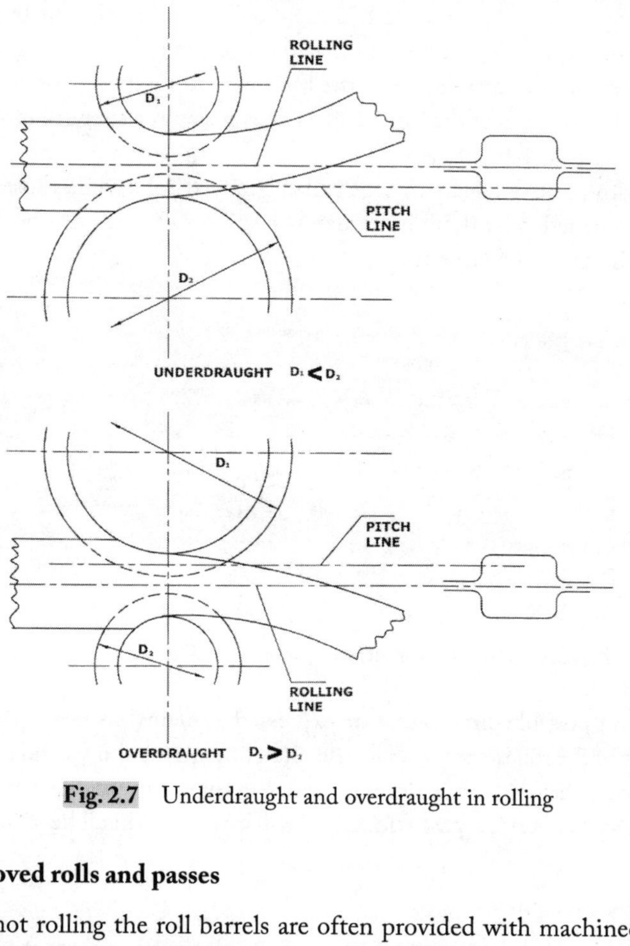

Fig. 2.7 Underdraught and overdraught in rolling

2.1.9 Grooved rolls and passes

In case of hot rolling the roll barrels are often provided with machined grooves. Between two rolls (top and bottom), these grooves create the *'passes'*, also referred as *'holes'*, through which the rolled stock passes by which the shape and dimensions are changed to comply with the shape of the hole. Even during hot rolling of thick rectangular stock, to restrict lateral spread, grooved rolls may be used.

For such a pair of grooved rolls, a roll drawing represents a section through the vertical plane of the roll centers and shows the passes (or holes) separated by *collars*. Each pass is separated from its neighbor by collars and if the collar protrudes beyond the boundary of the pass, it may be called a positive collar or simply a collar. If it does not, it is called a collar hole or negative collar, as illustrated in the hypothetical example for rolling flats by the tongue and groove method in Fig. 2.8. The collars between the outermost passes and the end face of the roll barrel are called end collars and the rest inner collars. The end face of the roll between the roll neck and the periphery of the outer collar is the shoulder of the roll. The breadth of the outer collars is often determined so that the stock may be manipulated into the outer pass without interfering with the housing and room may be required for a guide box. The breadth of the inner collars depends on the strength of the material of the rolls and the type of pass.

Rolls with positive collars may be called 'grooved' rolls while those with collar holes are 'tongued' rolls though frequently section rolls use both positive collars and collar holes on the same roll.

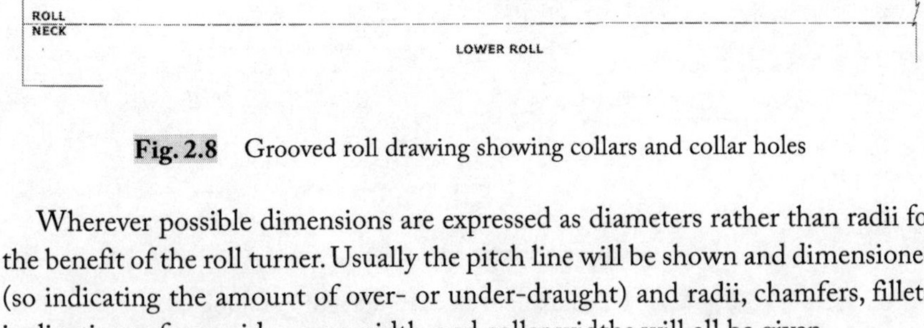

Fig. 2.8 Grooved roll drawing showing collars and collar holes

Wherever possible dimensions are expressed as diameters rather than radii for the benefit of the roll turner. Usually the pitch line will be shown and dimensioned (so indicating the amount of over- or under-draught) and radii, chamfers, fillets, inclinations of pass sides, pass width, and collar widths will all be given.

Open and closed passes

Definitions of open and closed passes differ from one plant to another. If the pass lies partly in one roll and partly in the other, the point where the boundary of the pass changes from one roll to the other is called the 'joint' or 'parting'. If this is formed by lines parallel to the roll axes, the pass may be called 'open', Fig. 2.9(a); but if it is formed by lines approximately perpendicular to the roll axes the pass may be called 'closed' as in a tongue and groove pass, Fig. 2.9(b). When the roll parting is inclined as in Fig. 2.9(c), the definition is not so clear. Tafel refers to the pass as closed if the inclination ∝ is more than 60^0 to the roll axis, and open, if it is less than 60^0.

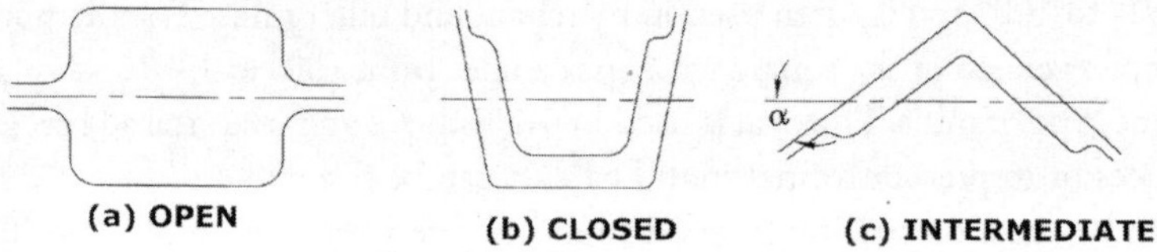

Fig. 2.9 Open and closed passes

End thrust

When hot metal is being reduced between two rolls the nature of the reduction may give rise to forces that attempt to push one roll axially in one direction and the other roll in the opposite direction. This axial force is called the 'end thrust' and results from unbalanced side thrust in the particular grooves forming the pass. Fig. 2.10 shows clearly the reasons for why end thrust is created during rolling of shapes with pass groove side inclined with horizontal and / or vertical axis.

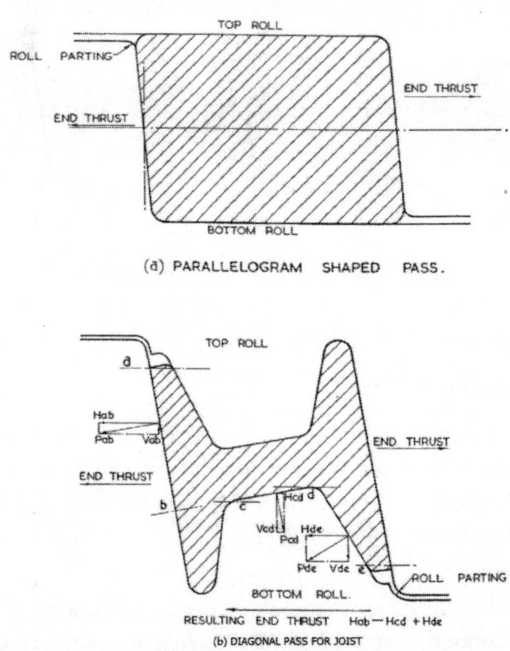

Fig. 2.10 Causes of end thrust

Types of passes

Figure 2.11 shows some types of passes which are frequently used in hot rolling of sections. The open and closed boxes are used frequently in breakdown (roughing) passes. The open square is a square set on the corner and with apex angles varying form 91^0 to 100^0 and is often used in merchant and billet mills. The diamond is a more open version of the square with apex angles from 100^0 to 130^0 and is used in the same type of mills. The oval is used in oval–square and oval–round breakdown sequences or to precede a final round pass. It can be fed into a round or a square and can be transformed into a good round in one pass when fed vertically. The diagonal pass is a less used type for billet rolling. The passes used in section rolling are discussed in greater details in sections 3.2.3–3.2.5.

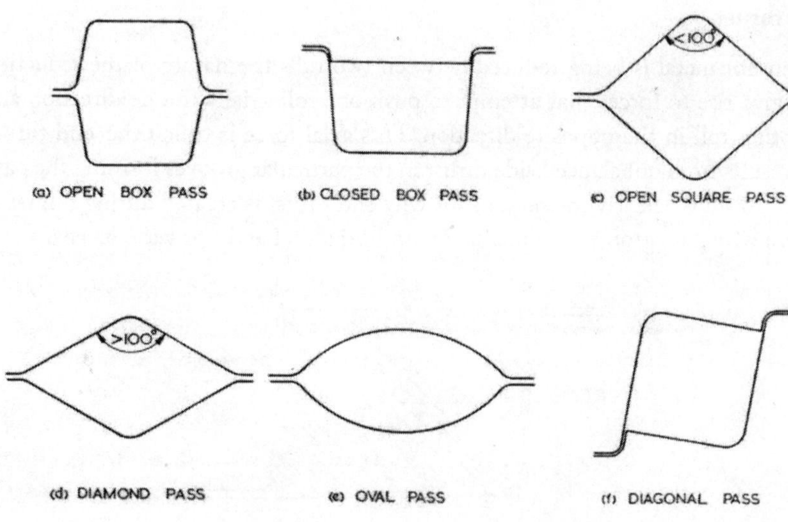

Fig. 2.11 Common types of passes

2.1.10 Spread in section rolling

The flow of material at right angles to the direction of compression and elongation is called *spread,* and it becomes a limitation on the extent of reduction possible in some types of rolling while in others it is beneficial to achieve the required shape, e.g., in section rolling. It is, therefore, necessary to estimate this spread with reasonable accuracy.

In flat rolling with unrestricted spread, a spread formula may be directly used. However, when diamond, square, oval passes, etc., are used the conditions have to be brought to the equivalent flat rolling conditions before the spread can be calculated. This may be done by the *equivalent rectangle* method that reduces the area of the ingoing stock and the pass, between the points of intersection of the two profiles, to rectangles of equivalent area between the same two points. Figure 2.12 shows an oval entering a square pass (the sharp corners have been retained for clarity in this example.) The profile intersects at points on vertical XX and YY and the distance between these is the width w_1 of the equivalent rectangles. The area of the oval between these lines gives a rectangle of height h_1 and the area of the square between these lines gives a rectangle of height h_2. A spread formula may now be used for the rolling of a rectangle of height h_1 and width equal to the actual width of the oval w_0 to one of height h_2 in rolls of diameter D'. The final width w_2 is found by using a spread formula and this is the width to which the square groove will fill. This can be used for other sequence, e.g., diamond-square-diamond, oval–square, oval–round, etc.

Most sections are rolled by controlling or inducing spread by balancing the reductions of various parts of the section, but in some cases spread formulae can be used. In the rolling of angles the growth of the leg under reduction can be calculated by considering the angle as a flat.

There are many formulae available and a few are mentioned below. In general, accuracy is only achieved with complex analysis but it is possible to adopt a simple formula based on experience to give good results.

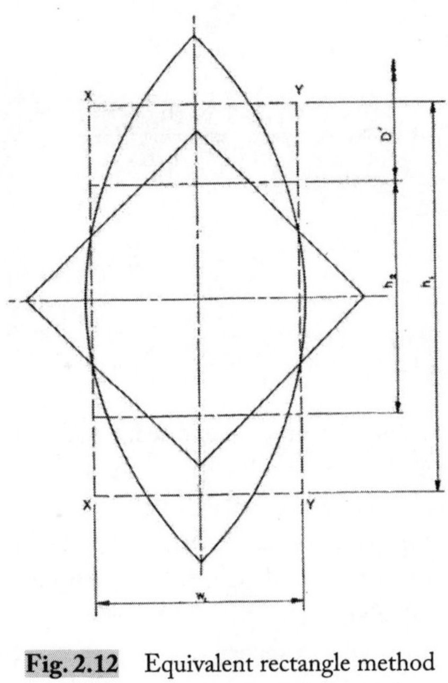

Fig. 2.12 Equivalent rectangle method

Spread in flat rolling is affected by many factors like temperature, roll diameter, stock height to width ratio, reduction, roll material, stock material, and speed of rolling. Many of the formulae do not account for all these factors, and recent work suggests that friction is one of the prime factors. Some of the spread formulae for steel materials are as follows:

(a) Siebel's formula

If $h_1 < w_1$, spread $\Delta_w = CR(h_1 - h_2)\dfrac{(h_1 - h_2)}{h_1}$

where C = 0.35 above 1000 ^{0}C
and C = 0.35–0.4 below 1000 ^{0}C.

With plain carbon steel C is reduced slightly and with alloy steel it is increased slightly and if $w_1 > 7\times\sqrt{R(h_1 - h_2)}$, C must be decreased as follows:

$\dfrac{w_1}{\sqrt{R(h_1 - h_2)}}$	7	8	10	12	14	16	18	20
Multiply C by:	1	.92	.75	.60	.45	.30	.15	0

(b) Sedlaczek formula

$$w_2 - w_1 = \frac{w_1\sqrt{w_1 R}\,(h_1 + h_2)}{3(w_1^2 + h_1 h_2)}$$

or in a modified version

$$w_2 - w_1 = \frac{h_1 - h_2}{6}\sqrt{\frac{R}{h_1}}$$

(c) Ekelund formula[1]

The Ekelund formula is complex and is not directly soluble but it has been stated in the alternative form of nomogram which can be used with ease. This is shown in Fig. 2.13, and the example given there explains the method. It is to be recommended as it accounts for all the major variables except the steel quality. In use, it has been found to slightly overestimate the spread in the rolling of carbon steels and underestimate it for those alloy steels that spread heavily.

The nomogram can be used in reverse by a trial and error method in cases when the required width w_2 is known and the ingoing width w_1 is to be calculated. A

R = Roll radius (INS)
Δh = Draught (INS)
W_0 = Initial Width (INS)
h_0 = Initial Height (INS)
μ = Coefficent of Friction
Δw = Spread (INS)
W_1 = Final width (INS)

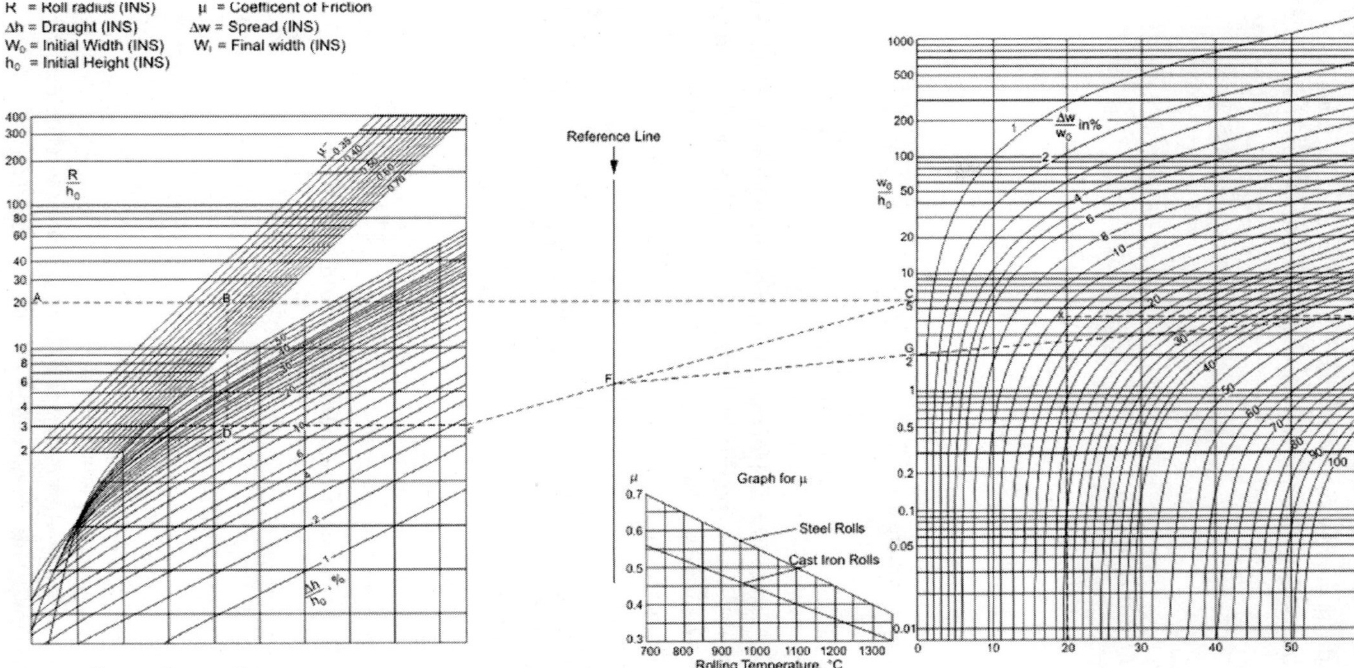

Example = $\frac{R}{h_0} = 20$, $\frac{\Delta h}{h_0} = 20\%$, $\frac{w_0}{h_0} = 2$, $\mu = 0.5$,

From the Diagram (Dotted lines) = $\frac{\Delta w}{w_0} = 13\%$, Therefore $W_1 = 1.13\ W_0$

Fig. 2.13 The Ekelund spread nomogram

value must be assumed for w_1, and w_2 found from the nomogram, and if this does not agree with the required value a corrected value of w_1 must be tried. If this also does not give the required value then interpolation or extrapolation will yield the answer for w_1.

(d) Hill Formula[1]

A recent formula is that of Hill.

$$\frac{w_2}{w_1} = \left(1 - \frac{h_1 - h_2}{h_1}\right)^{-q}$$

where $q = \frac{1}{2}e^{-x}$

and $x = \frac{\lambda w_1}{D(h_1 - h_2)}$ where $\lambda = 0.5$

This gives reliable results though like Ekelund it appears to overestimate the spread on plain carbon steels.

There are several other formulae for calculation of spread during flat rolling. Some of these namely (i) Wusatowski, (ii) McCrum are cited in the books mentioned at serial 16 and 4 respectively under "Suggested Teadings" given at the end of this book. Another simple formula, due to Bhakhtinov, which has been widely used in Russian literature, gives close results obtained by Hill, McCrum or Wusatowski. It is as follows:

(e) Bhakhtinov Formula

$$w_2 - w_1 = 1.15\frac{\Delta h}{2h_1}\left[\sqrt{R\Delta h} - \frac{\Delta h}{2\mu}\right]$$

2.1.11 Elastic deformation of components vis-à-vis set roll gap

Consider rolling of a rectangular stock, as shown in Fig. 2.1. During rolling, the rolling load works on the rolls, which in turn is transmitted to the mill housing. This rolling load (or roll separating force) causes the following deformations:

(a) Elastic deformation of the rolls. The arc of the roll which remains in contact under pressure with the stock tends to flatten. The net effect is that the instant radius of the roll in contact portion increases, and the gap set between the two rolls increases by the amount of total deformation of the two rolls. This phenomenon is called *roll flattening*.

(b) Elastic elongation of the total mill housing under the roll separating force. This also increases the set gap by the total elongation (or stretch) of the mill housing. This is known as *mill stretch*.

(c) The stock that comes out of mill bite has thickness under the compressive stress existing at the exit section. Therefore, as it comes out, the stress is relieved, and the thickness of the stock increases. This is called *elastic recovery* of stock.

Figure 2.14 illustrates that outgoing stock thickness, 'h_2', is equal to the roll gap, 'h_g', set between two rolls, plus the elastic elongation of the mill housing, 'Δf', the elastic deformation of each roll, 'Δ_2', and the elastic recovery of the stock per side, 'Δ_1'. Therefore, the final stock thickness, h_2, can be written as:

$$h_2 = h_g + \Delta f + 2\Delta_2 + 2\Delta_1$$

This relation is particularly applicable during cold rolling of thin flat product. In case of hot rolling, the elastic deformation of rolls and elastic recovery of the stock are small in comparison to the reduction (h_1-h_2), and hence may be neglected. So, for hot rolling, above relation becomes

$$h_2 = h_g + \Delta f$$

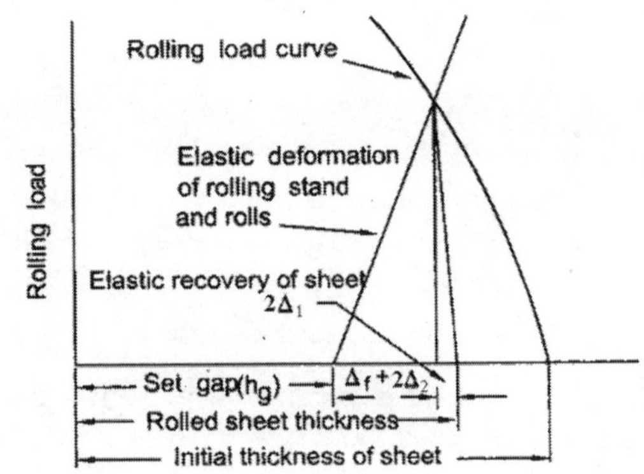

Fig. 2.14 Effect of elastic deformation on final stock thickness

2.1.12 Roll flattening

In cold rolling of stock (primarily strip or sheets) of steel and other hard metals, as a result of high specific pressure required for plastic deformation of the metal, the roll in contact with the stock gets elastically deformed. This is called *roll flattening.* The effect of roll flattening is that the set roll gap gets increased by twice the roll deformation of each roll, thus the draught is reduced to this extent. In order to achieve the set reduction, the roll gap has to be reduced sufficiently such that the final roll gap becomes equal to the desired gap after compensating for the roll deformation. This essentially means that due to roll flattening at the deformed position, the effective radius of the roll is increased, and the arc of contact also increases. Both these effects tend to increase rolling force.

The deformed roll radius can be estimated from Hitchcock's formula[2], which is given below:

$$R' = R + \frac{16(1-\gamma^2)PR}{w_m \pi E(h_1 - h_2)}$$

where, R= original roll radius; R′ = deformed roll radius; P = total rolling load; w_m = mean width of stock; E = Young's modulus for roll material; γ = Poisson ratio for roll material. For steel roll the values of E and γ are, $E \cong 215\times10^9$ N /M^2 and γ = 0.33

Figure 2.15 shows the roll flattening effect during rolling. The elastic deformation of each roll is denoted by Δ_2, while Δ_1 denotes the elastic deformation of rolled metal in each side.

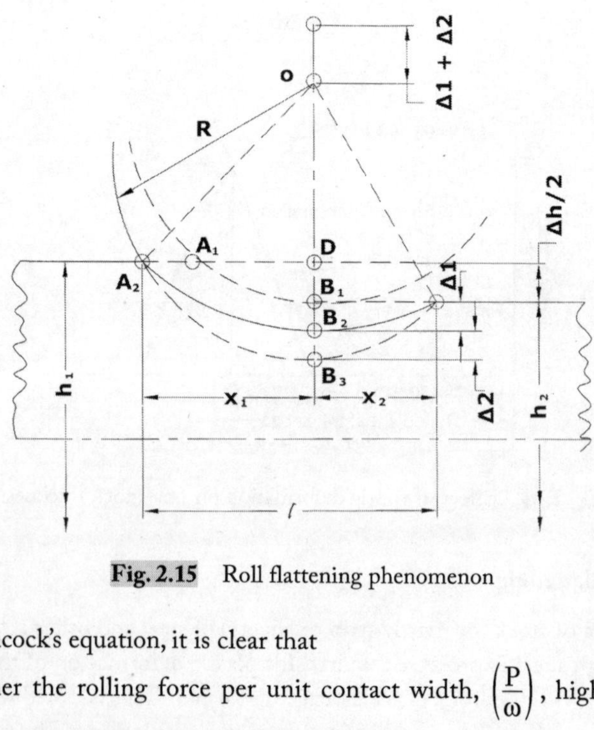

Fig. 2.15 Roll flattening phenomenon

From Hitchcock's equation, it is clear that

(a) higher the rolling force per unit contact width, $\left(\frac{P}{\omega}\right)$, higher is roll flattening effect;

(b) higher the strength of roll material, less will be the roll flattening effect, and less the draught, (h_1-h_2), of the pass, higher will be the roll flattening effect.

In cold rolling, rolling force per unit contact width being high, the work roll material is so chosen (generally hardened chrome alloy forged steel) that its strength is high. In case of very small sized work rolls (as in a Sendzimir cold strip mill), even tungsten carbide rolls are also used. While rolling thin strip, naturally draught is small, and effect of roll flattening becomes large and should be taken into account.

The estimation of increase of contact length under elastic deformation of rolls as well as stock has been further discussed under section 2.3.2.

2.1.13 Dark mode rolling

A peculiar situation is faced when rolling very thin strip which is termed foil (approximately 0.1 mm and below thickness). In such a case, with the foil in mill bite, if roll gap is tried to be reduced by operating the screw down, the deflection of the two rolls becomes such that the foil thickness is accommodated in this deflection and the barrels of the two rolls lying beyond the foil width touch each other. Under such a situation, any further attempt to reduce the roll gap sharply increases the rolling force but practically no reduction in the thickness takes place. The thickness of the output foil can be controlled by changing the rolling speed and input and output tension in the foil. This condition in foil rolling where the two rolls touch each other so that no light can pass through the edges of the rolls is called the *dark mode rolling*. Figure 2.16 illustrates the phenomenon of dark mode rolling.

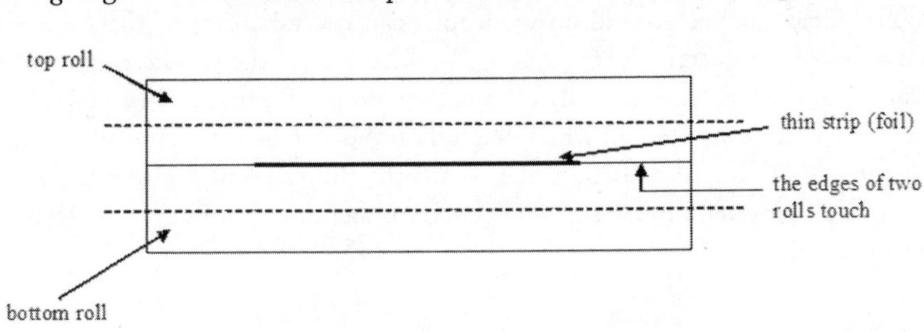

Fig. 2.16 Dark mode rolling

2.1.14 Roll bending and roll camber

During rolling of materials, the rolls in contact with stock experience roll separating force as a result of which the rolls supported at its journals at two ends bend away from the stock. This phenomenon is called *roll bending*. The effect of roll bending is particularly noticed during the rolling of flat material. The middle portion of the rolled material becomes thicker than the two sides, i.e., the shape of the rolled material becomes convex, and a *crown* is produced on the rolled material. Crown is the difference of thicknesses at the middle and edge of the rolled flat stock.

Roll bending under a given load may be minimized by increasing the roll diameter. But when the roll diameter is increased, the rolling load increases for the given reduction, and the roll bending also increases. Therefore, reduction of roll bending by increasing the roll diameter is rather limited in scope. In a 2-Hi mill, this bending can be calculated. In 4-Hi mills, this roll bending effect is much less. Smaller work rolls used results in less roll separating force, and the work rolls are backed up by much larger diameter rolls whose bending is quite small.

During rolling, particularly in hot rolling, a temperature differential gets developed between the edge and the middle of rolls. Temperature at center of the roll is more than that at the roll edges and experimental observations have shown that the distribution of temperature is of parabolic nature. Such a distribution of temperature adds to the camber of roll surface, which is called the *thermal camber*.

Roll bending and thermal camber have opposite effect on cross-section of rolled material and try to cancel each other.

To offset the resultant effect of thermal and roll bending effect, the rolls are generally given a barrel shape called positive *camber*. Since sheets of various widths may be rolled on the same pair of rolls and with different reductions, no one roll camber may be designed to suit all conditions of rolling. Therefore, the camber may be designed on the basis of the load expected in final pass, which delivers the finished flat products.

The fixed camber ground on work rolls can not take care of the different variations of parameters during rolling. Besides, one of the serious problems is wear of rolls, which continuously change the initial roll camber. Therefore, it is necessary to have a system in which the roll camber can be varied during rolling. Also it is desirable to distribute the wear over the roll length. The different methods employed for inducing variable roll camber during rolling are described in section 4.3.6.

2.1.15 Pass schedule

In order to produce a rolled product, the input material has to be passed several times between two work rolls of a rolling stand or successively through pair of work rolls of several rolling stands. Each such passage of the rolling stock through one set of work rolls, in which the stock is reduced in its cross-section, is called a *rolling pass* or simply a *pass*. If the stock passes through a mill stand without any change of cross-sectional area, then it is not a pass.

The reasons why a number of passes are required to obtain a desired product from a given input stock are:

(i) Maximum amount of reduction is limited by roll bite. In case of flat rolling, the incoming stock will not be taken inside the roll bite, if the draught is beyond a certain limit. Similarly in case of rolling in roll grooves, an input section beyond a certain size will not enter the pass groove.

(ii) Pass reduction is also kept restricted from the point of view of roll-separating force of the pass. Every mill stand is designed for a safe maximum roll-separating force, primarily based on stress developed in roll neck. Therefore, the maximum reduction of any pass is limited to a value such that the roll-separating force of that pass does not exceed the design limit.

(iii) In case of hot rolling of stock using grooved passes, the optimum reduction is often dictated by spread of the material. The input stock of a certain shape and size may enter the pass groove, but the material may not spread sufficiently to fill the entire roll groove. This is very important consideration of rolling sections like angle, channel, beams, and many other miscellaneous sections.

(iv) In case of cold rolling, the maximum pass reduction is also kept within different limits for different metals and alloys to avoid too much of crystal deformation giving rise to micro-cracks and other internal defects in the rolled material.

In case of hot rolling, pass schedule for producing a product from a given input stock in a rolling mill means determining the optimum number of passes with desired percentage reduction in each of these passes, and also the shape and size of each of these pass grooves (if groove rolling is involved) such that the mill productivity will be high, the finished product will be of acceptable quality (size and shape), the rolling load in each pass will remain within limit and the temperature of the stock, through the rolling, remains within a specified range for the specific material composition. In groove pass rolling, the shapes of the series of grooves often follow different pattern in sequence. Because of this sequence of the groove shapes, it is also known as *pass sequencing*.

Cold rolling is done to produce flat products like sheet, strip and foil. In case of cold rolling, reduction of the material thickness (draught) in each pass increases the strength of the material due to work hardening of the material. Therefore, with each cold pass it is important to calculate the increased strength of the material and accordingly design the reduction of the next pass. After a few cold passes, the strength of the material reaches such a high level that the percentage reduction at that stage has to be kept too small to be profitable. At such a stage, the cold rolled stock is annealed so that the material becomes soft and the percentage of reduction in next pass can again be increased. These annealing operations in the middle of a rolling schedule is called *inter anneal.* As the strength or hardness of cold rolled material depends on total cold working on the material from its annealed condition, it is customary to calculate not only the reduction of any pass but also the cumulative reduction on stock from its corresponding annealed thickness.

During cold rolling, the maximum reduction limit for a specific material is also kept into consideration. Typical examples of pass schedule for rolling different hot and cold rolled products have been discussed in further details in section 3.2.

2.2 Stresses in Rolling

2.2.1 Distribution of stress and strain across rolled strip thickness

All theoretical and experimental investigations show that stresses and strains across the thickness of rolled strip is not uniform. The distribution of such stress and strain arises out of different conditions, namely lateral spread, relative motion between the roll surface and the strip in contact, and sticking of the strip with roll.

In order to avoid effect of lateral spread, the width of the strip is considered to be large compared to the length of arc of contact and thickness of the strip, and it is assumed the rolling is done between smooth rolls. In such a condition, the stress and strain distribution may be considered as a two-dimensional problem.

The solution to this problem may be considered in two parts depending on the ratio of mean thickness 'h_m' to arc of contact length 'l'. The nature of the distribution of stress and strain varies considerably when the ratio $h_m : l$ is either <2–1 or >2.

When the ratio $h_m : l > 2$, the compressive strain does not penetrate the total thickness of the metal being rolled. Two cases may be considered separately:

(i) The case when h_m:l < 2–1. In this case the thickness being not very large in comparison to the length of arc of contact, it is assumed that the compressive strain extends over the entire thickness of the metal.

Referring to Fig. 2.17, if any point is considered on the surface of the roll subtending an angle ϕ with the line joining center of two rolls, where ϕ is within the limits $\phi = \gamma$ to α (where α is the contact angle and γ is the neutral point angle), the peripheral velocity of this point is greater than the mean velocity of motion of the strip, i.e.,

$$v_r \cos\phi > v_{xm}$$

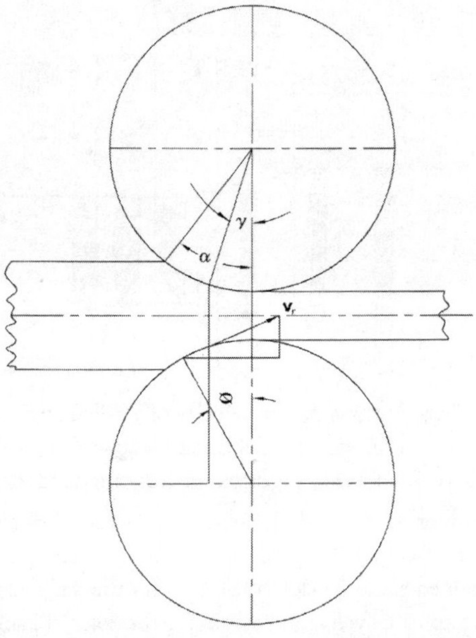

Fig. 2.17 Horizontal projection of the peripheral roll velocity

Consequently, the friction forces arising between the roll surface and strip, impart higher velocity to the zones of the strip between sections $\phi = \gamma$ to α, than to the middle of the strip. Because of this contact friction forces, a non-uniformity in the stress distribution arises across the strip thickness with the consequence of non-uniformity in strain and velocity distribution.

In the zone of forward slip, i.e., when ϕ is between 0 to γ, the peripheral velocity of the rolls is smaller than the mean velocity of the metal, i.e.,

$$v_r \cos\phi < v_{xm}$$

Figure 2.18 exhibits the velocities of rolled strip and velocity distribution at different strip thickness at different zones.

(ii) The second case is when h_m:l > 2. As this ratio increases, the effect of deformation at the outer zone of the strip becomes more active, and the degree of non-uniformity increases in the distribution of stresses across the thickness of the strip. Observations and investigations have shown that at higher ratios, the compressive stress does not penetrate the entire cross-section of the strip, but is localized in the zones adjacent to contact areas.

Figure 2.19 shows a typical velocity distribution of the metal being rolled under the condition (ii).

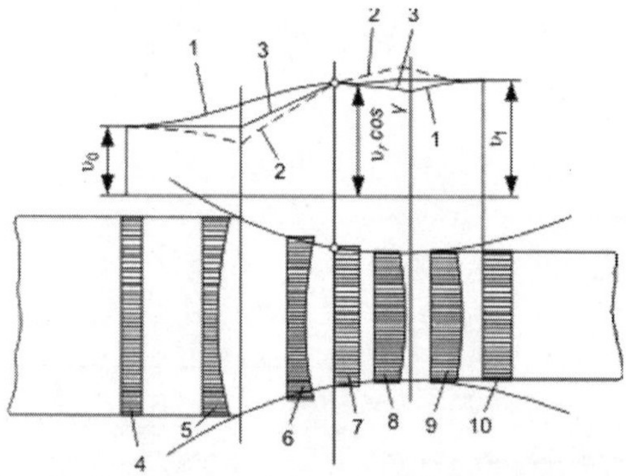

Fig. 2.18 Above: diagram showing the velocity of a rolled strip at different points in its cross-section as it passes through the deformation zone. Below: diagram showing the velocity distribution for different cross-sectional depths

1- the velocity of the outer portions of the strip cross-section; 2-the velocity of the middle portions of the strip cross-section; 3- the mean velocity of the strip cross-section; 4- velocity diagram for the unstrained zone; 5-velocity diagram for the deformation zone at the entry, away from the contact zone; 6- velocity diagram for the zone of backward slip; 7- velocity diagram for the neutral zone; 8-velocity diagram for the zone of forward slip; 9-velocity diagram for the deformation zone at the exit, away from the contact zone; 10-velocity diagram for the zone at the exit.

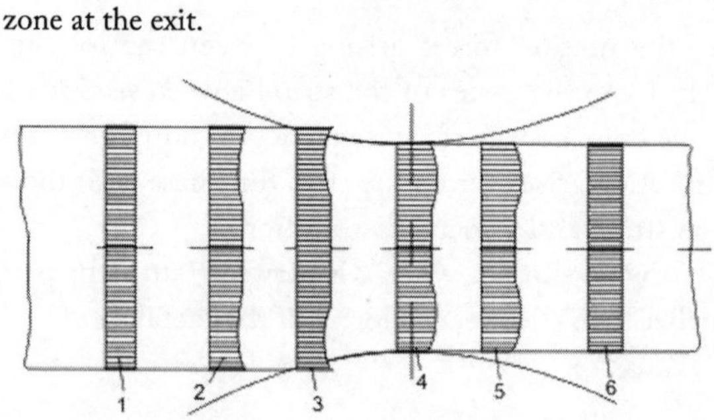

Fig. 2.19 The variation of velocity with depth in the rolled strip shown at various points along its length, with hm:l > 2 and D cos α > hn where D = roll diameter and hn is strip thickness at neutral zone

1- in the unstrained zone; 2- in the deformation zone at the entry, away from the contact zone; 3- in the zone of backward slip; 4- in the zone of forward slip; 5- in the zone of deformation at the exit, away from the contact zone; and 6- the steady zone at exit.

2.2.2 Contact stresses in two-dimensional deformation

Assuming flat stock rolling, it is possible to frame a differential equation, with suitable assumption and approximation, to derive an analytical solution for the stress distribution along the roll-strip contact surface under different rolling conditions. Such an analysis gives a theoretical basis for explaining the effects of different rolling parameters and conditions like (i) roll diameter, (ii) coefficient of friction, (iii) percentage of reduction, (iv) strip tension etc., on contact stress distribution and, therefore, on total rolling load.

For such an analysis it is assumed that the mean thickness of strip is less than the length of arc of contact, i.e., h_m:l < 1. It is further assumed that rolls are cylindrical and width of strip is many times more than arc length. With these assumptions it is considered to be a two-dimensional stress–strain problem.

Let us consider an element of strip **abcd**, as shown in Fig. 2.20, at a distance x from exit bounded by the cylindrical surface of the two rolls and by two planes perpendicular to the direction of rolling, having a very small gap '**dx**' between them. Consider the conditions of equilibrium of this element of unit width. It is further assumed that the element has no acceleration in the direction of rolling, i.e., resultant of forces acting on the element in the direction of rolling is zero.

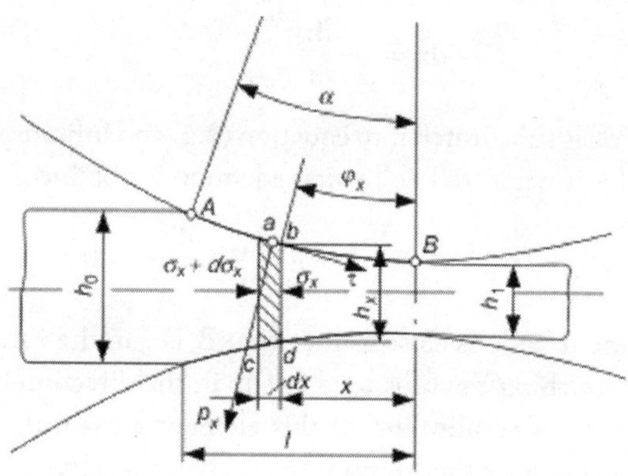

Fig. 2.20 Elementary forces acting on the rolled metal in the zone of backward slip

On the right-hand side of the strip, the element of unit width is acted on by the force:

$$\sigma_x h_x \quad \dots (2.1)$$

where σ_x is the mean normal compressive stress arising in the plane 'bd' of the strip being rolled, and 'h_x' is the height of the cross-section.

Suppose that in the plane 'ac' the mean normal stress is $\sigma_x + d\sigma_x$, and the height of the cross-section is $h_x + dh_x$. Then on the left-hand side the element is acted on by the force

$$(\sigma_x + d\sigma_x)(h_x + dh_x) \quad \dots (2.2)$$

In the first instance consider the equilibrium of the element when it is closer to the point A than the point B, i.e., this element is located in the zone of backward slip, and its particles, touching the rolls, tend to slip along them in the direction opposite to the rotation of the rolls. The horizontal projection of the force acting on the element form the direction of the rolls can be expressed as follows:

$$2\left(p_x \frac{dx}{\cos \phi_x} \sin \phi_x - \tau_x \frac{dx}{\cos \phi_x} \cos \phi_x\right) \quad \ldots (2.3)$$

where,

p_x is the specific roll pressure on the metal being rolled,

φ_x is the angle between the tangent to arc ab and the horizontal plane,

τ_x is the shear stress acting on the contact surface which is caused by the friction forces between the strip and the surface of the rolls.

The sum of the horizontal projections of all the forces acting on the element is

$$\Sigma X = (\sigma_x + d\sigma_x)(h_x + dh_x) - \sigma_x h_x - 2p_x \tan\phi_x dx + 2\tau_x dx = 0 \quad \ldots (2.4)$$

The relation between dx and tan φ_x can be expressed as

$$dx = \frac{dh_x}{2 \tan \phi_x} \quad \ldots (2.5)$$

If this value of dx is substituted into equation (2.4) and infinitely small quantities of the second order is neglected, following equation is obtained.

$$d\sigma_x - (p_x - \sigma_x)\frac{dh_x}{h_x} + \frac{\tau_x}{\tan \phi_x}\frac{dh_x}{h_x} = 0 \quad \ldots (2.6a)$$

When the element 'abdc' is close to the point B, i.e., in the zone of forward slip, then the surfaces touching the rolls tend to slip in the direction of rotation of the rolls. The conditions of equilibrium of this element remain the same as for the zone of backward slip, only the friction forces change direction. The equation of equilibrium for the zone of forward slip then can be expressed as follows:

$$d\sigma_x - (p_x - \sigma_x)\frac{dh_x}{h_x} - \frac{\tau_x}{\tan \phi_x}\frac{dh_x}{h_x} = 0 \quad \ldots (2.6b)$$

For solving equations (2.6a) and (2.6b), it is necessary to find the relationship between the specific pressure, $p_{x,}$ and the stress, σ_x. For this purpose the equation of plasticity for a two-dimensional deformation is used:

$$\left(\frac{\sigma_x - \sigma_y}{2}\right)^2 + \tau^2{}_{xy} = k^2$$

where 'k' is the shear yield stress. For a brief discussion on yield criteria and equations of plastic deformation, see the matter in the box at the end of this section.

Taking the vertical and horizontal stresses σ_1 and σ_3 as the principal stresses, we can write

$$\sigma_1 - \sigma_3 = 2k \qquad \dots (2.7)$$

where,

$$\sigma_1 = \left(p_x \frac{dx}{\cos \phi_x} \cos \phi_x \pm \tau \frac{dx}{\cos \phi_x} \sin \phi_x \right) \frac{1}{dx}$$

The second term of the right-hand side of this equation can be neglected being small in comparison with the first term. Then

$$\sigma_1 = p_x \text{ and } \sigma_3 = \sigma_x$$

From this, according to equation (2.7),

$$p_x - \sigma_x = 2k \qquad \dots (2.8)$$

Substituting this value of $p_x - \sigma_x$ into equations (2.6a) and (2.6b) the fundamental differential equation for the specific pressure is obtained:

$$d(p_x - 2k) = \left(2k \mp \frac{\tau_x}{\tan \phi_x} \right) \frac{dh_x}{h_x} \qquad \dots (2.9)$$

The minus sign in front of the second term on the right-hand side of this equation corresponds to the zone of backward slip, and the plus sign corresponds to the zone of forward slip.

The constants obtained as a result of integrating equation (2.9) are determined from the initial conditions. For this the specific pressure at the points A and B are obtained. Consider the more general case, when the strip being rolled is subjected to tension both at the entry to the rolls and at the exit from the rolls, which is often practiced in cold rolling, and also when rolling is carried out in continuous mills. The tensile stresses arising in the element as a result of its tensioning are denoted by σ_A at the entry to the rolls, i.e., when x = l, and by σ_B at the exit from the rolls, when x = 0. Then, in accordance with equation (2.8), the specific pressures at A and B are:

$$\left.\begin{aligned} P_A &= 2k - \sigma_A = \xi_0 2k \\ P_B &= 2k - \sigma_B = \xi_1 2k \end{aligned}\right\} \qquad \dots (2.10)$$

When tension is absent the contact pressures at the points A and B are equal to the quantity 2k.

The subsequent solution of equation (2.9) may be carried out in two ways: (1) considering the value of k varies along the arc of contact, and (2) the value of this quantity is constant. A variation of the quantity k along the contact surface during the rolling is possible owing to the strain hardening of the metal, the different strain rates at the beginning and the end of the arc of contact, or due to the variation of the temperature of the rolled metal in the direction towards the point B.

Different specialists have obtained different solutions to equation (2.9) by assuming

(i) k varying from A to B by a linear law;
(ii) considering variations in k due to strain hardening and variations of deformation rate.

V.S. Smirnov,[3] a Russian scientist and rolling mill specialist, analyzed the solution to equation (2.9) by assuming that the k varies due to strain hardening according to the equation:

$$p_x - \sigma_x \approx v2k\left(\frac{h_0}{h_x}\right)^n$$

where v and n are coefficient depending on the hardening characteristics of the metal.

Having made a detailed analysis of the results of his solution, V. Smirnov arrived at the conclusion that if instead of the variable value of k, the average value is taken, i.e.,

$$k = \frac{k_A + k_B}{2} \qquad \text{... (2.11)}$$

where k_A and k_B are the values of k at the beginning and end of the arc of contact, then the accuracy of the solution of the equation is not much altered.

Taking this into consideration, equation (2.9) can be solved with a constant value of k along the arc of contact as given by equation (2.11). Then in accordance with equation (2.8),

$$d\sigma_x = dp_x$$

The differential equation of specific pressure (equation 2.9) can be written as:

$$dp_x = \left(2k \mp \frac{\tau_x}{\tan \phi_x}\right)\frac{dh_x}{h_x} \qquad \text{... (2.12)}$$

The general solution of this equation in exact form is difficult owing to the inadequate definition of the relationship between p_x and τ_x.

We can now proceed to the next section (2.2.3) where some characteristic cases are analyzed based on the assumed nature of friction between the stock and the surface of the rolls.

Plastic Deformation and Yield Criteria

Plastic deformation in a body takes place when the applied stress exceeds the yield stress for the material. In case of uniaxial stress, say for tension, the tensile yield stress 'σ_y' can be determined experimentally. However, in most of the forming processes, the forces do not act uniaxially, and multiaxial stresses are set up on the body under deformation. It is necessary to consider the criterion for yielding to take place under such conditions. Different researchers have proposed different criteria, but only two of the most used ones are discussed here.

Tresca's maximum shear stress criterion

Plastic deformation in a material takes place through slip (dislocation) which is essentially a shearing process. Based on this fact, Tresca proposed in 1865 that plastic flow (i.e., deformation) will start when the induced shear stress will reach the *shear yield stress 'K'* for the material. If the principal stresses at any point in the material are σ_1, σ_2 and σ_3 ($\sigma_1 > \sigma_2 > \sigma_3$), then the maximum shear stress τ_{max} developed at that point is given by $\tau_{max} = ½ (\sigma_1 - \sigma_3)$. Now the plastic deformation will start if $\tau_{max} = K$. So, Tresca's criterion for elastic deformation becomes,

$$½ (\sigma_1 - \sigma_3) = K \qquad \ldots 2(i)$$

It is clear from the equation above that the yielding does not depend on the intermediate principal stress σ_2. In case of uniaxial loading of the material, say in case of tensile stress only, the ultimate shear stress K becomes,

$$k = \frac{1}{2}\sigma_y \qquad \ldots 2(ii)$$

Von Mises' maximum distortion energy criterion

Von Mises suggested in 1913 that plastic deformation takes place when the total shear strain energy in the material under applied stresses reaches a critical value. The total shear strain energy per unit volume '$\in$', can be expressed in terms of the three principal stresses as,

$$\in = \frac{1}{6G}[(\sigma_1 - \sigma_2)^2 + (\sigma_2 - \sigma_3)^2 + (\sigma_3 - \sigma_1)^2] \qquad \ldots 2(iii)$$

where G is the shear modulus of the material. As per this criterion, the plastic flow initiates when the $\in$ reaches a particular value, say S. Thus the Von Mises' criterion takes the form,

$$(\sigma_1 - \sigma_2)^2 + (\sigma_2 - \sigma_3)^2 + (\sigma_3 - \sigma_1)^2 = 6GS = C \text{ (constant.)} \qquad \ldots 2(iv)$$

According to this criterion, all the three principal stresses contribute to plastic deformation.

In order to apply these two criteria, it is necessary to ascertain the values of right hand sides of equations 2(i) and 2(iv), for a material. The tensile yield stress σ_y of different materials have been found out quite accurately by tensile tests.

By applying Tresca's yield criterion to pure tensile loading, we have already found out in equation 2(ii), The value of K in terms of σ_y as:

$$\mathbf{k} = \sigma_Y/2$$

Under pure tensile load yielding, $\sigma_1 = \sigma_y, \sigma_2 = \sigma_3 = 0$. The Von Mises' criterion [equation 2(iv)] can be written as

$$C = (\sigma_Y - 0)^2 + 0 + (0 - \sigma_Y)^2 = 2\sigma_Y^2 \quad \ldots 2(v)$$

If we consider yielding under pure torsion, the state of stresses in a material for a two dimensional situation is shown by the help of the following Mohr's circle diagram. It is clearly seen from the diagram that σ_1=k, σ_3=-k and $\sigma_2 = 0$. If we put these values in eqn. 2(iv), we get,

$$C = (K - 0)^2 + (0 - K)^2 + (-K - K)^2 = 6K^2 \quad \ldots 2(vi)$$

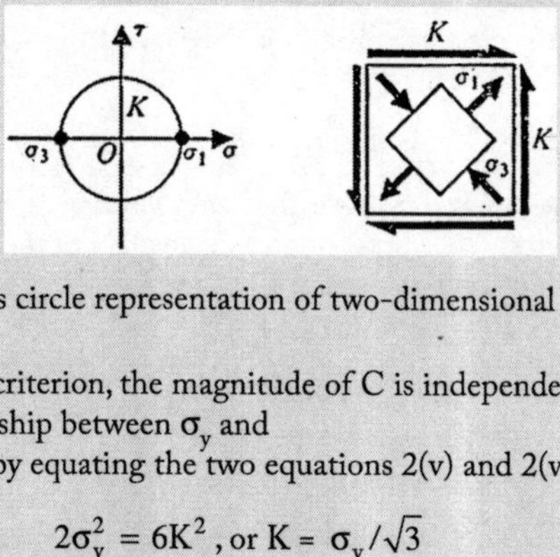

Fig. Mohr's circle representation of two-dimensional state of stress

In the Von Mises' criterion, the magnitude of C is independent of mode of loading. Therefore the relationship between σ_y and
k can be obtained by equating the two equations 2(v) and 2(vi), which becomes,

$$2\sigma_y^2 = 6K^2\text{, or } K = \sigma_y/\sqrt{3} \quad \ldots 2(vii)$$

From experiments, Von Mises' criterion has been found to be more realistic, and thus the value of K= $\sigma_y/\sqrt{3}$ is normally used.

2.2.3 Specific pressure in the case of slip with a constant coefficient of friction (dry friction)

In the case when the metal being rolled slips along the surface of the rolls with dry friction the contact shear stress τ_x can be expressed as follows:

$$\tau_x = \mu p_x$$

After substituting this value of τ_x into equation (2.12) we obtain what is known as the equation of Von Karman:

$$dp_x = \left(2k \mp \frac{\mu p_x}{\tan \phi_x}\right)\frac{dh_x}{h_x} \quad \ldots (2.13)$$

The variable quantities h_x and φ_x appearing in this equation are determined by the coordinates of the arc of contact (Fig. 2.21). The equation of a circle which represents the cross-section of the roll is

$$x^2 + (y - \bar{y})^2 = r^2$$

where y and $\bar{y}$ are the distance along the vertical, from the axis of the rolled strip to any point on the surface of the roll, and from the axis of rolled strip to the center of roll, respectively (Fig. 2.21), and r is the radius of the roll.

From this we find

$$x = \sqrt{r^2 - (y - \bar{y})^2}$$

$$dx = \frac{(y - \bar{y})dy}{\sqrt{r^2 - (y - \bar{y})^2}}$$

$$\tan \phi_x = \frac{dy}{dx} = \frac{\sqrt{r^2 - (y - \bar{y})^2}}{y - \bar{y}}$$

After substituting this value of tan φ_x into equation (2.13), and noting that $\frac{h_x}{2} = y$, we obtain

$$dp_x = \frac{2k\,dy}{y} \mp \frac{\mu p_x (y - \bar{y})dy}{y\sqrt{r^2 - (y - \bar{y})^2}} \qquad \text{... (2.14)}$$

Solution of this equation is cumbersome and not convenient for practical use.

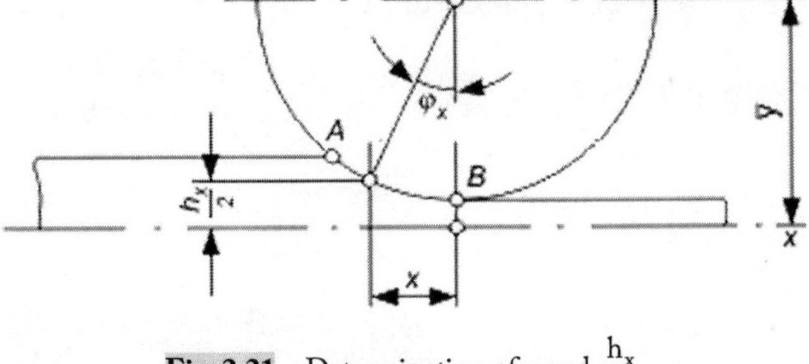

Fig. 2.21 Determination of x and $\frac{h_x}{2}$

Since the angle of contact of the metal in most cases does not exceed 20^0 to 30^0 for hot rolling and 4^0 to 8^0 for cold rolling, the arc of contact can be equated to a curve for which the solution of equation (2.13) is simplified.

This method of solution is quite justified particularly in cold rolling, where due to the high specific pressure the rolls are subjected to a considerable strain, with the result that in practice the arc of contact does not remain an arc of a circle.

In the section below, the two simplest solutions of this equation will be considered: in the first case equate the arc of contact to a parabola; in the second, equate it to a chord.

For all practical purposes, both of these solutions may be considered as sufficiently accurate.

In this first case, the arc of contact is equated to a parabola whose vertex is located at point B and whose side passes through point A (see Fig. 2.21).

The equation of this parabola has the form:

$$h_x = ax^2+b$$

When the curve passes through points A and B, the constant coefficient a and b will be:

$$a = \frac{\Delta h}{l^2} \text{ and } b = h_1$$

where l is the horizontal projection of the arc of contact, and Δh is the reduction, in one side of the strip.

It is further found that

$$dh_x = 2axdx = \frac{2\Delta h}{l^2}xdx$$

$$\tan \varphi_x = \frac{dh_x}{2dx} = ax$$

Then substitute these values of h_x, dh_x, and $\tan \varphi_x$ into equation (2.13):

$$dp_x = \left(2akx \mp \mu p_x\right)\frac{2\,dx}{ax^2+b} \qquad \text{... (2.15)}$$

Introduce the new variable u:

$$x = \sqrt{\frac{b}{a}}\tan u$$

Then,

$$u = \tan^{-1} x\sqrt{\frac{a}{b}}\text{, and } du = \sqrt{\frac{a}{b}}\frac{dx}{1+\frac{a}{b}x^2}$$

from which one can write,

$$\frac{dx}{ax^2 + b} = \frac{du}{\sqrt{ab}}$$

After substituting the value of u into equation (2.15), one obtains

$$dp_x - 4k \tan u \, du \pm \frac{2\mu p_x}{\sqrt{ab}} du = 0 \qquad \text{... (2.16)}$$

Instead of p_x, introduce two new variables:

$$p_x = \nu t$$

Then,

$$dp_x = t \, d\nu + \nu \, dt$$

Substituting these values of p_x and dp_x into equation (2.16), and also denoting

$$\frac{2\mu}{\sqrt{ab}} = m$$

one obtains:

$$t \, d\nu + \nu \, d\nu - 4k \tan u \, du \pm m\nu t \, du = 0 \qquad \text{... (2.17)}$$

It is assumed that the variable quantity t just introduced satisfies the condition

$$dt + mt \, du = 0 \qquad \text{... (2.18)}$$

from which one can write,

$$\log t = -mu$$

or

$$t = e^{-mu} \qquad \text{... (2.19)}$$

Substituting the expression (2.18) and (2.19) into equation (2.17) one obtains

$$e^{-mu} d\nu = 4k \tan u \, du$$

After integration one gets

$$\nu = 4k \int e^{mu} \tan u \, du + C$$

or

$$p_x = e^{-mu}[4k \int e^{mu} \tan u \, du + C] \qquad \text{... (2.20)}$$

The integral in this equation can be calculated only approximately, as a result of expanding one of the integrands in a series.

For low and medium reductions, when Δh does not exceed h_1 and $u < 1$, one finds the convergent series for tan u:

$$\tan u = u + \frac{u^3}{3} + \frac{2u^5}{3 \times 5} + \ldots$$

Substituting this value of tan u into equation (2.20), p_x can be calculated by means of term by term integration with any accuracy, depending on the number of terms of the series used.

In the given case one can confine to a calculation of tan u for small reductions, when the convergence of the series obtained is so rapid that one can limit oneself to its first term. Then equation (2.20) for the zone of backward slip assumes the form

$$p_x = C_0 e^{-mu} - \frac{4k}{m^2}(1 - mu) \qquad \ldots (2.21)$$

while for the zone of forward slip this expression is

$$p_x = C_1 e^{mu} - \frac{4k}{m^2}(1 + mu) \qquad \ldots (2.22)$$

The constant quantities C_0 and C_1 in these equations can be determined from the initial conditions:

At the point A, where $x = l$,

$$p_x = 2k\text{-}\sigma_A$$

$$u = u_0 = \tan^{-1}\sqrt{\frac{\Delta h}{h_1}}$$

at the point B, where x=0,

$$p_x = 2k\text{-}\sigma_A$$

$$u = 0$$

Substituting these values of p_x and u into equations (2.21) and (2.22) one gets:

$$C_0 = 2k\left[\xi_0 + \frac{2}{m^2}(1 - mu_0)\right]^{emu_0}$$

$$C_1 = 2k\left(\xi_1 + \frac{2}{m^2}\right),$$

where $\xi_0 = 1 - \frac{\sigma_A}{2k}$ and $\xi_1 = 1 - \frac{\sigma_B}{2k}$

After substituting these values of C_0 and C_1 into equations (2.21) and (2.22), the final expression for the specific pressure in the zone of backward slip becomes

$$p_x = 2k\left\{\left[\xi_0 + \frac{2}{m^2}(1 - mu_0)\right]e^{m(u_0-u)} - \frac{2}{m^2}(1 - mu)\right\} \qquad \ldots (2.23)$$

whilst in the zone of forward slip:

$$p_x = 2k\left[\left(\xi_1 + \frac{2}{m^2}\right)e^{mu} - \frac{2}{m^2}(1 + mu)\right] \qquad \ldots (2.24)$$

where
$$m = \frac{2\mu}{\sqrt{ab}} = \frac{2\mu}{\sqrt{\Delta h/_{l^2}\, h_1}} = \frac{2\mu\, l}{\sqrt{h_1 \Delta h}}$$

$$u_0 = \tan^{-1}\sqrt{\frac{\Delta h}{h_1}}$$

and
$$u = \tan^{-1}\left(\sqrt{\frac{\Delta h}{h_1}}\, \frac{x}{l}\right)$$

According to these equations the minimum specific pressure is observed at the point A and B; this pressure increases towards the neutral section as indicated in Fig. 2.22. The maximum specific pressure occurs close to the neutral section where both the pressure curves expressed by equations (2.23) and (2.24) intersect.

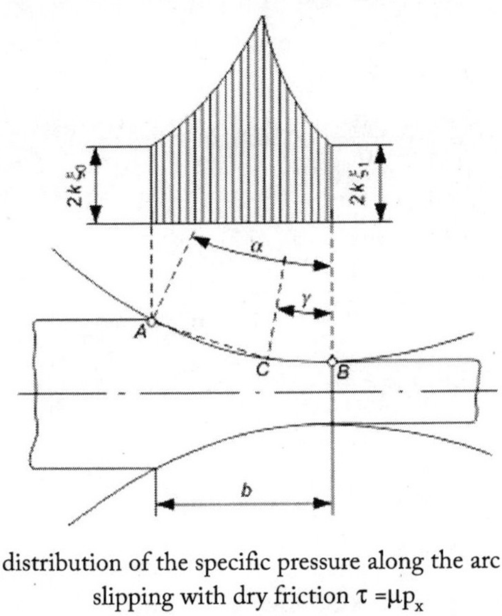

Fig. 2.22 The distribution of the specific pressure along the arc of contact during slipping with dry friction $\tau = \mu p_x$

Considering the second case, if the arc of contact is equated to a chord passing through the points A and B (Fig. 2.22); or to two chords (AC for the zone of backward slip, and BC for the zone of forward slip, the solution of equation (2.13) is considerably simplified. The final results obtained in this case, as proposed by *A. Tselikov,*[3] are simpler and hence more suitable for practical calculations. The discrepancies arising from this assumption, as comparative calculations have shown, are quite insignificant. In particular, for medium and high reductions, and also for angles of contact which are less than the angle of friction, this method of solution gives considerably more accurate results than calculations based on equations (2.23) and (2.24).

With this assumption, if following substitutes are made:

$$\tan\phi_x = \tan\frac{\alpha+\gamma}{2}\text{, for the segment AC, and}$$

$$\tan\phi_x = \tan\frac{\gamma}{2}\text{, for the segment BC}$$

in equation (2.13), one obtains

$$dp_x = \left(2k - \delta_0 p_x\right)\frac{dh_x}{h_x} \quad \text{... (2.25)}$$

for the zone of backward slip, and

$$dp_x = \left(2k + \delta_1 p_x\right)\frac{dh_x}{h_x} \quad \text{... (2.26)}$$

for the zone of forward slip, where

$$\delta_0 = \frac{\mu}{\tan\frac{\alpha+\gamma}{2}} \text{ and } \delta_1 = \frac{\mu}{\tan\frac{\gamma}{2}} \quad \text{... (2.27)}$$

After integration one obtains

$$p_x = C_0 h_x^{-\delta_0} + \frac{2k}{\delta_0} \quad \text{... (2.28)}$$

for the zone of backward slip, and

$$p_x = C_0 h_x^{\delta_1} - \frac{2k}{\delta_1} \quad \text{... (2.29)}$$

for the zone of forward slip.

The initial conditions are:

$$p_x = \xi_0 2k\text{, when } h_x = h_0\text{, and}$$

$$p_x = \xi_1 2k\text{, when } h_x = h_1$$

From this one finds C_0 and C_1:

$$C_0 = 2k\left(\xi_0 - \frac{1}{\delta_0}\right)h_0^{\delta_0} \qquad \text{... (2.30a)}$$

$$C_1 = 2k\left(\xi_1 + \frac{1}{\delta_1}\right)h_1^{-\delta_1} \qquad \text{... (2.30b)}$$

Substituting these values of C_0 and C_1 into equations (2.28) and (2.29) the final expressions for determining the specific pressure are obtained:

$$p_x = \frac{2k}{\delta_0}[(\xi_0\delta_0 - 1)\left(\frac{h_0}{h_x}\right)^{\delta_0} + 1] \qquad \text{... (2.31)}$$

for the zone of backward slip, and

$$p_x = \frac{2k}{\delta_1}\left[(\xi_1\delta_1 + 1)\left(\frac{h_x}{h_1}\right)^{\delta_1} - 1\right] \qquad \text{... (2.32)}$$

for the zone of forward slip,

where $$\delta_0 = \frac{\mu}{\tan\frac{\alpha+\gamma}{2}} \text{ and } \delta_1 = \frac{\mu}{\tan\frac{\gamma}{2}} \qquad \text{... (2.33)}$$

If the metal is rolled without tension and no external forces, with the exception of that of the rolls act on the strip, then the values of σ_A and σ_B, where $h_x = h_0$ and $h_x = h_1$, are zero. Then equations (2.31) and (2.32) assume the form:

$$p_x = \frac{2k}{\delta_0}\left[(\delta_0 - 1)\left(\frac{h_0}{h_x}\right)^{\delta_0} + 1\right] \qquad \text{... (2.34)}$$

for the zone of backward slip, and

$$p_x = \frac{2k}{\delta_1}\left[(\delta_1 + 1)\left(\frac{h_x}{h_1}\right)^{\delta_1} - 1\right] \qquad \text{... (2.35)}$$

for the zone of forward slip.

It follows from the analysis of the equations describing the distribution of the specific pressure along the arc of contact that this pressure depends on a number of factors: the coefficient of external friction, the thickness of the strip being rolled, pass reduction, the diameter of the rolls and, finally, the tension of the strip at the entry to the rolls and at the exit from them. Figures 2.23, 2.24, 2.25 and 2.26 illustrate the nature of the influence of these factors on the specific pressure. These figures have been plotted based on the data calculated from equations (2.31) and (2.32), (2.34) and (2.35).

Figure 2.23 shows the theoretical distribution curve of the specific pressure along the arc of contact (under its horizontal projection l) in the case of rolling of broad strips with the same reduction of 30% but with different coefficients of friction (0.075, 0.1, 0.15, 0.2, 0.3 and 0.4). As is seen from this curve, at higher values of frictional coefficient, the contact pressure increases rapidly near the neutral section, and consequently rolling force increases.

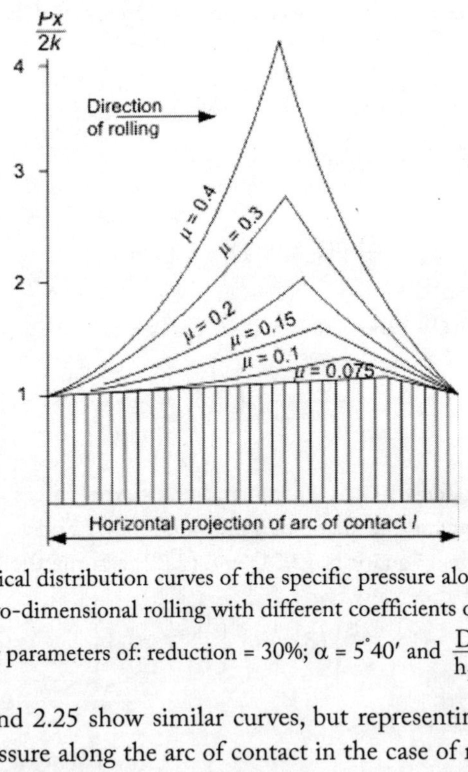

Fig. 2.23 Theoretical distribution curves of the specific pressure along the arc of contact (with $\tau = \mu p_x$) for two-dimensional rolling with different coefficients of friction with other rolling parameters of: reduction = 30%; $\alpha = 5^{\circ}40'$ and $\frac{D}{h_1} = 86$.

Figures 2.24 and 2.25 show similar curves, but representing the distribution of the specific pressure along the arc of contact in the case of rolling a wide strip with different reductions and with rolls of different diameter (i.e., with different D/h_1 ratios) under otherwise identical conditions. If these two curves are carefully analyzed, one can easily conclude that the mean specific pressure increases with the length of arc of contact. The arc of contact length increases under the following conditions:

(i) when reduction increases, all other conditions remaining the same;
(ii) when the roll diameter increases, other conditions remaining the same;
(iii) when the thickness of the strip diminishes, other conditions remaining the same. In this case when the output thickness, h_1, reduces with same pass reduction, basically the ratio D/h_1, increases. This has the same effect as that of increasing the roll diameter.

The rolling force increases not only as a result of increase in the area of contact between the stock and the rolls, but also as a result of an increased specific pressure.

These theoretical equations for the distribution of contact stress (i.e., specific pressure) along the arc of contact also explain the effect of back and front tension of the strip under rolling. Figure 2.26 illustrates the theoretical contact pressure distribution curves with different tensile stresses of magnitude 0, 0.2(2k) and 0.5 (2k). Formulas (2.31) and (2.32) have been used for calculating the data for plotting of the two curves. Figure 2.26a shows the effect of having only front tension in the strip, while Fig. 2.26b depicts effect of both front and back tension acting together on the strip under rolling.

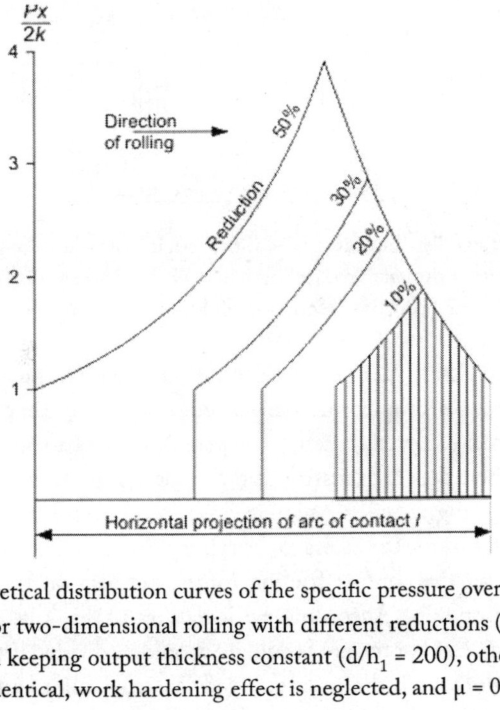

Fig. 2.24 Theoretical distribution curves of the specific pressure over the arc of contact (with $\tau = \mu p_x$) for two-dimensional rolling with different reductions ($\Delta h/h0 = 0.1, 0.2, 0.3$ and 0.5) and keeping output thickness constant ($d/h_1 = 200$), other conditions are identical, work hardening effect is neglected, and $\mu = 0.2$

By comparing the two curves it is noticed that when the rolled strip is subjected to tension, the average specific pressure is reduced and this reduction increases with increase in strip tension. It is interesting to note that the reduction in contact stresses at the beginning of contact (point of roll bite) and at the end of contact (point of release from roll bite) is exactly same as the magnitude of stresses caused by back tension and front tension respectively.

In reality (particularly, when hard alloys are rolled) the reduction in the pressure of the metal on the rolls due to tension will be considerably more than that shown in Fig. 2.26, owing to the reduction in the local elastic compression of the rolls (contact compression) due to tension, and consequent shortening of the actual length of the arc of contact.

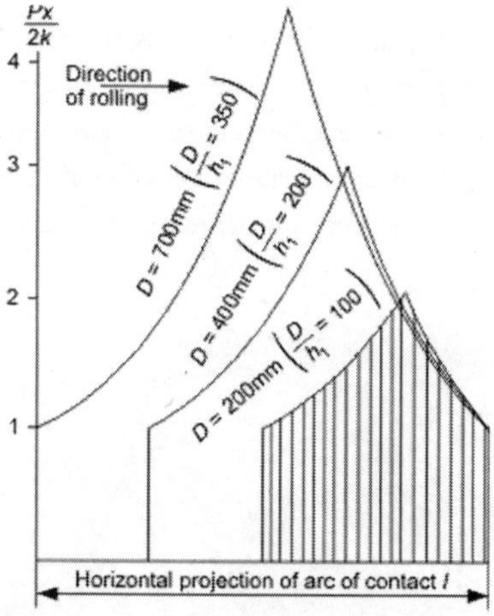

Fig. 2.25 Theoretical distribution curves of the specific pressure along the arc of contact (with $\tau = \mu p_x$) for two-dimensional rolling with rolls of different diameters:$D/h_1 = 100$, 200, 350; reduction of 30% and $\mu=0.3$

Equations (2.31) and (2.32), in the derivation of which the arc of contact was equated to a chord, are simpler and hence more convenient than equations (2.23) and (2.24). The error incurred thereby is very small and has no practical importance, except in cases where specific pressure and the specific friction forces are analyzed for various contact angles whose value is close to the angle of friction or exceeds it. Under these conditions the stress, $\sigma_{x,}$ varying along the rolled strip will strongly depend not only on the contact friction forces, but also on the variation of the horizontal projection of p_x. Accordingly calculations of this kind should be carried out using equation (2.23) or the arc of contact should be divided into several sections (say 4–6), on each of which the arc is equated to a small chord and the calculation is carried out using equations (2.31) and (2.32), with different values of δ for each section. The curves of the distribution of the specific pressure along the arc of contact thus calculated—when the angles of contact are greater than the angle of friction—have an interesting feature over the segment where

$$\tan \phi_x > \mu$$

the coefficient δ is negative and accordingly the curve of the specific pressure on the section from point A in the direction of rolling are characterized not by a rise, as usual, but by a fall (Fig. 2.27).

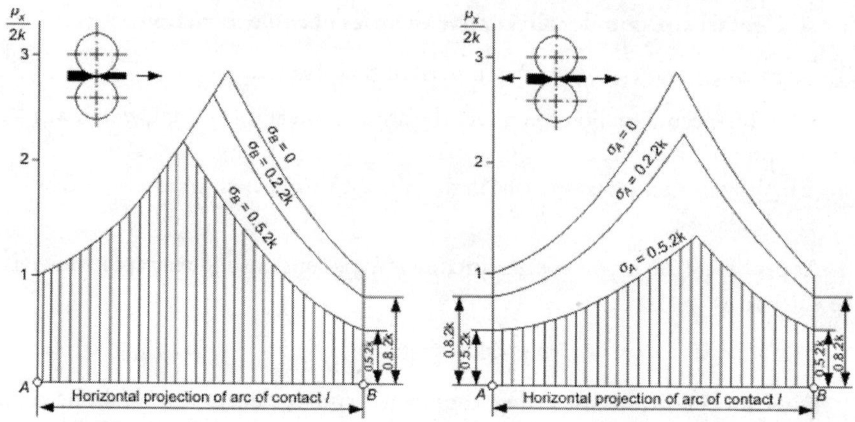

(a) tension at front end only

(b) tension at both front and back ends.

Fig. 2.26 Theoretical distribution curves of the specific pressure along the arc of contact (with $\tau = \mu p_x$) in the case of two-dimensional rolling with different tension $\sigma = 0$; 0.2 (2k) and 0.5 (2k) and a reduction of 30% $\alpha = 30^0 5'$; $\mu = 0.2$; $D/h_1 = 200$ and $\delta = 6$

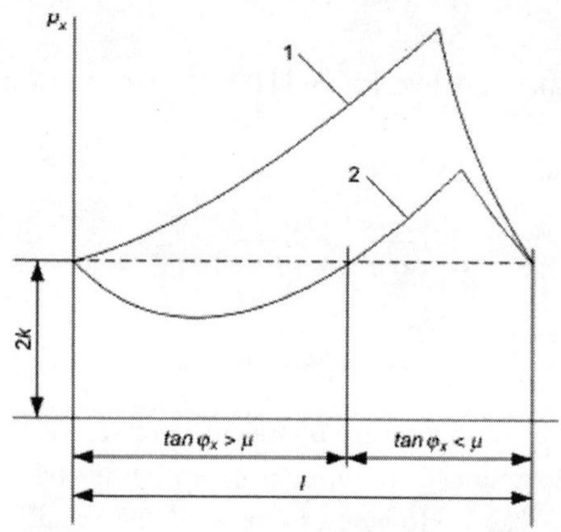

Fig. 2.27 Theoretical distribution curves of the specific pressure along the arc of contact (with $\tau = \mu p_x$) in the case of two-dimensional rolling with different arcs of contact but with the same horizontal projection of the arc of contact:

Curve 1: $\tan\phi_x < \mu$ along the entire arc of contact;

Curve 2: $\tan \phi_x > \mu$ at the beginning of the arc of contact

From this it can be concluded that if the angle of contact is greater than the angle of friction or close to it, the specific pressure is less than in the case where for the same extent of the arc of contact the rolling takes place for angles ϕ_x which are less than the angle of friction.

2.2.4 Contact stress under alternative theories of contact friction

Among many theories, two of the important theories are:

(i) The specific friction force over the arc of contact is constant, which was proposed by Siebel.

(ii) Viscous friction takes place in the contact zone, which was suggested by Nadai[4].

As per Sibel's theory, the specific friction force is constant and may approximately be equated as,

$$\tau = \mu 2k \text{ (constant)}$$

For this condition, assuming the arc of contact becomes parabolic, i.e., $h_x = ax^2 + b$, the solution of the basic equation (2.12) is greatly simplified. The specific pressure p_x becomes,

$$p_x = 2k\left(\xi_0 - \log_e \frac{z_0^2 + 1}{z^2 + 1}\right) + 2\tau\sqrt{\frac{r}{h_1}} \tan^{-1} \frac{z_0 - z}{1 + z_0 z} \quad \text{... (2.36)}$$

for the zone of backward slip, and

$$p_x = 2k\left[\xi_1 + \log_e\left(z^2 + 1\right)\right] + 2\tau\sqrt{r/h_1} \tan^{-1} z \quad \text{... (2.37)}$$

for the zone of forward slip, where

$$z = \sqrt{a/b}\, x = \sqrt{\Delta h/h_1}\, \frac{x}{l}$$

and

$$z_0 = \sqrt{\Delta h/h_1}$$

As per Siebel's theory, the nature of specific pressure curve along the contact length remains approximately the same as given by dry friction theory ($\tau = \mu p_x$) considered in section 2.2.3. However, the rise of the specific pressure curves are less steep and maximum value of specific pressure at the neutral zone will also be less. This is because the friction force is assumed to be less in this theory than that considered by the earlier theory.

As per the Nadai's viscous friction theory, according to Newton's law of viscosity, the specific friction force can be expressed as,

$$\tau = \eta \frac{dv}{dy},$$

where

η = coefficient of viscosity

$\frac{dv}{dy}$ = velocity gradient in the direction perpendicular to the plane of slip.

With the viscous friction condition and assumption of a parabolic arc of contact, it can be shown that the specific pressure distribution for both backward and forward slip zone can be expressed by a single equation of the form:

$$p_x = 2k\left[\xi_1 + \log_e(1+z^2) + \frac{A\tau_0}{2}\left(\frac{z}{1+z^2} - B\tan^{-1}z\right)\right] \quad \text{... (2.38)}$$

where, $A = 2l/\sqrt{h_1}$

$$B = \frac{Ak}{A\tau_0 \tan^{-1} z_0}\left[\xi_1 - \xi_0 + \log_e(1+z_0^2)\right] + \frac{z_0}{(1+z_0^2)\tan^{-1} z_0}$$

$$z = \frac{x}{l}\sqrt{\Delta h / h_1}$$

In this case, the neutral section does not coincide with the point of maximum specific pressure. Also the shape of specific pressure curve is different from the other two cases. Instead of having a pointed peak, the curve is domed shape.

Figure 2.28 shows the distribution of specific pressures and specific friction forces along the arc of contact according to the theories of Von Karman, Siebel and Nadai.

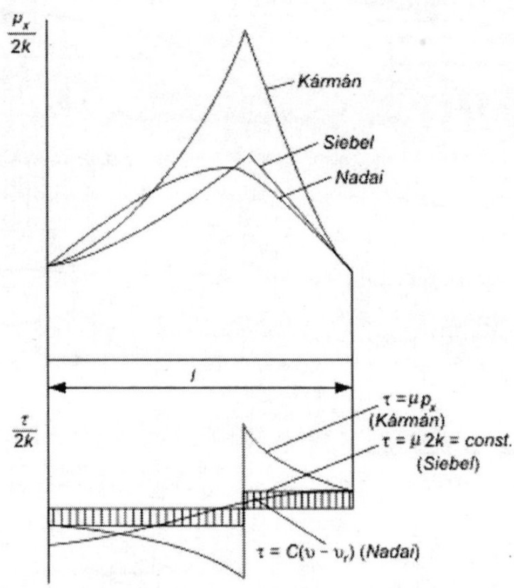

Fig. 2.28 The distribution of specific pressure and specific friction forces along the arc of contact during two-dimensional rolling according to different theories: dry friction (Von Karman), constant friction force (Siebel), and viscous friction (Nadai)

Another interesting theory based on zone of sticking was put forward in 1933 by S. Sobolevsky[3]. As per his theory, between the backward and forward slip zone, there is a zone where there is no slip between the metal and rolls, called zone of sticking. In this zone, the condition of deformation and hence the distribution of specific friction forces are different from those in zones of slipping.

If rolling of wide strips are divided into four categories of rolling based on ratio of length of arc of contact to the mean strip depth (categories being $l/h_m \geq 5$, = 2 to 5, =0.5 to 2 and <0.5), then the distribution of specific pressure and specific friction forces along arc of contact for the above four categories of strip rolling will be as shown in Fig. 2.29.

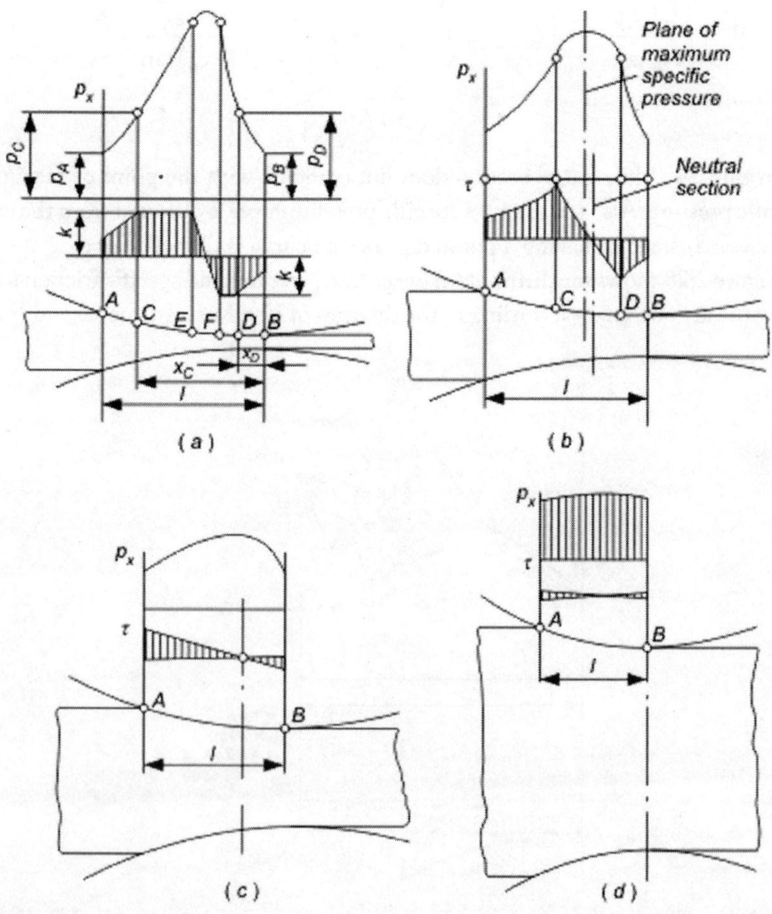

Fig. 2.29 Variation of the specific pressure and specific friction forces along the arc of contact for two-dimensional rolling with different l: h_m ratios: (a) l:h_m > 5; (b) l:h_m (2–5; (c) l:hm (0.5–2; (d) l:hm <0.5

It is interesting to note that various experiments conducted by different researchers for measurement of actual pressure between metal and the roll indicate that nature of distribution of rolling pressure follow closely to those indicated in Fig. 2.29.

2.3 Rolling Force

The force exerted by the metal being rolled on the rolls, trying to separate the two rolls away from each other, is called roll separating force or rolling force in short.

The design of a rolling stand, particularly the roll diameter (work rolls for a 2-Hi mill or the back-up rolls for a 4-Hi mill), the roll neck size, selection of roll neck bearings, screw down design as well as the cross-section of roll stand housings depends on the maximum rolling load a roll stand will be subjected to. Therefore, it is very important to determine the roll separating force for rolling an input stock to the desired size of the output.

2.3.1 Direction of forces in simple rolling

Consider a *simple rolling process*, which means no external force other than the forces from the rolls act on the metal; both rolls are of uniform circular cross-section and having equal diameter; both are driven at same peripheral velocity. This ensures the rolling process is completely symmetrical with respect to the rolls.

As discussed in the earlier section, it can be stated that the forces applied by each roll on the metal being rolled, can be represented by three resultant forces as shown in Fig. 2.30.

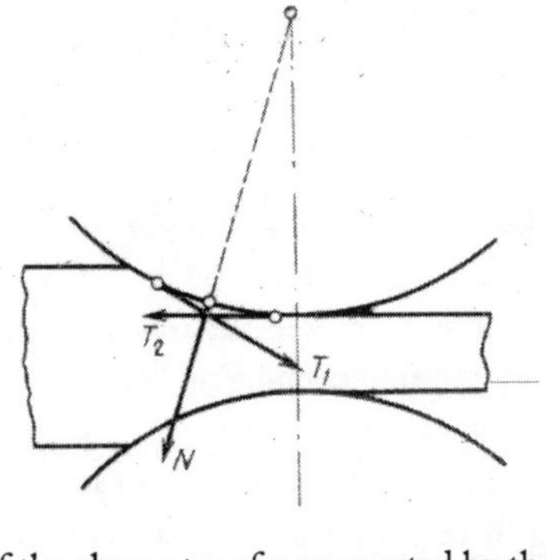

Fig. 2.30 Resultants of the elementary forces exerted by the roll on the rolled metal

The forces are N which is the resultant of pressure exerted over area of contact, and two resultant friction forces T_1 and T_2 one from the zone of backward slip and the other from the zone of forward slip, acting tangentially with arc of contact, opposite to each other.

As the rolling process is symmetrical with respect to the rolls, the forces from the other roll can also be expressed as three similar forces.

As per assumption of simple rolling, there is no external force acting on the metal, and the velocity of the metal is uniform at the entry and exit sections. The portion of metal in the deformation zone (under area of contact with rolls) undergoes the change in velocity, thus has an inertial force acting in the direction of rolling. However, this force is very small compared to force N and hence can be neglected. Then, we can conclude that the forces acting on the metal form the two rolls must balance out. This is possible when the resultant of N, T_1 and T_2 from the top roll is equal, opposite and collinear to the resultant of the corresponding forces from the bottom roll. This situation is shown in Fig. 2.31, where resultant forces P_1 and P_2 are equal and opposite, acting parallel to the line joining the centers of two rolls. Their points of application A and B are equidistant from points C and D respectively.

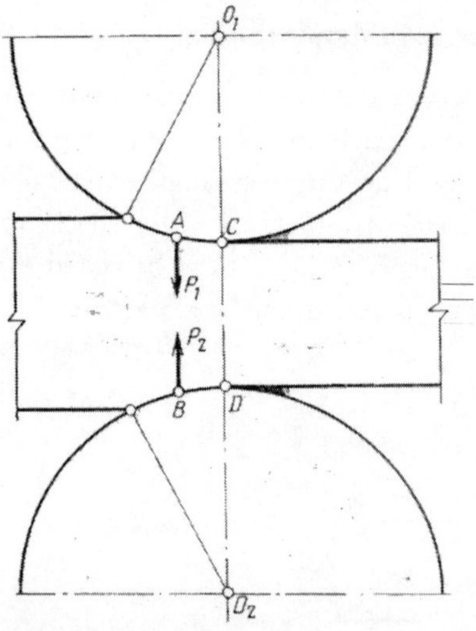

Fig. 2.31 Direction of overall resultant forces exerted by the roll on the rolled metal in a simple rolling process

Considering the force applied by the metal on rolls is same as applied by the rolls on metal, the rolling force P on the rolls can be shown as in Fig. 2.32.

From this figure, one can calculate the torque required at the roll to rotate each roll which is:

$$M_1 = Pa \qquad \ldots (2.39)$$

or $$M_1 = P\,D/2\ \text{Sin}\beta \quad \dots (2.40)$$

where,

D = roll diameter, and

β = is the angle characterizing the point of application of resultant force.

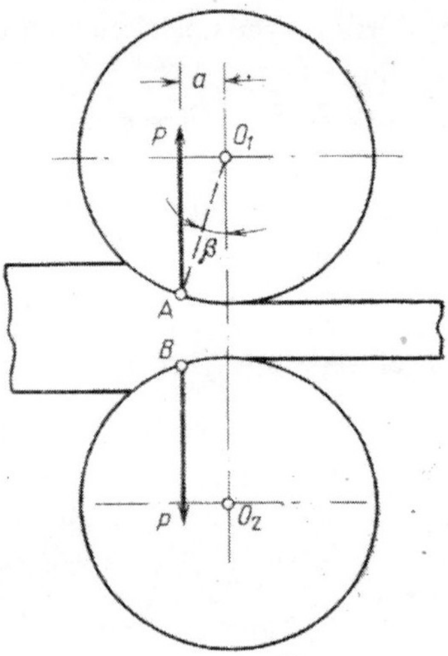

Fig. 2.32 Direction of the resultant forces applied to the rolls in a simple rolling process

The total torque required for rotation of both rolls, neglecting roll journal friction is $2M_1 = 2Pa$.

If 'd' is the roll neck (journal) diameter, and μ is the coefficient of bearing friction, then journal friction due to rolling load P will be = Pμ and the torque will be Pμd/2.

Therefore, total torque considering bearing friction, M becomes,

$$M = P(D \sin \beta + d\mu) \quad \dots (2.41)$$

It may be noted that the conditions for simple rolling is not always satisfied, and therefore, the separating force does not act vertically. In such cases the rolls are subjected to lateral forces. When rolling process is unsymmetrical the torque required for the two rolls may be unequal. Possible reason for such variation may be one or combination of a few of the following:

(i) Back and front tension applied to the metal under rolling, particularly when the rolling is performed in continuous stands or during unwinding and rewinding by coilers in cold rolling of strip.

(ii) The motion of the material is non-uniform. This happens when a reversible mill changes direction with the material in the roll bite.

(iii) When only one roll is driven, as in a thin strip mill, the symmetrical rolling condition is not maintained.

(iv) The peripheral speed of the two rolls is different. This is practiced deliberately in section mills for creating underdraught or overdraught condition.

(v) The rolls are not equal as in a Lauth mill.

(vi) The mechanical properties of the stock is not uniform (unevenly heated) or during rolling of bimetals.

(vii) The two rolls do not have same surface finish thus having different coefficient of friction.

(viii) The width of the stock is different on two sides which are in contact with the two rolls respectively.

2.3.2 Determination of roll separating force

If it is assumed that the contact pressure remains constant over the width of the rolled strip, then taking a unit width of the strip, the rolling force 'P' may be expressed by the following equation,

$$P = \int_o^\alpha p_x \frac{dx}{\cos\alpha_x} \cos\alpha_x + \int_\gamma^\alpha \tau_x \frac{dx}{\cos\alpha_x} \sin\alpha_x - \int_0^\gamma \tau_x \frac{dx}{\cos\alpha_x} \sin\alpha_x$$

All the terms in right hand side of the equation are as shown in Fig. 2.33. γ is the angle characterizing the neutral point.

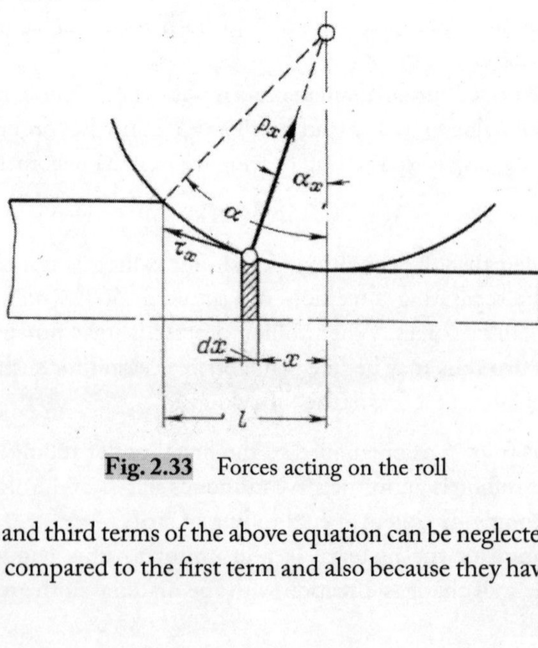

Fig. 2.33 Forces acting on the roll

The second and third terms of the above equation can be neglected considering to their smallness compared to the first term and also because they have opposite signs (plus and minus) and hence their contribution becomes further small. Therefore, the rolling force may be expressed as:

$$P = \int_o^{\alpha} p_x dx \qquad \dots (2.42)$$

In practice, this quantity is calculated as a product of two quantities:

$$P = Fp_m \qquad \dots (2.43)$$

where F = the projection of the total contact area between metal and roll, at right angles to the direction of force P. Therefore, $F = lb_m$, where b_m is the average width of metal under roll bite, and p_m is the average or mean specific stress between the metal and roll, which can be expressed as

$$p_m = \frac{1}{l}\int_0^l p_x d_x \qquad \dots (2.44)$$

Determination of roll separating force, thus, reduces to finding out two basic quantities:

(i) the total contact area between the metal and the rolls and its projection to a plane normal to the roll axis, denoted by F;

(ii) the mean specific pressure on the rolls during rolling, denoted by p_m.

The quantity F can be determined comparatively easily, particularly for flat rolling, form the geometry of the roll bite considering the roll diameter and thickness of the strip at entry and exit form the roll bite. For shape rolling through grooves, different methods are employed for estimating the contact area, which have been discussed later in this section.

Determination of p_m, the average pressure at the roll, can be more complicated, depending on different factors. These factors can broadly be divided into two groups. The first group consists of factors influencing the mechanical properties of the material. The factors which affect the mechanical properties, i.e., the resistance to linear deformation, σ_a, or the resistance to pure shear, k, include factors like temperature of rolled material, work hardening and strain rate. The second group is made up of factors that govern the state of stress of the rolled material, i.e., contact friction, the outer zone, external tensions, etc.

(A) Determination of contact area

Horizontal projection F of the contact area is given by the equation

$$F = l\,b_m$$

where, l is the horizontal projection of arc of contact, and b_m is the mean width of the rolled strip. As determined earlier (section 2.1.3), the horizontal projection of arc of contact, in case of flat to flat rolling can be closely approximated as:

$$l = \sqrt{r\Delta h} \qquad \dots (2.45)$$

where, r = roll radius,

$$\Delta h = h_1 - h_0 = \text{reduction in thickness.}$$

So contact area

$$F = b_m \sqrt{r\Delta h} \qquad \dots (2.46)$$

If the radii of the two rolls differ considerably, l can be written as:

$$l = \sqrt{\frac{2r_1 r_2}{r_1 + r_2}\Delta h} \qquad \dots (2.47)$$

This equation is based on the assumption that the contact areas of both rolls is equal (and thus contact pressure is also equal) by virtue of different reductions on both sides of the metal. If the reductions are z_1 and z_2 for rolls with radius r_1 and r_2 respectively, then from equality of l on both sides, one can write,

$$l = \sqrt{2r_1 z_1} = \sqrt{2r_2 z_2} \qquad \dots (2.48)$$

so that

$$z_1 = \frac{r_2}{r_1} z_2$$

since $z_1 + z_2 = \Delta h$, one can write

$$z_1 = \frac{r_2}{r_1 + r_2}\Delta h \text{ and } z_2 = \frac{r_1}{r_1 + r_2}\Delta h$$

from equation (2.48);

$$l = \sqrt{2r_1 \times \frac{r_2}{r_1 + r_2}\Delta h} = \sqrt{\frac{2r_1 r_2}{r_1 + r_2}\Delta h}$$

When sections are rolled in grooved rolls, the contact area is determined either graphically or analytically. In graphical method the lines of intersection of the roll and entering stock is plotted in three projections from which the contact area is determined.

There are various analytical relationships, proposed by various experts for determining the Δh while rolling different sections.

One simple method is to take Δh equal to the mean linear reduction over the width of section, when rolling flat but non-rectangular section, i.e.,

$$\Delta h = \frac{A_0}{b_0} - \frac{A_1}{b_1} \qquad \text{... (2.49)}$$

where, A_0 and A_1 are the cross-sectional areas of the profile before and after reduction, and b_0 and b_1 are the corresponding mean widths.

In case of section rolling in grooved rolls, following relations can be used:

For a rhombus rolled from a rhombus (Fig. 2.34a)

$$\Delta h = (0.55 \text{ to } 0.6)\ (h_0 - h_1)$$

For an oval rolled from a square (Fig. 2.34b)

$$\Delta h = h_0 - 0.7h_1 \qquad \text{(for a shallow oval)}$$

and

$$\Delta h = h_0 - 0.85h_1 \qquad \text{(for a round oval)}$$

For a square rolled from an oval (Fig. 2.34c)

$$\Delta h = (0.65 \text{ to } 0.7)h_0 - (0.55 \text{ to } 0.6)\ h_1$$

For a circle rolled from an oval (Fig. 2.34d)

$$\Delta h = 0.85h_0 - 0.79h_1$$

where, h_0 and h_1 refer to the depth of the cross-section before and after the pass.

In order to calculate the contact surface in section mills the following formulae can also be recommended:

For a square rolled from an oval

$$F = 0.75b_1\sqrt{r_1\ (h_0 - h_1)}$$

For an oval rolled from a square

$$F = 0.54(b_0 + b_1)\sqrt{r_1(h_0 - h_1)}$$

For a rhombus or square rolled from a rhombus

$$F = 0.67b_1\sqrt{r_1\ (h_0 - h_1)}$$

where, h_0 and h_1 are the depths at the middle of the cross-section of the rolled stock before and after the pass, respectively (see Fig. 2.34), and r_1 is the roll radius at the middle of the section.

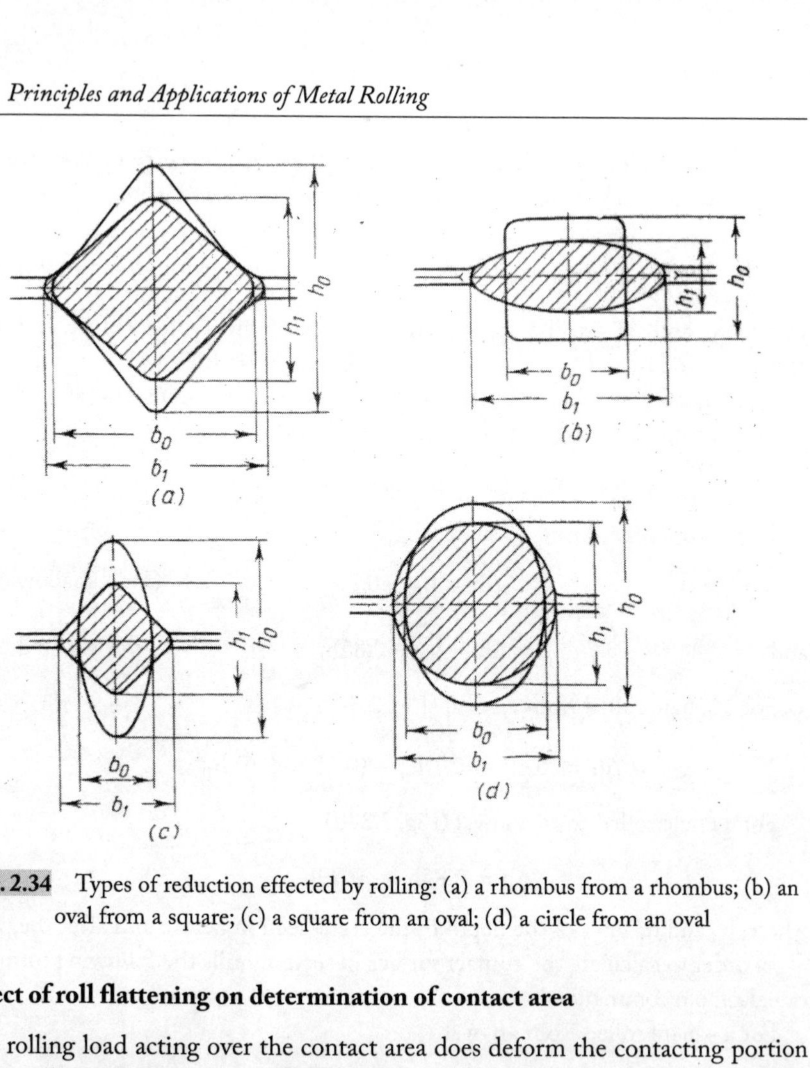

Fig. 2.34 Types of reduction effected by rolling: (a) a rhombus from a rhombus; (b) an oval from a square; (c) a square from an oval; (d) a circle from an oval

Effect of roll flattening on determination of contact area

The rolling load acting over the contact area does deform the contacting portion of the roll elastically, however small it may be. During cold rolling of high strength materials, because of high specific pressure, such deformations may be appreciable. The phenomenon of roll flattening has already been discussed in section 2.1.12. The net effect of roll flattening is that the effective roll radius is increased and consequently, for the same pass reduction, the arc of contact gets elongated and so is the area of contact.

The deformed roll radius, r′, as given by the famous Hitchcock's formula, can be written as:

$$r' = r + \frac{16(1-\gamma^2)\,Pr}{b_m \pi E(h_0 - h_1)} \quad \ldots (2.50)$$

Putting this new value of flattened roll radius r′ in equation (2.46), we can estimate the quantity F under roll flattening as

$$F = b_m \sqrt{r'\Delta h}$$

$$= b_m \sqrt{r\Delta h + \frac{16\left(1\text{-}\gamma^2\right)Pr}{b_m \pi E}} \qquad \ldots (2.51)$$

It is worth noticing here that we need to find out F to estimate the rolling load P, but during roll flattening, the quantity P is required to estimate F. Therefore, to start with, calculation of F starts considering no roll flattening and calculates first approximate value of P. Then putting this value of P in eqn. (2.51), next value of F is estimated. If a substantial difference between two successive values of P exists, the iteration is continued.

The effect of roll flattening is particularly noticeable when the factor $\left(\frac{16(1-\gamma^2)P}{b_m \pi E}\right)$ is not negligible compared to reduction Δh, i.e., when P/b_m is large, E is small or Δh itself is small.

(B) Determination of average pressure of rolling

It has already been stated that the mean specific pressure 'p_m' at which rolling takes place depends on many factors. Theoretical analysis of these factors is difficult and as yet not extensively studied. The practice, therefore, is to consider the effects of these factors by introducing corresponding coefficients.

When the first group of factors like temperature of rolled material, work hardening and strain rate are considered, which modify the resistance to linear deformation (or pure shear), the actual yield stress 'σ_a' can be expressed as:

$$\sigma_a = n_t n_w n_v \sigma_s \qquad \ldots (2.52)$$

where, σ_s is the yield stress for the annealed metal when tested under static conditions at room temperature by means of the conventional testing machine. n_t, n_w and n_v are the coefficients accounting for the effect of temperature, work hardening, and strain rate, respectively.

The second group of factors comprises external factors which affect the nature of the state of stress developed in the metal under roll bite and thus influence value of the mean specific pressure, p_m, which can be expressed as:

$$p_m = \frac{2}{\sqrt{3+\xi^2}} n_\sigma \sigma_a \qquad \ldots (2.53)$$

The first factor, $2/\sqrt{3+\xi^2}$, takes care of mean normal stress, σ_2, while the second factor, n_σ, takes care of the second group of factors mentioned above.

For a two-dimensional deformation (i.e., rolling of a flat stock without side spread), the value of $\xi = 0$, hence first factor becomes, $2/\sqrt{3} = 1.15$.

The second factor, n_σ, can be expressed by $n_\sigma = n'_\sigma \, n''_\sigma \, n'''_\sigma$ to take care of effect of external friction, external tension, and width of the metal determining side spread respectively.

Thus, we may finally write,

$$p_m = 1.15 \, n_\sigma \sigma_a$$

where, $\sigma_a = n_t n_w n_v \sigma_s$ and, $n_\sigma = n'_\sigma n''_\sigma n'''_\sigma$

(a) The first group of factors influences the mechanical properties, mainly yield stress of the materials being rolled, factors like temperature, work hardening, and strain rate.

The yield stress, σ_a, may be expressed by the equation (2.52) as:

$$\sigma_a = n_t n_w n_v \sigma_s$$

where, n_t, n_w and n_v are the coefficients accounting for the effect of temperature, work hardening, and strain rate, respectively, and σ_s is the yield stress of the annealed material determined in tensile testing under static condition and at room temperature.

As rolling of rectangular piece without side spread has been considered, for the sake of analysis, it has implicitly considered two-dimensional strain. In this case, the resistance to deformation should not be determined from tensile yield stress σ_a, but should preferably be considered from resistance to pure shear k.

where, $$k = \frac{\sigma_a}{\sqrt{3}} = 0.57 \, n_t \, n_w \, n_v \, \sigma_s \qquad \ldots (2.54)$$

It is also to be noted that rolling involves compression. It has been established form experiments that yield stress for steel in compression is approximately 10% higher than that in tension, while for aluminum and copper it is almost the same. Depending on the material under consideration the value of σ_s in equation (2.54) should accordingly to be adjusted.

(i) Effect of temperature of the metal

It has been found out through experimental investigations that the mechanical properties like ultimate strength, hardness, etc., reduces with increase in temperature exponentially following the equation:

$$P = M \, e^{-mT}$$

where,

P is respective property,
M and m are constants depending on the metal
T is the absolute temperature

However, instead of using such a theoretical equation, in practice, laboratory test results are used for estimation of value of n_t in equation (2.52), that is, for the yield stress data for different metals and alloys at different temperatures.

The property of reduction of ultimate yield stress of a material is applicable for hot rolling operation. The temperature of the metal during rolling goes on changing due to:

(a) loss of heat due to radiation
(b) loss of heat by convection
(c) loss of heat by conduction to the rolls, guides, rollers of roller table and other equipment to which the hot metal comes into contact.
(d) Gain of heat resulting from the work done to cause plastic deformation of the metal.

The loss due to convection and conduction is very small compared to that due to radiation at high rolling temperature, and hence is generally neglected in the heat balance.

Mathematical models are set up for calculation of temperature of the metal during rolling. Rollers follow a suitable rolling practice to ensure that the metal temperature does not reduce much so that the advantage of hot rolling is not lost. In continuous form of rolling, the temperature of the metal actually tends to increase and requires water spray cooling to keep the metal temperature within the limit.

For steel, loss of temperature, ΔT_l, in °C due to radiation is given by the equation:

$$\Delta T_l^* = 0.0067\, A_s/W' \left(T_m/100\right)^4 \Delta t \qquad \ldots (2.55)$$

where,

A_s = surface area in sq. meter of 1 meter length of rolled section
W' = weight in kg of 1 meter of rolled section
T_m = mean value of absolute temperature of metal over the time interval Δt in second

Similarly the temperature gain, ΔT_g, in the metal per pass is given by the equation:

$$\Delta T_g^* = A(1-\alpha)/427 Ws \qquad \ldots (2.56)$$

---

* refer to stress and strain in metal rolling by A. Tselikov, Mir publishers, for detailed deductions of the equations.

where,
A = the energy required to deform the metal, kg-m
α = the coefficient denoting part of deformation energy lost in heating rolls and surrounding medium
W = weight of metal in kg.
s = the specific heat, cal/kg°C

value of A is obtained from experimental curves of specific energy expended during rolling of different metals and alloys under different temperatures.

(ii) Effect of work hardening

During cold rolling of metals, the metal gets work hardened and the resistance to deformation is increased.

The value of yield stress, σ_s, as a function of % of reduction in rolling is determined form suitable experiments conducted by different researchers as well as rolling mill manufacturers in experimental mills of their laboratories. Figure 2.35 shows a typical yield stress vs. % reduction curve

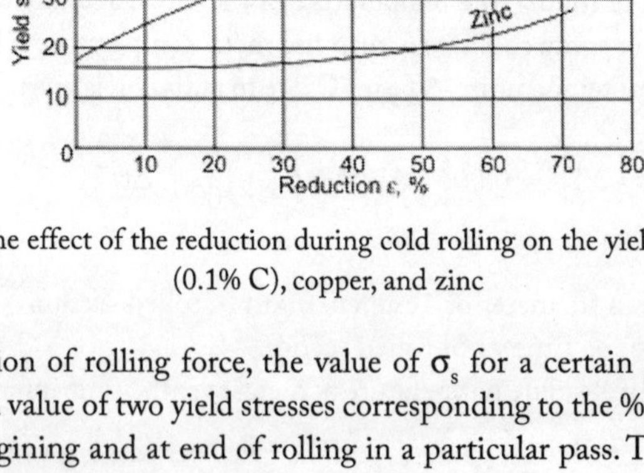

Fig. 2.35 The effect of the reduction during cold rolling on the yield point of steel (0.1% C), copper, and zinc

For calculation of rolling force, the value of σ_s for a certain pass is generally chosen as mean value of two yield stresses corresponding to the % reduction of the metal at the begining and at end of rolling in a particular pass. This takes care of the coefficient, n_w, in equation (2.52) or (2.54).

(iii) Effect of strain rate

The mean strain rate, u_m, in rolling operation can be expressed as

$$u_m = \frac{v_1}{l} \frac{\Delta h}{h_0} \qquad \dots (2.57)$$

where,

v_1 = exit velocity of the metal

l = horizontal projection of the arc of contact

h_0 = entry thickness of the strip

Δh = reduction of the pass.

Thus it can be assumed that rate of strain is directly proportional to the rolling speed that determines the exit velocity of the metal.

Experimental results by various researchers (Nadai and Manjone, Cook, Dinnik, Alder and Philips and others) through both tensile and compressive tests in laboratories show that in hot rolling, a considerable effect of strain rate on the resistance to deformation exists. Figure 2.36 shows the variation of ultimate strength of mild steel with rate of deformation at different temperatures. This figure clearly shows that the effect of strain rate is more at higher temperatures of the metal. The ultimate tensile stress has an increasing trend with increasing strain rate, and the ultimate stress value may increase roughly 5–7 times than static values. In hot rolling strain rate varies over a wide range. For a breakdown mill, it may range from 0.5 to 5 per second, while for a continuous strip mill or wire rod mill, it can vary between 100–400 sec^{-1} or even higher.

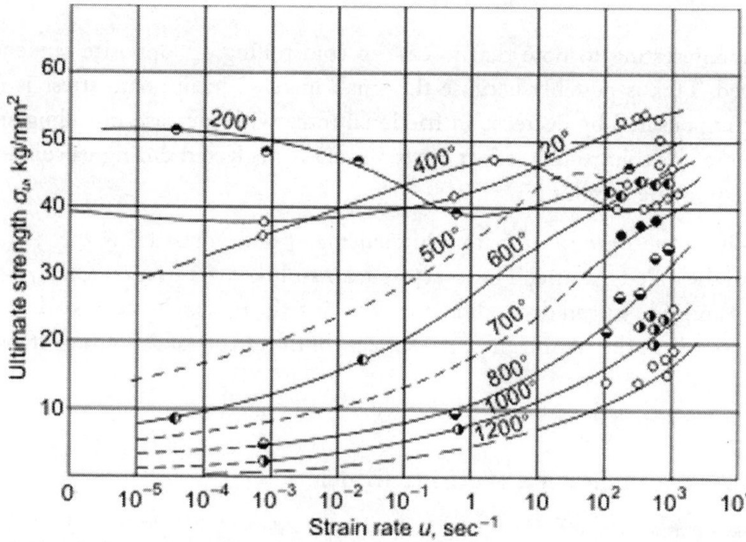

Fig. 2.36 Variation of the ultimate strength of mild steel with strain rate at various temperatures in °C (A. Nandai and M. Manjone)

The experimental work by P. Cook, on different grades of steel further establishes that the effect of strain rate on resistance to deformation depends to a large extent on the amount of deformation. For small strains up to 20–30%, the yield stress strongly increases with the strain. For strains more than 30% the rate of increase in the yield stress is reduced, and in many cases the yield stress decreases when strain is increased still further. Figure 2.37 shows the results of his tests for a true strain of 30%.

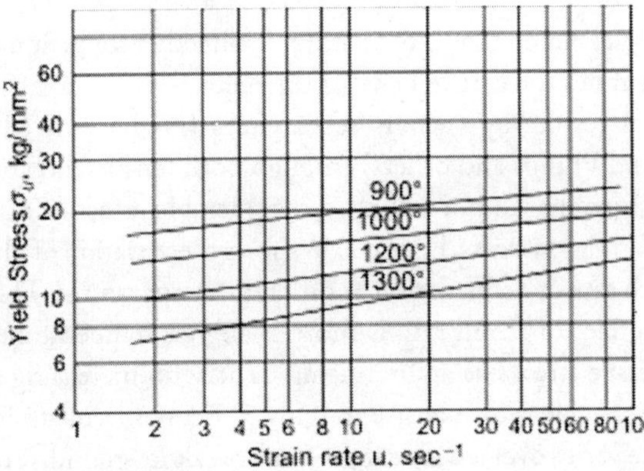

Fig. 2.37 Variation of yield stress of mild steel with 0.15% C with the strain rate when

$$\log_e \frac{h_0}{h_1} = 0.3 \text{ for various temperature}$$

It is interesting to note that in case of cold rolling, an opposite tendency is observed. This is possibly because the small increase in ultimate stress is more than compensated by decrease in frictional force with increase in rolling speed. Therefore, for cold rolling, effect of strain rate is neglected during calculation of average pressure of rolling.

(*b*) The second group of factors influencing specific pressure is the nature of stress of the rolled material. These factors are mainly contact friction forces, tension applied and side spread depending on width of strip. p_m can be expressed as: $p_m = 1.15\, n_\sigma \sigma_a$ where the coefficient, n_σ, takes care of the effects of this group of factors.

Let,
$$n_\sigma = n'_\sigma\, n''_\sigma\, n'''_\sigma;$$

where,

n'_σ takes care of the effect of external friction,

n''_σ takes care of the effect of tension, and

n'''_σ takes care of the effect of strip width.

(i) Friction between the roll surfaces and roll material and their effect

It has already been discussed (see section 2.1.5) the importance of friction in gripping and feeding of the stock in the roll bite. With increased coefficient of friction, higher reduction is possible. On the other hand, the specific pressure between rolls and stock increases.

Because of this reason, in hot rolling, increased coefficient of friction is favored to achieve a high percentage of reduction. The associated increase in rolling force is, in general, not of much concern. However, for cold rolling generally rolling force is more important than reduction in the pass. Therefore, in cold rolling the coefficient of friction is kept low by using ground/polished roll surface and applying lubricant during rolling.

It is important to distinguish between the three kinds of coefficient of friction in rolling:

(a) during gripping process, designated as 'μg'
(b) when rolls slip over the entire contact surface, 'μ_{sl}'
(c) during the steady state rolling when slipping of metal takes place on opposite directions about the neutral section, 'μ_{sl}'.

As per S. Ekelund, the μ_g during hot rolling of 0.15% C steel is given by the formula,

with steel rolls, $$\mu_g = 1.05 - 0.0005t \qquad \text{... (2.58)}$$

with C.I. rolls, $$\mu_g = 0.8\,(1.05 - 0.0005t) \qquad \text{... (2.59)}$$

where t = temperature in °C of the stock.

The gripping coefficient of friction during hot rolling of a few non-ferrous metals and alloys, as experimentally determined by A. Presnyakov is given in Table 2.1.

Table 2.1 The coefficient of friction during hot rolling of non-ferrous metals at the instant of gripping

Metal	*Temperature, °C*	*Coefficient of friction, μ_g*
Copper	900	0.52
Brass		
JI62	800	0.45
JI68	800	0.38
German silver		
HM-81	950	0.4
Nickel	1,100	0.4

Source: Stress and strain in rolling by A. Tselikov

The coefficient of friction during slipping of metal over rolls μ_{sl} is always found to be less than μ_g. As per investigation by N. Get for hot rolling of steel containing 0.5–0.8% C, the value of μ_{sl} for t > 700°C, can be represented by the equation:

$$\mu_{sl} = 0.55 - 0.00024\, t \qquad \text{... (2.60)}$$

The coefficient of friction during cold rolling under steady slipping condition has been measured by many researchers following different methods. One of the methods proposed by D. Bland and H. Ford is to estimate frictional force by measuring on the roll while rolling with the zone of forward slip coinciding to zero, (i.e., neutral point is coincident with exit point of roll bite). This condition is achieved by applying large back tension. In this condition, μ_{sl} can be estimated from the equation:

$$\mu_{sl} = \frac{M}{P \times r} \qquad \text{... (2.61)}$$

where,
M = rolling torque/roll
P = rolling force
r = roll radius

The experiments were carried out on a two-high laboratory mill. Rolls of 100 mm diameter were made of steel, polished and velocity of rolling was 9 mps. Following table shows some of the results for carbon steel, annealed copper, annealed 60:30 brass and annealed aluminum under different % reduction and a few lubricants.

Table 2.2 Coefficient of friction under cold rolling of different materials under different conditions

Lubricants	*Carbon steel*		*Annealed copper*		*Annealed brass*		*Annealed aluminum*	
	% *redu-ction*	μ_{sl}	% *redu-ction*	μ_{sl}	% *redu-ction*	μ_{sl}	% *redu-ction*	μ_{sl}
1. dry rolls and strip	15.0	0.085	29.0	0.093	15.1	0.093	22.5	0.092
2. Paraffin	16.5	0.080	23.9	0.067	15.0	0.067	24.5	0.081
3. Paraffin + 5% sodium oleate	23.0	0.049	-	-	-	-	27.9	0.059
4. 615 Graphite in oil	15.5	0.072	36.8	0.054	27.8	0.049	33.5	0.055
5. Oil, olive	18.1	0.057	30.2	0.058	33.8	0.057	-	-
6. Oil, castor	23.0	0.045	34.1	0.046	-	-	21.6	0.057
7. Lanoline	26.5	0.041	-	-	28.0	0.043	25.4	0.025

The effect of the friction forces between rolls and metal at the contact surface, on the specific pressure is taken care by the factor, n'_σ.

For the cases of rolling where $5 \geq l/hm > 2$ (particularly for cold rolling), the variation of this factor n'_σ with δ, where $\delta = 2\mu_{sl}\ l/\Delta h$, for different reductions $\Delta h/h_0$ is shown in Fig. 2.38.

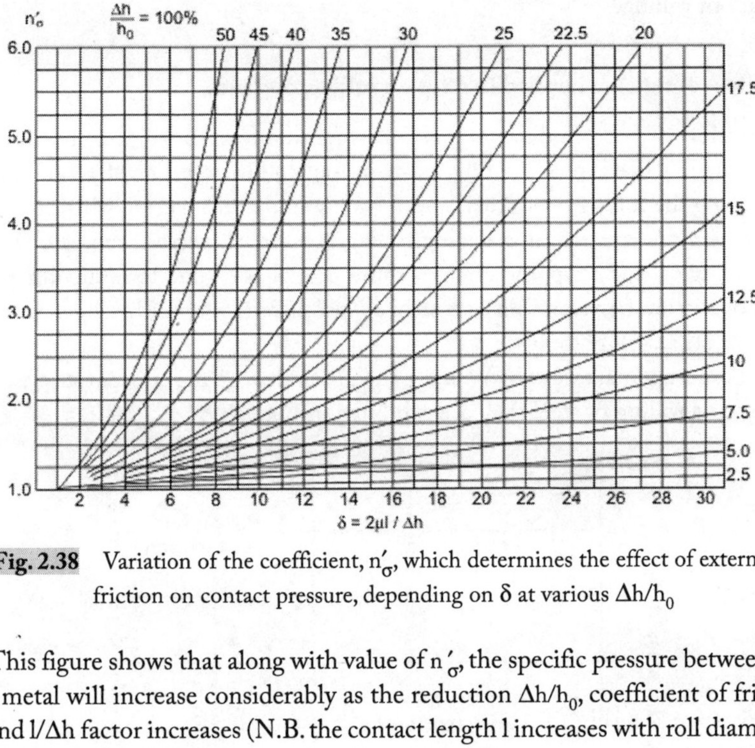

Fig. 2.38 Variation of the coefficient, n'_σ, which determines the effect of external friction on contact pressure, depending on δ at various $\Delta h/h_0$

This figure shows that along with value of n'_σ, the specific pressure between roll and metal will increase considerably as the reduction $\Delta h/h_0$, coefficient of friction μ_{sl} and $l/\Delta h$ factor increases (N.B. the contact length l increases with roll diameter).

For example, if the following data for a cold rolling operation are considered:

$$\delta h/h_0 \times 100 = 15\%,$$

$$\mu_{sl} = 0.085,$$

$$1/\Delta h = 50$$

Then $$\delta = 2\mu\ l/\Delta h = 8.5$$

From above figure, $$n'_\sigma \cong 1.4$$

By using proper lubrication, if μ_{sl} can be brought down to the value of 0.05; the δ becomes equal to 5 and the corresponding n'_σ becomes equal to 1.15 approximately. This clearly shows the importance of keeping the coefficient of friction low in cold rolling operation, for keeping the rolling pressure low.

When the ratio l/h_m remains less than 2, the factor n'_σ is approximated by the equation

$$n'_\sigma = 1 + \frac{\sqrt{r\Delta h}}{3(h_0 + h_1)} = 1 + \frac{\sqrt{r\Delta h}}{3 \times 2h_m} = 1 + \frac{l}{6h_m} \quad \ldots (2.62)$$

which is independent of coefficient of friction. This equation is generally used in case of hot rolling.

(ii) Effect of tension applied to strip on rolling pressure

Application of back and front tension is common in continuous rolling and essentially practiced in cold rolling of strip from coil to coil. The effect of tension in reducing the rolling pressure has been widely studied. It is established that back tension is more effective in reducing rolling pressure than front tension. During cold rolling of thin strip, the effect is more pronounced due to reduction of the effect of friction between rolls and strip as well as reduction of local deformation of the roll (i.e., roll flattening).

The experimental results by W. Lueg and F. Schultze on the effect of strip tension during cold rolling of steel strip are exhibited in Fig. 2.39.

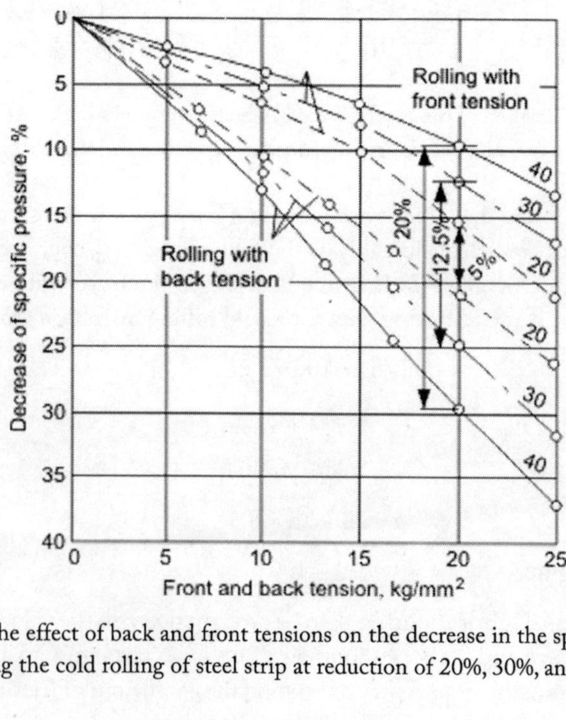

Fig. 2.39 The effect of back and front tensions on the decrease in the specific pressure during the cold rolling of steel strip at reduction of 20%, 30%, and 40%

It is interesting to note that difference of effect with back tension and with front tension increases with increase in amount of reduction.

According to W. Hessenberg and R. Sims, the specific pressure of metal on the rolls, p_m, can be expressed as shown below, when effect of tension is taken into consideration:

$$p_m = p'_m\left(1 - \frac{\sigma_0 + \sigma_1}{4k}\right)$$

where, p'_m is the mean specific pressure when tension is absent.

Therefore, the effect of tension expressed by the factor n''_σ can be approximated as:

$$n''_\sigma = \frac{p_m}{p'_m} = \left(1 - \frac{\sigma_0 + \sigma_1}{2p'_m}\right) \quad \ldots (2.63)$$

(iii) Effect of the width of the rolled strip on pressure

It has already been discussed that when the average width 'b_m' is more than 5 times the length of arc of contact 'l' or more than 12 times the thickness of the strip with the condition $l/h_o \geq 2.5$, the side spread of the strip is restricted. However, when width $b_m \leq 5l$, and free spreading in a smooth barreled or grooved roll is allowed, the mean specific pressure tends to get reduced. This is due to reduction in the stress σ_2 in the direction of the width of the metal, and as a result of the influence of external friction. The value of n'''_σ therefore is equal to or less than 1.

2.3.3 Formulae for calculation of mean rolling pressure

In the following, a few well known formulae for calculation of mean rolling pressure are reproduced, suggested by some of the pioneers in the field of rolling mill technology.

(i) Ekelund's formula[5]

Way back in 1927, Swedish designer S. Ekelund suggested a formula for calculation of rolling force in the hot rolling of steel, considering different factors.

The formula* put forward by Ekelund is:

$$P = b_m k_m \sqrt{R\Delta h}\left[1 + \left\{1.6\mu\sqrt{R\Delta h} - 1.2\,(\Delta h)\right\}\frac{1}{h_0 + h_1}\right] \quad \ldots (2.64)$$

* This is really a modified version of original Ekelund's formula. The original formula had a factor added to k_m, to take care effect of strain rate and viscosity of rolled metal on the value of resistance to deformation.

The effect of external friction represented by the factor n'_{σ}, as per this formula becomes:

$$1 + \frac{1}{h_0 + h_1}\left\{1.6\mu\sqrt{R\Delta h} - 1.2(\Delta h)\right\}$$

The formula recommended by Ekelund for value of k_m, the specific resistance to compressive deformation, for steels containing manganese and chromium (not more than 2-3%) besides carbon is

$$k_m = (14\text{-}0.01t)(1.4\text{+}C\text{+}M_n\text{+}0.3Cr),\ \text{kg/mm}^2 \qquad \ldots (2.65)$$

where, C, M_n, and C_r indicates the % content of the three elements in the steel.

Ekelund's formulas for calculation of μ have already been mentioned in equations 2.58 and 2.59.

Though Ekelund's formula was suggested for hot rolling, it has been established by later researchers (cook and Lake, and Ford) that this equation can be readily modified for application to cold rolling conditions, which takes the form

$$P = p_m b_m \sqrt{R\Delta h}\left[1 + \left\{1.6\mu\sqrt{R\Delta h} - 1.2\Delta h\right\}\frac{1}{h_0 + h_1}\right]$$

where, $$p_m = \frac{2}{\sqrt{3}}\sigma_a = 1.15\, n_t n_w n_v \sigma_s$$

(ii) Geleji's formula*

The formula for estimating mean specific rolling pressure, suggested by A. Geleji of Hungary, is reproduced below:

$$k_m = k_f\left(1 + C\mu\frac{l_d}{h}\sqrt[n]{v}\right) \qquad \ldots (2.66)$$

where,

k_m = the mean specific pressure, kg/mm²

k_f = the resistance to linear deformation (the quantity corresponding to σ_a), kg/mm²

C = an experimental coefficient, depending on the l_d / h ratio, and determined from a graph (Fig. 2.40)

μ = the coefficient of friction

$\sqrt{r\Delta h} = l_d$ is the length of the arc of contact

n = an index, taken to be equal to four

v = the peripheral velocity of the rolls, m/s

* The formula was developed basically for hot rolling of flat product, but may be used for cold rolling also.

The value of the coefficient C, shown in Fig. 2.40, was obtained by A. Geleji by analyzing a series of test results, principally from the data of O. Emicke.

For the determination of the value k_f for hot rolling of plain carbon steels, A. Geleji recommends the formula

$$k_f = 0.015(1{,}400 - t)\ \text{kg/mm}^2 \qquad \ldots (2.67)$$

or the formula recommended by Ekelund as per equation (2.65). For a more exact determination of k_f, Geleji also recommends that the graph shown in Fig. 2.41 be used.

A. Geleji recommends that the coefficient of friction be determined from the equation:

$$\mu = 1.05 - 0.0005t - 0.056v \qquad \ldots (2.68)$$

for steel rolls,

$$\mu = 0.94 - 0.0005t - 0.056v \qquad \ldots (2.69)$$

for hardened cast iron roll, and

$$\mu = 0.82 - 0.0005t - 0.056v \qquad \ldots (2.70)$$

for polished steel or hardened cast iron rolls,
where, v = the peripheral velocity of the rolls, m/s
t = the rolling temperature, ^{0}C

These formulae are valid for t > 700 $^\circ$C and v < 5 m/s.

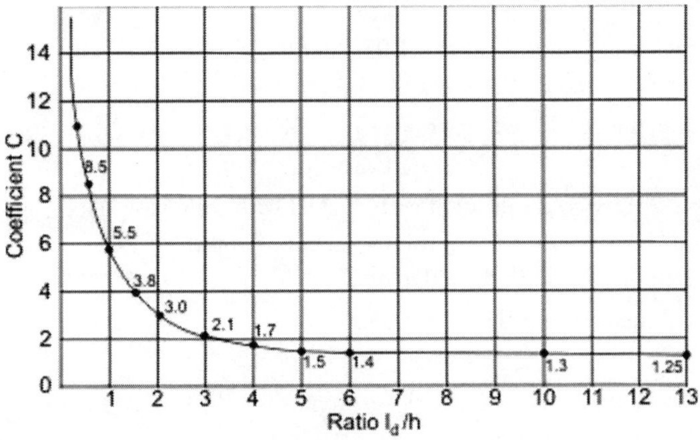

Fig. 2.40 Variation of the value of the coefficient C in formula 2.66 with ratio l_d/h (A. Geleji)

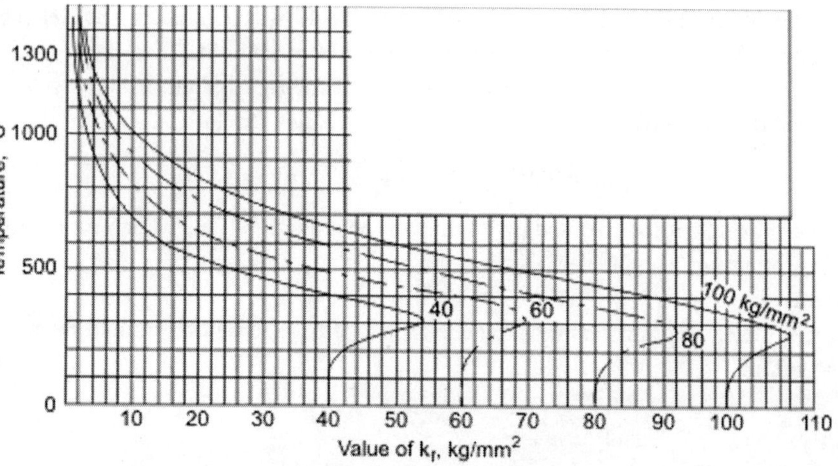

Fig. 2.41 Variation with temperature of k_f of a plain carbon steels (C < 0.6%; Si < 0.5% and Mn < 0.8%), characterized by ultimate strengths of 40, 60, 80, and 100 kg/mm^2 in the cold state (A.Geleji)

As per Geleji's formula, combination of strain rate factor n_v and the combined stress factor 'n_σ', the resistance to linear deformation 'σ_a' is increased by the combined factor expressed as:

$$n_v\, n_\sigma = 1 + C\mu\, l/h_m\, \sqrt[n]{v}$$

Let us consider two examples of hot and cold rolling, respectively.

Example 1: Hot rolling with following data:

(a) Steel roll, (b) rolling temperature = 700 °C, (c) l/h_m = 2, (d) rolling speed = 2 mps

Then coefficient of friction, μ = 1.05-0.0005×700-0.056×2 = 0.588

From graph in Fig. 2.40, value of C = 3 at l/h = 2,

Therefore, the combined factor = $1+3 \times 0.588 \times 2 \times \sqrt[4]{2}$

$= 1+4.196 = 5.196$

Example 2: Dry cold rolling of carbon steel with following data:

(a) μ = 0.085, (b) l/h_m = 5, (c) rolling speed = 5 mpm

from graph in Fig. 2.40, the value of C = 1.5

The combined factor = $1+1.5 \times .085 \times 5 \times \sqrt[4]{5}$

$= 1+.953 = 1.953$

It is to be noted that in Geleji's formula, the effect of strain hardening and external tension is not included.

(iii) Sims' formula[6] (for calculation of hot rolling load)

The formula proposed by Sims in 1954, for estimation of mean rolling pressure is derived by solving the differential equation (2.4) of equilibrium of an element of metal under rolling as shown in Fig. 2.20.

The respective equations for zones of backward and forward slip, respectively, are as follows:

$$\frac{p_x}{2k} = \frac{\pi}{4}\log_e\frac{h_x}{h_o} + \frac{\pi}{4} + \sqrt{\frac{r}{h_1}}\tan^{-1}\left(\sqrt{\frac{r}{h_1}}\alpha\right) - \sqrt{\frac{r}{h_1}}\tan^{-1}\left(\sqrt{\frac{r}{h_1}}\varphi\right) \quad \ldots (2.71)$$

$$\frac{p_x}{2k} = \frac{\pi}{4}\log_e\frac{h_x}{h_1} + \frac{\pi}{4} + \sqrt{\frac{r}{h_1}}\tan^{-1}\left(\sqrt{\frac{r}{h_1}}\varphi\right) \quad \ldots (2.72)$$

According to above two formulae of Sims, the stress factor represented by $p_m/2k$ depends only on the % reduction of the pass and the ratio r/h_1. Figure 2.42 shows the values of the stress factor $p_m/2k = n_\sigma$ calculated from the formula of Sims.

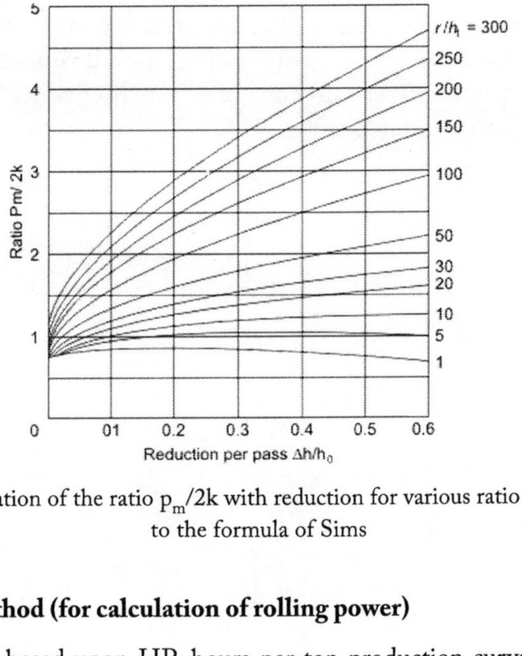

Fig. 2.42 Variation of the ratio $p_m/2k$ with reduction for various ratio r / h_1 according to the formula of Sims

(iv) Energy method (for calculation of rolling power)

This method is based upon HP–hours per ton production curves derived from experimental results and can be used with some degree of confidence if it is known that the mill to be considered is of similar type and rating to the mill from which the results have been obtained. When reliable curves are available, the maximum rolling power HP_{max} (in horse power) can readily be calculated from the simple expression:

HP_{max} = HP–hours per ton × maximum production rate of the mill in any pass (in tons/hr).

This method is particularly applicable for hot blooming, billet, slab and strip mill.

2.4 Rolling Torque and Power

The torque required to rotate the rolls for rolling a material is called *rolling torque.* This is calculated either from the forces acting on rolls multiplied by a constant called lever arm, estimated through experiments, or from the energy consumption data obtained through experiments. The first method gives more accurate results during both hot and cold rolling of rectangular sections, i.e., during rolling of blooms, slabs, strips, etc. The second method is generally used for estimating rolling torque during section rolling.

2.4.1 Lever arm method

It is already determined in section 2.2.2. that the rolling pressure distribution varies along the arc of contact. In case of simple rolling, the roll separating force may be considered to be parallel to the axis joining the centers of two rolls, and the force distribution will be as indicated in Fig. 2.43.

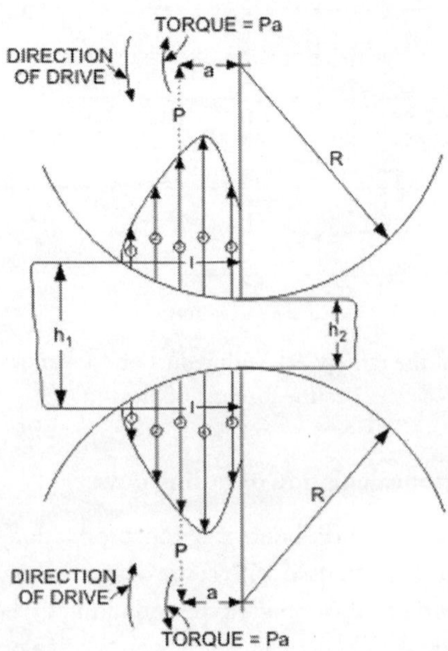

Fig. 2.43 Diagram illustrating the meaning of lever arm and torque

If P is considered to be the resultant force acting on each roll, at a distance of 'a' from the roll axis, then the torque exerted by this resultant force on each roll against the direction of roll rotation will be given by:

$$T' = Pa$$

and total torque required from the drive to overcome torques on both the rolls will be:

$$T = 2Pa \text{ tonnes-mm} \qquad \ldots (2.73)$$

where,
a = length of the lever arm in mm
P = rolling force in tonnes
T = total basic torque required for reduction of the material (without considering drive efficiency) in tonnes-mm

From the geometry of rolling condition it can easily be said that the lever arm 'a' depends on the length of the projected arc of contact. This can be expressed by the relation:

$$a = \lambda\sqrt{R\Delta h} \qquad \ldots (2.74)$$

where, 'λ' is called lever arm coefficient.

Ford along with colleague Bland has done exhaustive cold rolling experiments[7],[8],[9] (nearly 1000 tests) on different material under different rolling conditions in an experimental mill where rolling force P and rolling torque T could be accurately measured, to find out the average values of λ. The results are shown in Table 2.3 below:

Table 2.3 Average values of λ for all results determined by Ford

Material	*Thickness in.*	*Roll finish*	*Average value of λ*	*Percentage of each value of λ differing from the average by not more than*		
				5%	*10%*	*20%*
0.2% C steel	0.100	Mirror	0.40	59	86	100
H.C Copper	0.100	Mirror	0.40	54	79	96
H.C Copper	0.050	Matt	0.31	31	41	75
H.C Copper	0.075	Matt	0.32	40	68	88
H.C Copper	0.100	Matt	0.33	35	61	91
0.2% C steel	0.100	Matt	0.32	42	61	97
0.2% C steel	0.100	Matt*	0.33	11	53	100
0.11% C Steel	0.074	Mirror	0.36	21	50	79
0.07% C Steel	0.065	Mirror	0.35	58	84	100

*Rolls dry. All other results refer to rolls lubricated with Vacuum 40A oil.

Each of the λ values is an average of many results obtained during rolling of a particular material in a particular finish of the roll under different percentage reductions and strain hardening. These individual values show large variation as indicated by the data given under last three columns of the table.

These values of λ, however, are based on undeformed radius R of the roll. Ford and Bland showed further that if roll flattening effect is taken into consideration, the fluctuation of λ for any one particular set of data is reduced considerably.

The modified λ′ corresponding to roll flattening, can be expressed, as derived by Ford and Bland, as:

$$\lambda' = \lambda\sqrt{\frac{R}{R'}} + \frac{1}{2}\left(1 - \frac{R}{R'}\right) \quad \ldots (2.75)$$

The average value of λ′ corresponding to the same set of experiments show, as given in Table 2.4, that average values of λ′ are more consistent.

Based on these findings, Ford concluded that for all practical purposes, two values of λ′ will suffice namely,

λ′ = 0.43 for rolls with matt finish

λ′ = 0.48 for rolls with polished surface

Table 2.4 Average values of λ′ for all results determined by Ford

Material	*Thickness in.*	*Roll finish*	*Average value of λ′*	*Percentage of each value of λ′ differing from the average by not more than*		
				5%	*10%*	*20%*
0.2% C steel	0.100	Mirror	0.47	55	93	97
H.C Copper	0.100	Mirror	0.49	56	88	100
H.C Copper	0.050	Matt	0.43	38	66	97
H.C Copper	0.075	Matt	0.42	44	68	100
H.C Copper	0.100	Matt	0.41	70	91	100
0.2% C steel	0.100	Matt	0.43	90	100	100
0.2% C steel	0.100	Matt*	0.44	79	100	100
0.11% C Steel	0.074	Mirror	0.48	46	71	88
0.07% C Steel	0.065	Mirror	0.47	53	95	100

*Rolls dry. All other results refer to rolls lubricated with Vacuum 40A oil.

From equation (2.75), one gets,

$$\lambda = 0.5\sqrt{R/R'} - (0.5 - \lambda')\sqrt{R'/R} \quad \ldots (2.76)$$

and R′ to be calculated form Hitchcock's formula given in section 2.1.12.

2.4.2 Rolling power calculation

The total power required during rolling can be divided into three main groups, namely:

(i) that needed to deform the material as it passes through the roll bite, this being called the basic power of rolling;

(ii) that needed to overcome the frictional resistance in the roll neck bearings;

(iii) that required to balance the frictional losses in the pinions, gear box and losses in electric motors and various electrical circuits, which are grouped into one called drive efficiency.

The above definition of rolling power does not include the combined power requirement of a coiler and decoiler during coil to coil rolling. This is somewhat complex because while coiler motor consumes power, decoiler motor acts as a generator and generates power, which is fed back to grid. Moreover, due to the effect of front and back tension from coiler and decoiler, the basic power of rolling gets reduced.

Basic rolling power

As discussed in an earlier section, if P is the roll separating force in tonnes (1000 kg), and 'a' is the lever arm in mm, then in each revolution of the roll, the work done per roll is,

$$\text{Work done} = 2\pi a\,P \text{ mm-tonnes}$$

The total work done considering both rolls will be,

$$\text{Total work done} = 2\,(2\pi aP) \text{ mm-tonnes.}$$

During a pass, if the rolls are rotating at N rpm, then total work done per minute is

$$4\pi aPN, \text{ mm-tonnes per min}$$

So the basic kilo-watt power needed for getting above work per minute is,

$$\text{kw power (basic)} = \frac{1000 \times 9.81 \times 4\pi a\,PN}{60 \times 1000 \times 1000} = 0.002\ aPN \text{ kw} \quad \ldots (2.77)$$

Using equation (2.74) for the value of a,

i.e., $$a = \lambda\sqrt{R\Delta h},$$

then $$\text{kw power (basic)} = 0.002\lambda PN\sqrt{R\Delta h} \quad \ldots (2.78)$$

where value of λ can be determined by the equation (2.76).

Roll neck bearing power

Each loaded roll neck bearing, i.e., each four bearings of a 2-Hi mill, each 4 bearings of the back-up roll in a 4-Hi mill or each four bearings of the outer rolls of a 3-Hi mill, experiences a load equal to half of the roll separating force, assuming that the metal is centrally positioned with respect to the roll barrel. If μ is the value of coefficient of friction associated with roll neck bearings, and 'd' is the journal diameter, then work to be done per revolution of the rolls to overcome the frictional forces developed in the four bearings due to the rolling load will be:

$$\text{Work done} = 4 \times \frac{2\pi P}{2}\;\frac{\mu d}{2} \times \frac{1000 \times 9.81}{1000} \text{ meter-Newton}$$

where,
P = roll separating force, in tonnes
d = journal dia in mm of respective loaded roll

If N is the rpm of the respective roll, then

$$\text{Kilo-watt power of bearings} = \frac{1000 \times 9.81}{1000 \times 60 \times 1000} \times 2\pi P\mu dN$$

$$= 0.001\ P\mu dN. \qquad \ldots (2.79)$$

In order to find out the value of the bearing power, numerical value of μ is to be used. Table 2.5 shows the value of μ for different types of bearings, as published by L.R. Underwood[10] in 1943.

Table 2.5 Values of the coefficient of friction for various types of roll-neck bearings

Types of bearing	*Coefficient of friction* μ	*Types of lubrication*
Roller	0.002 to 0.003	Oil or grease
Totally enclosed fluid film	0.001 to 0.004	Oil
Laminated synthetic resin	0.005 to 0.010	Water or soluble oil solution
Plain white- metal	0.01 to 0.04	Pressure grease or oil
	0.05 to 0.08	Imperfect grease lubrication
Bronze	0.05 to 0.08	Pressure grease
	0.08 to 0.15	Grease brick

References

1. Helmi and Alexander, "Geometric Factors Affecting Spread in Hot Flat Rolling of Steel"; J.Iron Steel Institute, N.206, 1968.
2. "Elastic Deformation of Rolls during Cold Rolling", A.S.M.E. Research Publication "Roll Neck Bearings", published by Amer.Soc.Mech.Engg.

3. Tselikov, A.I. "Stress and Strain in Metal Rolling", MIR Publishers, Moscow, 1967.
4. Nadai, A., "Theory of Flow and Fracture of Solids", 1950.
5. Ekelund, S., "The Analysis of Factors Influencing Rolling Pressure and Power Consumption in the Hot Rolling of Steel", Steel, 1933 (translated from Jernkontorets Ann, 1927).
6. Sims, R.B., "Calculation of Roll Force and Torque in Hot Rolling Mills", Proc. Inst. Mech. Eng., 1954.
7. Ford,H. "Experimental Research in Cold Rolling of Metals", J.West Scot. Iron and Steel Inst., 1944-45.
8. Ford, H, "Research into the Deformation of Metals by Cold Rolling", Proc. Inst. Mech.Engg., 1948.
9. Ford, H, "Cold Rolling Technique", Sheet Met.Ind., 1949.
10. Underwood, L.R., "Roll Neck Bearings", Published by Iron an Steel Ind. Res. Council, 1943.

CHAPTER 3

Rolling Practices

In chapter 1, the process of rolling has been discussed along with the definitions and meaning of different terms used in the field of rolling technology. The different types of rolling stands that have evolved over time with their specific design features and uses have also been discussed.

In chapter 2, the mechanics of the rolling process has been discussed. Theoretical analysis of the rolling process in roll bite has been carried out to estimate the mean specific pressure necessary to be exerted on the metal in order to have the rolling with desired reduction possible. The role of the static yield stress of the metal and the effect of various other rolling conditions on this yield stress have also been discussed so that the actual rolling force involved can be worked out. Estimation of the rolling force along with torque and power necessary for rolling are the basic data for selection of the rolling stand to do a particular rolling operation.

In this chapter, the rolling practices generally followed by rollers have been discussed. For producing one or a group of finished rolled products (of different metals/alloys, shapes, sizes, quantities, etc.) the choice of mill stand design, its layout, rolling methods and techniques, pass sequence and schedule, etc., followed by a roller, is generally the meaning of the term *rolling practice.*

It is important to understand that the various rolling practices have been developed based on practical experiences of many rolling practitioners over ages. Still there exists a lot of variation between practices followed by different rollers and rolling plants. For producing same or similar product mix, different plants may be having different rolling equipment and / or different equipment layout. To suit the shape and size of the input materials and the type and layout of equipment, the plants are following different alternative rolling methods and pass schedules.

In this chapter effort has been made to explain some of the widely accepted practices for rolling of common products and provide background information on the principles and applications of roll pass design for the products.

3.1 Layout of Stands

A rolling mill may have one or more mill stands depending upon the number of passes required for rolling one or a set of products and also upon the required output from the mill. Depending on the positional arrangement of the stands with respect to each other and corresponding path of the rolling stock through these stands, the different arrangement of stands in a mill is called *layout of stands.* When number of stands is more than one, layout of stands are also referred as *mill train.* Each of the layouts is given a specific name as detailed below.

3.1.1 Single stand

This is the simplest and the most common type of layout. This consists of one mill stand connected with its drive consisting of motor, gear box, pinion stand, and spindle-couplings.

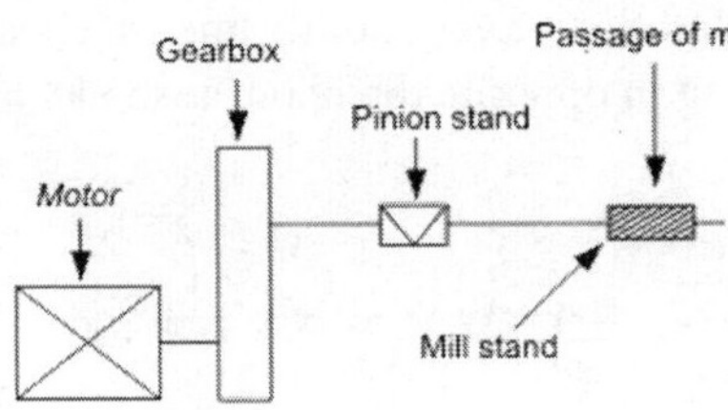

Fig. 3.1 Single stand layout

This layout is used as : i) reversing blooming, slabbing, plate mills; (ii) non-reversing 2-Hi mills for rolling tubes and for cold rolling of sheet and strip; (iii) 4-Hi or multi-roll mill for cold rolling narrow and wide strip and also for hot rolling wide strips; (iv) 3-Hi mill for rolling sections and sheet, etc.

3.1.2 Multi stands side by side

This layout is also called *open train,* which offers the facility of access to the mill stands from both sides. All the stands in such a layout are driven by one motor through a common gear box and pinion stand.

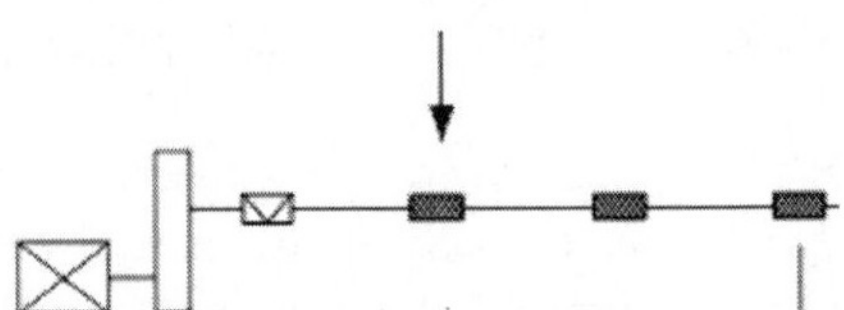

Fig. 3.2 Three stands side by side

The number of stands in such a layout depends on the specific function of the mill. Generally two to five stands are used. These stands are used for rolling large and small sections and also for plate and wide sheets. The individual stands are 3-Hi stands.

The major drawbacks of this layout are (i) the speeds of the train are fixed and (ii) it becomes necessary to transfer the rolled stocks laterally between one stand to the next.

3.1.3 Looping train

This consists of several alternate 2-Hi stands arranged side by side, in which the stock coming out from one stand after one single pass is looped to the next stand. For this to happen, the adjacent stands must operate in the opposite directions. This is achieved by the *alternate 2-Hi drive* system as illustrated in Fig. 3.3(a) utilizing one single motor, G.Box and 3-Hi pinion stand. Alternatively, the alternate 2-Hi stands may be displaced slightly to form another line which is driven through two 2-Hi pinion stands rotating in opposite direction, as shown in Fig. 3.3(b).

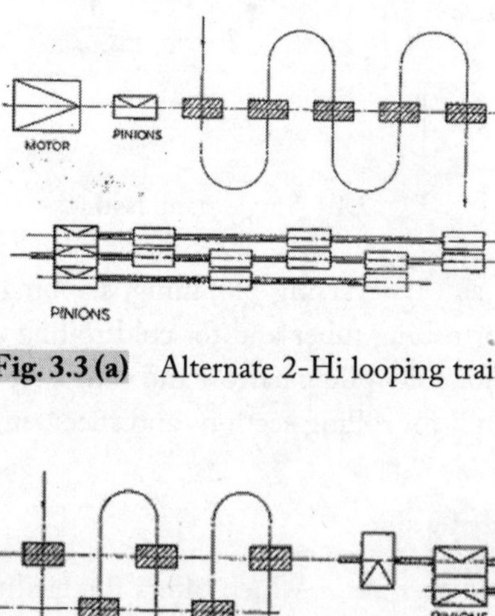

Fig. 3.3 (a) Alternate 2-Hi looping train

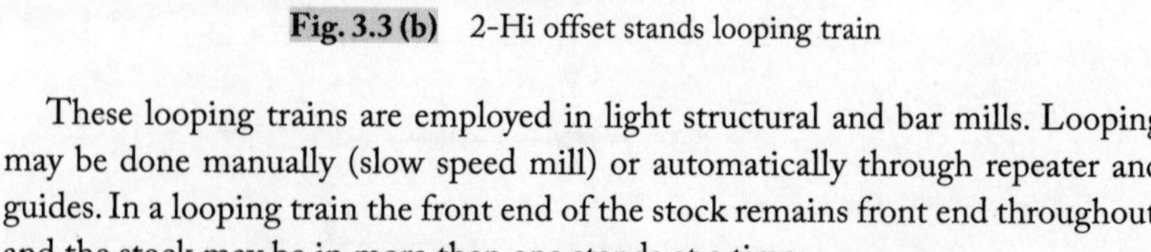

Fig. 3.3 (b) 2-Hi offset stands looping train

These looping trains are employed in light structural and bar mills. Looping may be done manually (slow speed mill) or automatically through repeater and guides. In a looping train the front end of the stock remains front end throughout, and the stock may be in more than one stands at a time.

3.1.4 Two-stand tandem mill

In this layout, two single stands are laid one after another in two lines, as shown in Fig. 3.4. The stands may be 2-Hi, 3-Hi or 4-Hi and are reversible.

The input stock is rolled in several passes, first in the first stand and then in the second stand. This type of mill layout is generally used for rolling plates.

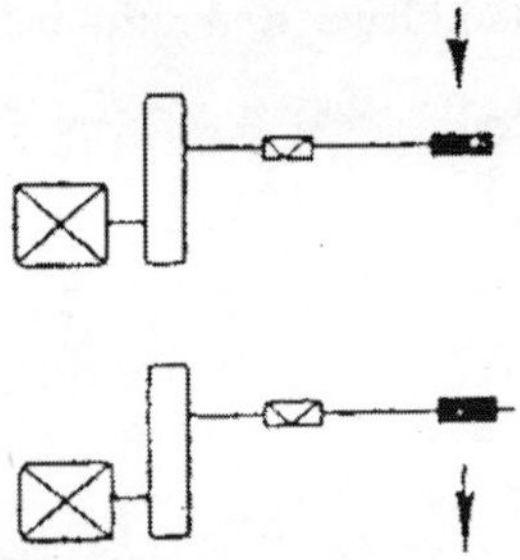

Fig. 3.4 Two stand tandem mill

3.1.5 Stepped trains of side by side stands

For rolling of sections, a two stand tandem mill may not provide requisite number of passes. In such a case this type of layout, also called *stepped open train*, may be used where side by side stands may be used. The layout is shown in Fig. 3.5. Each of the trains in stepped layout having more than one stand is in itself an alternate 2-Hi train.

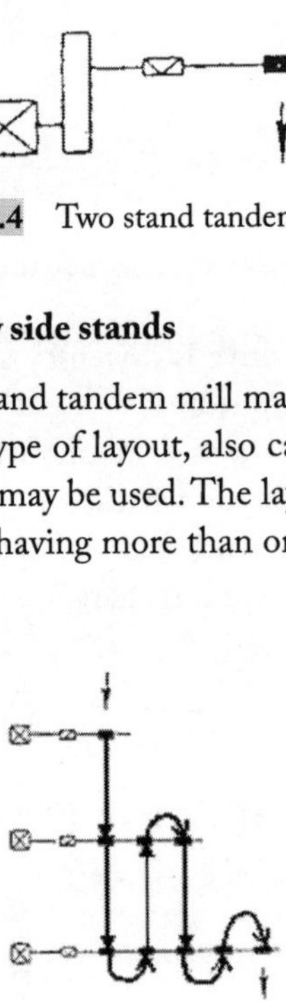

Fig. 3.5 Stepped open train

3.1.6 Tandem continuous stands

In order to avoid the time loss in reversing the direction of rolling, a number of stands are put in different lines close together. Each of the stands offer single pass and the stock undergoes rolling in all or several stands simultaneously. The diameter and revolutions of the rolls in each stands are so chosen that the volume of metal passing through each stand per unit time is equal, i.e., $a_1v_1 = a_2v_2 = a_3v_3 = a_nv_n =$ constant, where, a_n and v_n are the cross sectional area and velocity of the stock at the output of the nth stand, respectively.

Continuous stands may have a *group drive* or *individual drive*, as shown in Fig. 3.6(a) and 3.6(b), respectively. In group drive, a group of stands is driven from one motor but through series of gearings having different fixed ratios. Thus the revolutions of rolls of each stand have fixed ratios between them.

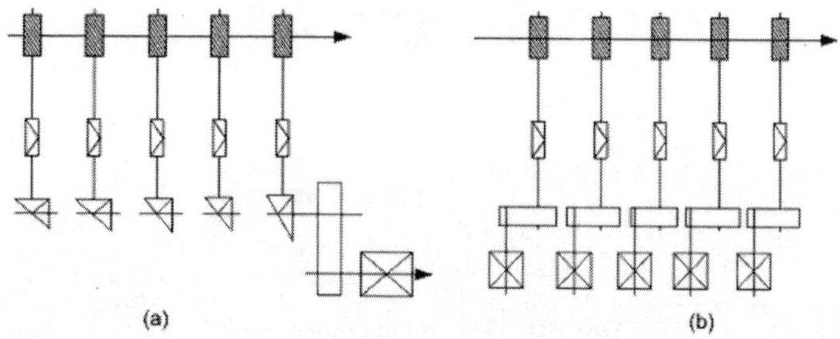

Fig. 3.6 Continuous stands with (a) group drive and (b) individual drive

Each stand with individual drive is driven by separate drive. Hence, their roll speeds can be individually controlled.

Tandem continuous stand layout is economically employed for large production of hot or cold strip in coils. The drawbacks of continuous mill are difficulty in adjusting the stands. Hence they are not preferred for rolling different types of shaped products involving frequent roll charges.

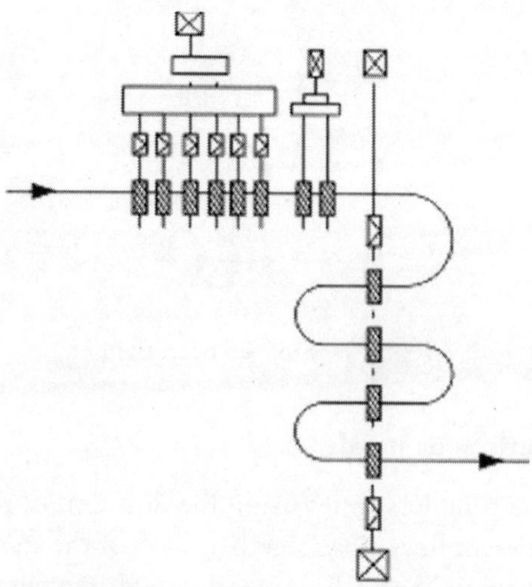

Fig. 3.7 Semi-continuous mill layout

3.1.7 Semi-continuous mill layout

For rolling of complicated sections (angles, channels, etc.) continuous mills have not been used owing to the trouble of setting the rolls in all stands. Semi-continuous layouts have been widely used for these products. Figure 3.7 shows the layout. The first part consists of a few stands in continuous layout that is used for primary roughing passes of the input stock. The intermediate and finishing passes are taken in the second set of stands which are laid out as looping train. In these looping train stands frequent adjustment of roll is required due to wear of the grooves.

3.1.8 Cross-country mill layout

Cross-country mills are very popular for rolling medium and heavy sections. In this layout the stands are placed side by side and the stock is transferred from one stand to the next, and the direction of rolling is reversed in each successive stand as shown in Fig. 3.8(a). This type of layout is also known as *zigzag layout*. Alternatively, the stock may pass through more than one stands in one direction, then, in order to reduce the total length of the mill, the stock is transferred and returned back through another set of mill stands. This layout is shown in Fig. 3.8(b). Such a layout is characterized by the facts that the stock is never in more than one stand at any time, separate stands have separate rolling speed for each pass, and after each transfer the front end of the stock becomes back end until the next transfer. Any one or more stands at the beginning of the mill may be a multi-pass 3-Hi or 2-Hi reversing stand. The term "cross country" is often applied to any arrangement of stands which makes use of to and fro (or zigzag) motion of the stock at any stage of the layout.

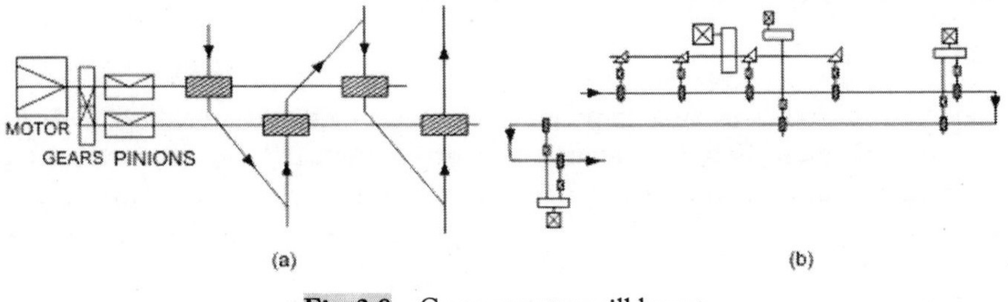

Fig. 3.8 Cross-country mill layout

3.1.9 Chequerboard mill layout

Chequerboard layout is a special variety of cross-country layout, in which the last stands are laid out in staggered (zigzag) formation in order to save space. This is shown in Fig. 3.9.

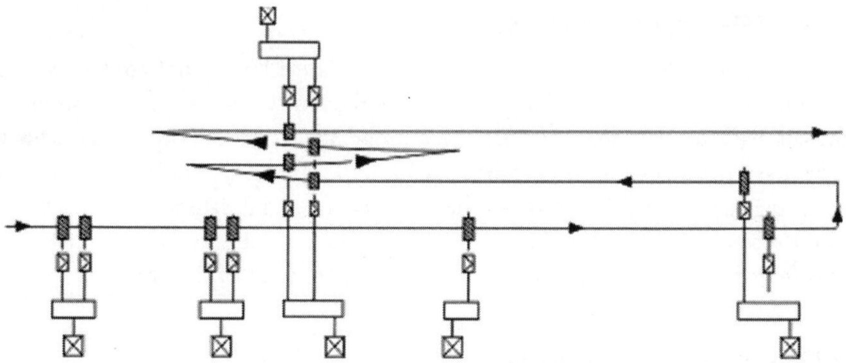

Fig. 3.9 Chequerboard mill layout

3.2 Rolling Applications

3.2.1 Introduction

Rolling is used for manufacture of different types of finished or semi-finished products. For manufacturing these different types of products, the rolling practices being followed are also different from various aspects like (i) method of rolling; (ii) rolling techniques; (iii) types of mill stands; (iv) mill layout; (v) rolling equipment other than the mill stands; (vi) rolling pass schedule, etc. These aspects considered together for rolling one or a group of similar products is termed as *rolling practices.* These practices constitute the practical application of rolling principles. The rolling practices are briefly discussed, groups of product-wise, in this chapter.

3.2.2 Rolling of steel blooms, slabs and billets

(i) Definitions

Blooming (also called Cogging) mills are primary rolling mills to reduce reheated cast *ingots* weighing from 1 ton (small size) up to 20 tons, to blooms of cross section from 125×125 mm up to 450×450 mm. Slabs of sizes ranging from 75 to 250 mm in thickness and 400–1600 mm in width may also be rolled in a blooming mill by using suitable rolls.

These are generally reversing 2-Hi single stand mills, categorized as large, medium and small depending on roll diameters of 1150 to 1300 mm, 900–1000 mm and 750–850 mm respectively. The barrel lengths are between 2800 mm to 3000 mm. Blooming mills with two 2-Hi stands in tandem and continuous blooming mills consisting of several non-reversing 2-Hi stands are also used. The ingots used are generally with square cross section and with small taper from one end to another. This taper is necessary for easy removal of the cast ingot from their metallic moulds.

Slabbing mills are specially designed for reducing rectangular section ingots weighting from 7 tons to 22.5 tons to slabs from 75 to 250 mm thickness and 700 to 1600 mm or more width. The most common are two stand universal slabbing mills. The first stand is a vertical stand with about 900 mm roll diameter and the second is a 2-Hi horizontal mill with 1100–1150 mm roll diameter and about 2100 mm barrel length or more.

Billets are rolled products having square cross sections normally ranging from 50×50 mm up to 200×200 mm. The billets are rolled from the blooms produced in a blooming mill. It is, therefore, customary to locate the billet mill just after a blooming mill, which avoids reheating the blooms. Billet mills are continuous vertical and horizontal stands.

Figure 3.10 below gives a schematic layout of a continuous billet mill, which is placed behind a blooming mill. The billet mill consists of two continuous vertical and horizontal trains of 6 stands in each. In the first group horizontal roll diameters are 730 mm and in the second 530 mm. Starting from 300×300 mm blooms, the first train can produce down to 120×120 mm billets while up to 60×60 mm billets can be produced at the end of the second train.

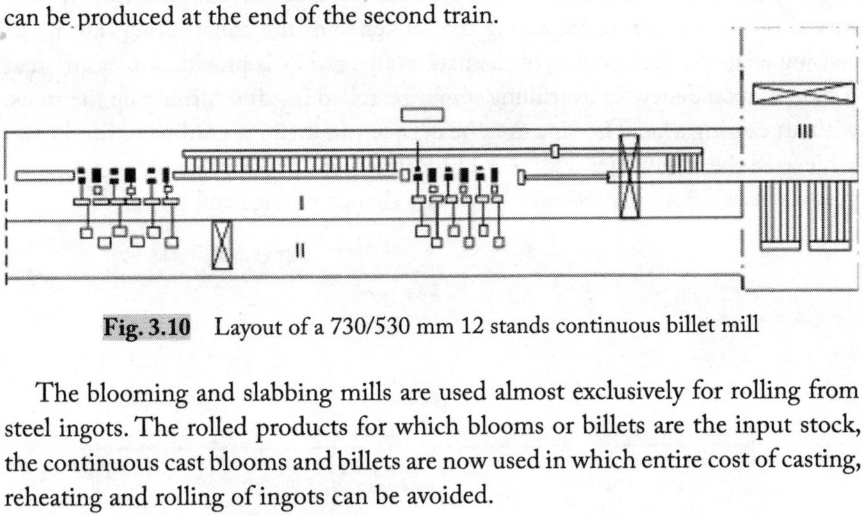

Fig. 3.10 Layout of a 730/530 mm 12 stands continuous billet mill

The blooming and slabbing mills are used almost exclusively for rolling from steel ingots. The rolled products for which blooms or billets are the input stock, the continuous cast blooms and billets are now used in which entire cost of casting, reheating and rolling of ingots can be avoided.

(ii) Method of rolling an ingot

The cast ingot is tapered from one end to the other; the cast structure is generally coarse with marked trans-crystallization zones and somewhat porous due to blow holes in some areas. In the beginning, the reheated (nearly 1250 °C for low carbon steel) ingot is given some light early passes with reduction between 10–15% to eliminate the taper, break down cast structure and close down internal pores. The middle passes are to give heavy reduction up to 25% without causing defects like laps from overfills, corner cracking, surface cracking and surface seams. The last two passes are with little draught of less than 10% to obtain the exact dimension and bloom shape.

If the ingot is rolled for several passes without turning through 90°, tension is set up in the side surfaces due to spreading and differential elongation of the surface and center. This causes cracks on the faces, and it is advisable to turn the stock at intervals of every two passes to work all sides in succession, particularly at the beginning. While rolling in grooves, turning is not critical, hence it is advisable to put the stock into grooved passes as soon as the initial passes are over, to protect the corners from cooling consequently cracking. Later, as the stock becomes smaller, the rolling deformation penetrates deeper and the spread and elongation tends to occur more at the center. If the side spread is too large, heavy wear of the collars (side walls of the grooves) results and there is possibility of overfilling at the roll gap. Two precautions are taken to avoid this. First, the grooved passes are given a convex shape at the bottom, so that when the stock is turned up, it will have hollow sides permitting considerable spread without overfilling the roll gap. Secondly, the collars should be tapered to give a grooved pass an inclined side permitting a little spread of the stock and decreasing the tendency of the collar biting into metal causing a lap. At the opening of the pass a large radius is provided to ensure that if there is a tendency for overfilling it may be rolled in, after turning up the stock, without causing a lap. The taper may be between 10° to 20° according to the duty to achieve all the advantages. Figure 3.11 illustrates the above principle of designing grooved pass for rolling rectangle or square shaped blooms and ingots.

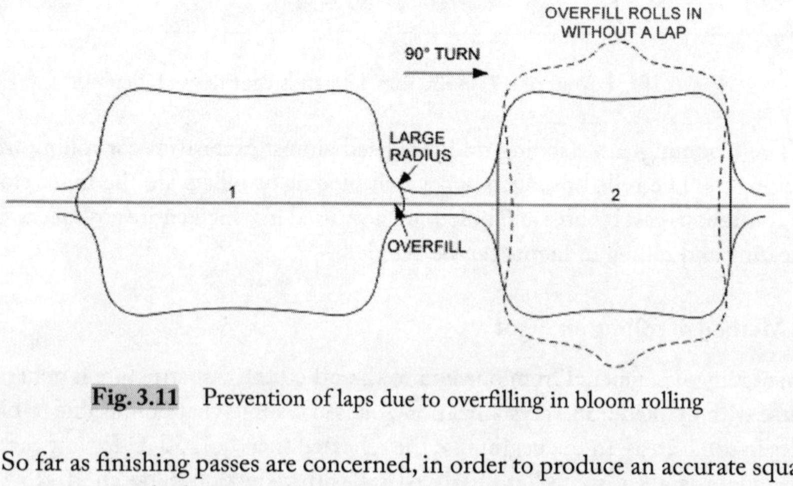

Fig. 3.11 Prevention of laps due to overfilling in bloom rolling

So far as finishing passes are concerned, in order to produce an accurate square, the convexity in the last pass cannot be used and that the penultimate pass should be such that the final reduction will result in a flat sided square by spreading. The radii in the pass corners should also correspond to those desired in the product.

(iii) Pass schedule in a 2-Hi reversing blooming mill

2-Hi mill with sufficient vertical adjustment of top roll through screw down arrangement ensures complete flexibility of choosing draught in any pass. Two different techniques have been developed for designing the pass grooves in the roll. According to the duty required of the mill, the first makes extensive use of a bullhead pass (rolling between plain cylindrical surfaces of two rolls) together with several edging grooves and the second makes use of grooved, i.e. open box, passes throughout. A bullhead pass consists of a portion of the barrel which is plain cylindrical and of a width greater than any stock which is to be rolled in it, and though it may have collars at its ends they are never used to restrict the spread of the stock so that the bullhead gives parallel vertical reduction with free spread. Sketches of these two types of design are shown in Figs. 3.12 and 3.13. The main advantage of the first method is that it economizes on roll barrel space as many passes are taken in the bullhead. If the bullhead pass is placed in the middle of the barrel the majority of the earlier passes are carried out with equal loads and equal wear on the bearings. The main disadvantage is that the stock is not subjected to side work or supported by collars during the bullhead rolling and if the heating is not uniform a parallelogram section may result. The corners and edges are not protected and cracking may occur particularly on "difficult" steels and in the earlier passes, as the spread takes place nearer the surface, some lapping may occur at the corners after turning up the stock for edging.

The general technique with the first method, as shown in Fig. 3.12, is to break down the ingot in the bullhead pass with progressively increasing reductions, after which the stock is alternatively edged in a grooved pass and reduced in the bullhead until the required dimensions are reached. Alternatively, after breaking down and entering the grooved passes the stock is not returned to the bullhead if sufficient grooves are available for the completion of the sequence. This alternative is the one used with the rolls illustrated in Fig. 3.12 and the approximate sequence for obtaining an 8 in. (200 mm) square bloom from a 21 in. (533 mm) square ingot is shown in the figure. Extensive use of the bull head leads to high production due to the decrease in manipulative time, because if a large bullhead is used the need for accurate positioning of the ingot is eliminated. Where very high production is not the prime requirement, a bullhead may be used, in place of a grooved pass, to give flexibility of ingot size, but the first edging pass is made as large as is permitted by the available roll barrel, so that the stock may be worked in a groove as early in the sequence as possible. The bullhead may also be exploited to increase the range of finished sizes which are otherwise limited by the grooves available.

It should be noted that consideration as to whether to use a bullhead or not may be influenced by mill design factors as well as mill duty. Slabs may be rolled by using the bullhead, together with the grooves as edgers, but the surface of these slabs will suffer unless the bullhead has a good surface and it is preferable, if space permits, to have a separate finishing flat pass for slab.

The second type of design as illustrated in Fig. 3.13, requires a long barreled roll as the ingot is placed into grooved pass right from the beginning, so that the spread is restricted and the corners are protected throughout the sequence.

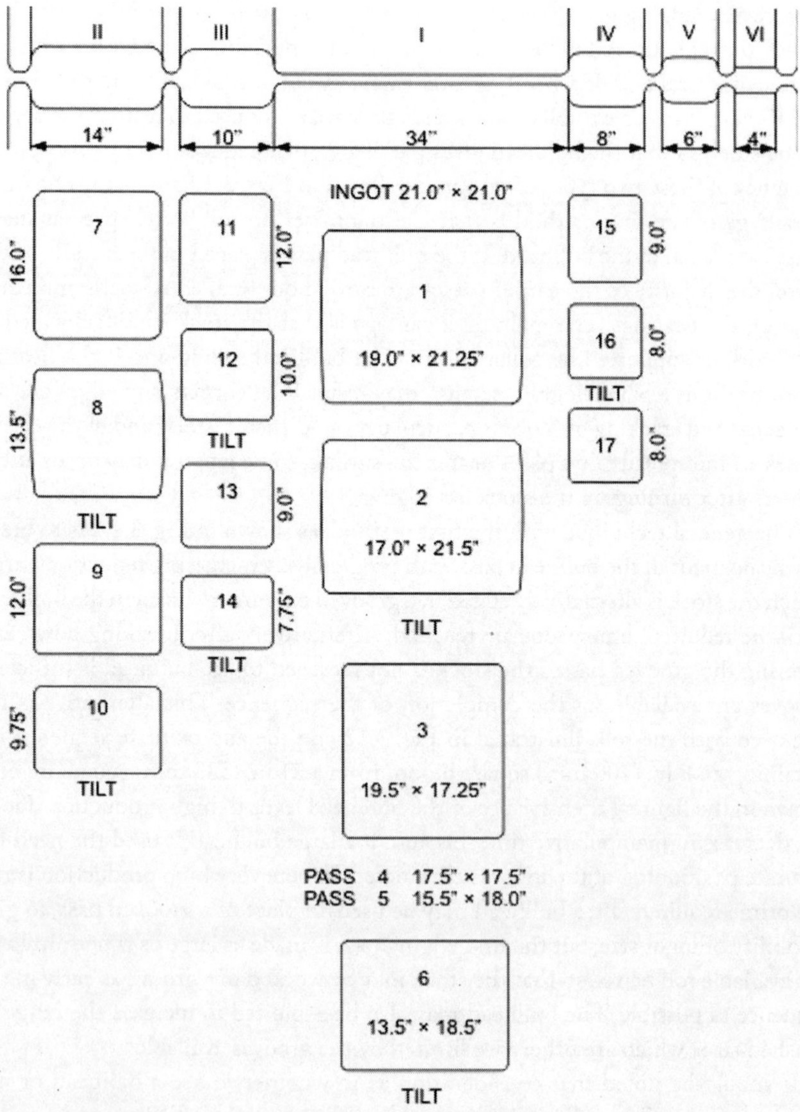

Fig. 3.12 2-Hi blooming rolls and pass sequence- using a bullhead

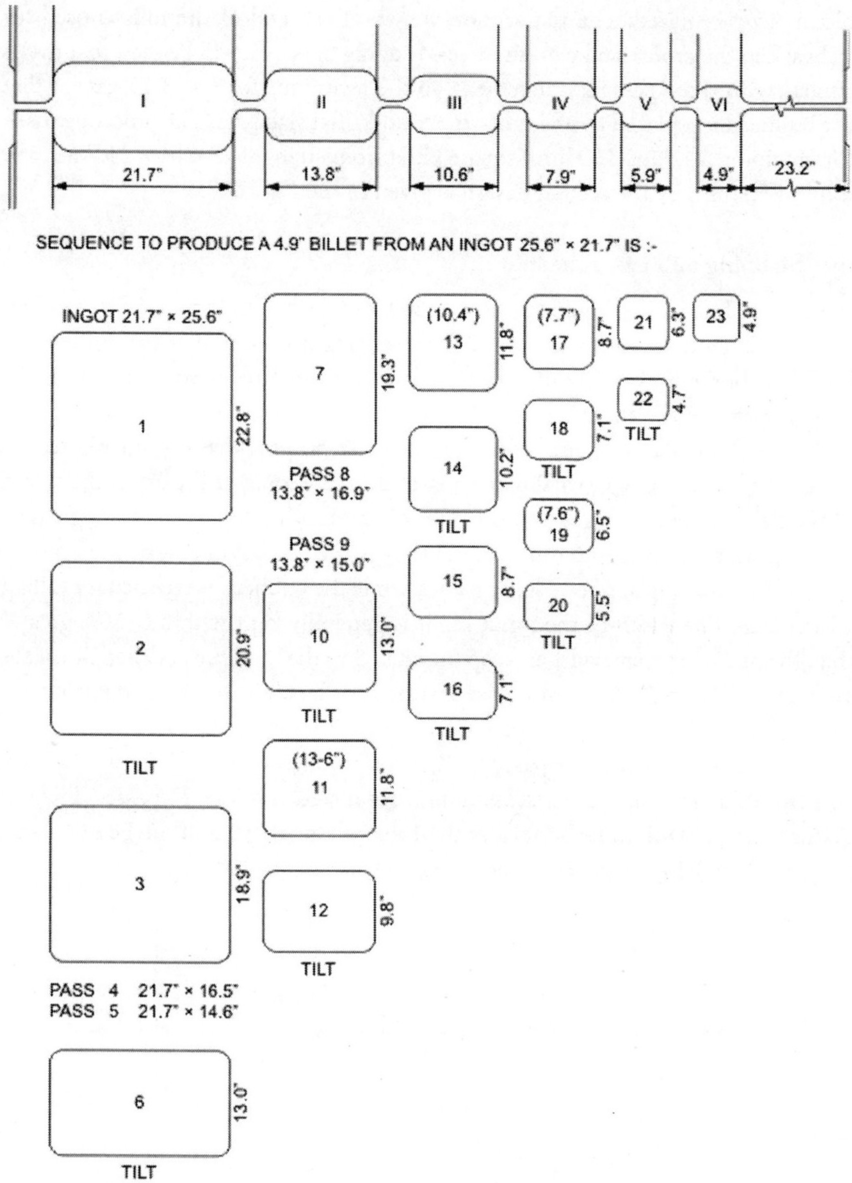

Fig. 3.13 Blooming rolls and pass sequence- using grooved passes

The ingot is reduced in the ingot pass until, with turning, it is of a suitable size to enter the second groove and the process is continued with frequent tilting until the required bloom size is reached. Grooves must be available for all the standard bloom sizes required. The flat section shown at the end of the rolls is used for slabs when the grooves take on the role of edging passes. It will be seen that this is similar, with some pass rearrangement, to the technique for Fig. 3.12, except that the bullhead is replaced by an ingot groove and a flat slab pass. The procedure used for obtaining a 4.9 in. (120 mm) square billet from an ingot 25.6 in. × 21.7 in. (650 mm × 550 mm) by the second design is given in the Fig. 3.13.

(iv) Slabbing mill pass schedule

In many rolling plants, liquid steel is cast as large slabs. Otherwise, slabs are rolled from rectangular ingots either in a 2-Hi reversing mill or a universal mill which has plain horizontal rolls and a pair of driven vertical rolls (edger rolls) on the outgoing side of the stand.

In the 2-Hi slabbing mill the top roll must be provided with a higher lift compared to that on a normal blooming mill. The degree of lift limits the width of the slab that can be edged and the slab thickness is limited by the edging passes available. In a slabbing mill roll, some of the grooved passes of a normal blooming roll are therefore replaced by edging passes, while the bullhead is retained for rolling widest slab. The width of the input ingot is generally kept within 6–10% greater than the maximum slab width to keep the edging to the minimum. When thin slabs are required, the bullhead is often used without collars. The design and setting out of the grooves follow the principles set out for blooming mill.

Table 3.1 shows a realistic pass schedule for production of a 1025 mm×225 mm slab from a 1175 mm×700 mm medium carbon steel ingot in 21 passes. The roll diameter is 1050 mm. The barrel length of the roll for above mill can be calculated from the Fig. 3.14, which works out to be 2485 mm.

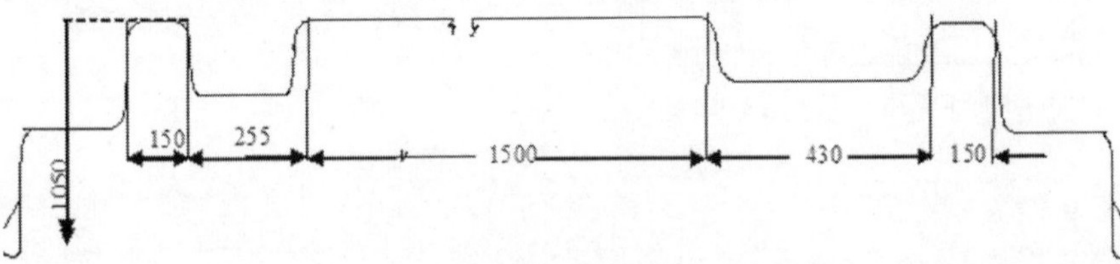

Fig. 3.14 Roll of a slabbing mill

Universal mills which have plain horizontal rolls and a coupled pair of vertical driven rolls at the outgoing side of the stand are also used for slab rolling. The horizontal rolls control the slab thickness while vertical rolls control the width. Large cast slabs are ideally suited as input to such a mill.

Table 3.1 Typical slabbing mill pass schedule

Rolling Groove	*Pass No.*	*Draught, mm*	*Height, mm*	*Width, mm*	*Manipulation*	*% Reduction*
Bullhead	Ingot	---	700	1175	---	---
	1	50	650			7.14
	2	38	612			5.85
	3	26	586			4.25
	4	38	548			6.48
	5	38	510			6.93
	6	38	472			7.45
	7	32	440			6.78
	8	30	410		Tilt	6.82
I (430 mm)	9	---	1125	(410+)		---
	10	38	1087			3.38
	11	38	1049			3.49
	12	25	1024		Tilt	2.38
Bullhead	13	..	400	(1024+)		..
	14	38	362			9.50
	15	38	324			10.50
	16	31	293			9.57
	17	31	262			10.58
	18	31	231		Tilt	11.83
II(255)	19	..	1050	(231+)		..
	20	38	1012		Tilt	3.62
Bullhead	21	..	225	1025		..

(v) Billet mill pass schedule

Billets are smaller version of blooms and are used for re-rolling to bars, rods, etc., and are also used for forging or direct machining. Billets are always rolled from blooms, unless continuous casting process is followed to produce billets directly.

It has been mentioned that primary mills (blooming mills) can be used for production of billets of size not less than 100 mm square and when the quantity of billet is low. However, for higher production, this is not an economic proposition. In the later case, the bloom at a suitable size is taken out and rolled in billet mill with smaller rolls and higher output speed. Frequently blooms of 175×175 mm size are used. The billet mills may be mainly of three types:

(a) Open train billet mill consisting of 2-Hi or 3-Hi stands
The open train with two 3-Hi stands is shown in Fig. 3.15(a). Due to the difficulty of manipulation, the bloom is cut into several small pieces and reheated before rolling.

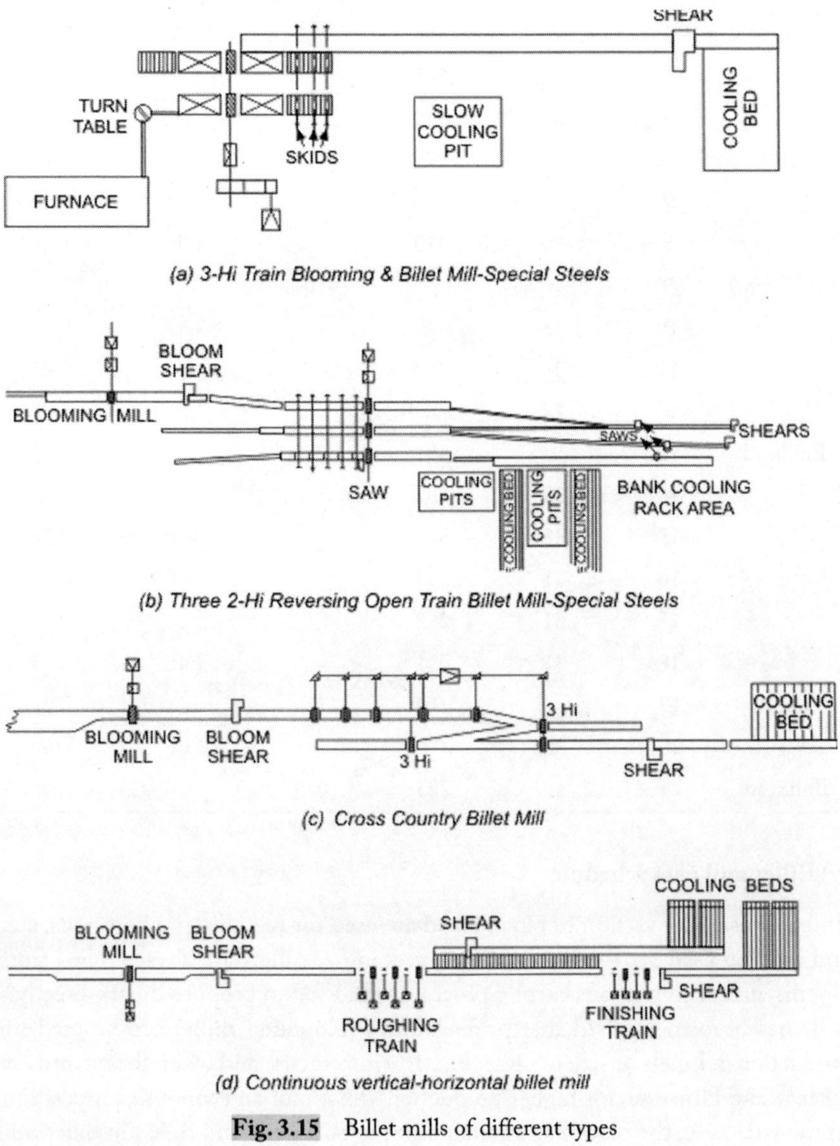

(a) 3-Hi Train Blooming & Billet Mill-Special Steels

(b) Three 2-Hi Reversing Open Train Billet Mill-Special Steels

(c) Cross Country Billet Mill

(d) Continuous vertical-horizontal billet mill

Fig. 3.15 Billet mills of different types

The open train may also consist of two or three 2-Hi reversing stands to match the tonnage from the blooming mill, without the need for reheating. The layout is shown in Fig. 3.15(b).

(b) Cross-country mill consisting of several 2-Hi and/or 3-Hi stands following the blooming mill, as shown in Fig. 3.15(c).

(c) Continuous mill consisting of alternate vertical horizontal stands in two sets, one for roughing passes and the other for finishing passes, as shown in Fig. 3.15(d). In this layout, blooms may pass from the cogging mill straight through the billet mill without reheating and without shearing into smaller length. This gives a high yield and high and uniform finishing rolling temperature; hence much better control of shape and quality of billets than with other layouts.

A flying shear installed at the output cuts the long billet to size. For production of billets from blooms, following four different grooved pass sequences may be employed. (Refer to section 2.1.9 for types of passes).

(a) Open box passes

These, whether they are in the 3-Hi, 2-Hi, cross-country or continuous mills, are similar in principle to the ones described under blooming mills. The basic principles are: inclined sides, radiused collars, and fillets in the bottom corners of the pass. Sometimes chamfers are used instead of fillets as some designers feel that the corners of the stock are worked better by this means. The percentage reduction taken in a box pass depends to some extent on the amount of spread restriction, but generally lies between 20–30%. If sufficient allowance is provided for spread, higher reduction within the power and bite limitations may be taken.

(b) Open square sequence

The open square is a square set on the diagonal with an apex angle of from 91°–98°. A rhombus with an apex angle greater than this may be considered as diamond. In this sequence, the stock issuing from one pass is turned up 90° and entered into the next square, the angle being gradually decreased towards the finishing pass. The last pass is usually given an angle of 91° or 92° as this permits a greater life, because as the pass wears the angle decreases and when it drops below 90° it must be restored by dressing.

The open square pass rolls all four sides of the billet but the draughts have to be light as there is little allowance for spread, and reductions of 20% may lead to overfilling. To obtain higher draughts the obvious way is to broaden the open square by increasing the apex angle so giving more room for spread. This gives a diamond–

diamond sequence and the obvious drawback is that not only entry difficulties may be encountered but the diamond is not stable and will turn down. To avoid this problem and still achieve good draughting, a diamond–square sequence may be used.

(c) Diamond–square sequence

This sequence makes use of Vee-shaped grooves with alternating large and small apex angles. The diamond may be from 100° to 125° while alternate passes will be squares with angles of 92° or 91°, if finished billets are required from that pass. Angle may be a little higher, e.g. 94° or 96°, if no finished product is required from the pass. The diamond is usually limited to 125° as above this limit the entering square tends to turn down through 45° and enter the pass on the flat as when entering an oval. Also a wide flat diamond is more difficult to enter into the next square groove due to the high angle of contact. Reduction up to 28% is possible in the diamond to square stage. The inclined sides of the square tend to restrict spread but the large vertical draught from a diamond to a square gives such a degree of spread and such an angle of contact that this limit of 28% reduction is difficult to exceed.

In the reduction from square to diamond the actual vertical draught is less than the above for a given reduction in area and the pass is wider so allowing more spread. However, the more open sides impose less restriction on spread and 30% is found to be the approximate limit of reduction.

These two steps give an overall reduction from square to square of about 47.5%. If a higher rate of reduction is required a square–oval–square sequence must be used. The bottoms of the grooves of both square and diamond are usually filleted or chamfered for ease of working. The fillet or chamfer ensures corner working but it must be of the correct dimensions. The radius in the passes should be large enough to enable the entering stock to fit the corners but must be small enough to hold the stock stable. The finishing pass will have the radius required on the finished billet and the earlier passes will be derived from this and form the need to prevent overfilling.

Another drawback of diamond–square sequence is that the corners cool fast and for alloy steel cracks may develop during rolling of the corners hence not recommended. Figure 3.16 shows a typical pass sequence for rolling billets in a continuous mill.

(d) Oval–square series

This is a heavy reduction series which is widely used for rolling products from billets but it is not so common for billet production. However, it may be used if there is a shortage of passes but no shortage of power.

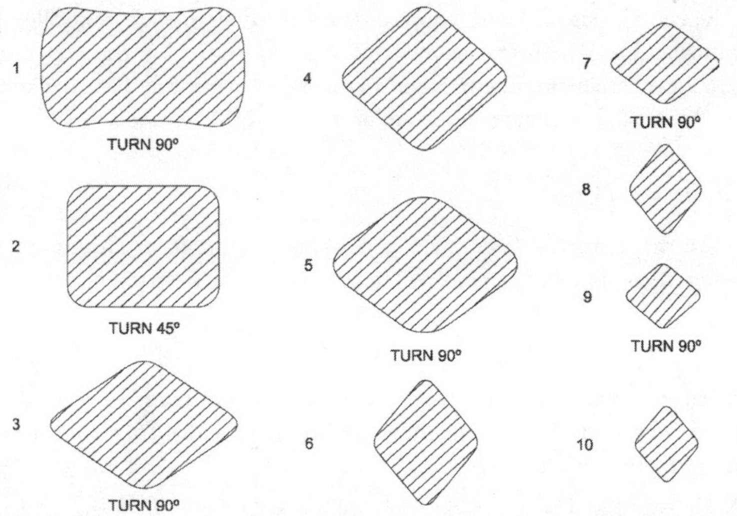

Fig. 3.16 Diamond–square pass sequence for a billet mill

Percent reductions from 1st to 10th passes are: 23.65%, 17.66%, 16.37%, 16.75%, 20.87%, 27.5%, 30%, 28.5%, 26.5%, and 25%, respectively

(e) Summary

The decision as to what pass sequence should be used for billet rolling depends on the overall reduction required and the number of passes and power available. If, for example, ten passes are available for getting from the bloom area A to the billet area B, then the overall reduction is $\frac{A-B}{A} \times 100$ percent, and the overall elongation is, $E' = \frac{A}{B}$. If alternate passes are to be squares and the elongation from square to square is constant throughout the sequence, this elongation equals $\sqrt[5]{\frac{A}{B}}$.

If this is not greater than 1.86, a diamond–square sequence may be considered, but if it is greater, then a sequence, of higher reducing ability must be used. If the square to square elongation calculated above is lower than 1.5, the open square sequence may be used. In a mill where the number of passes may be varied (e.g., with a 2-Hi reversing mill) the sequence will depend on the quality of steel being rolled, the output required and the roll barrel available. With alloy steels the oval square sequence would probably be avoided but with low and medium carbon steels it may be chosen if a high output is required and if the power is available. The reduction from square to square may then be suitably proportioned between the two passes in question, e.g., square–oval: oval–square = 1:0.7, and square–diamond: diamond–square = 1:0.83. The individual passes may then be designed to give the reduction in area required.

In the continuous mill with coupled drives the draughts are controlled by the gear ratios and roll diameters from stand to stand. Assuming the gear ratios are fixed then the variables are reduction in area and effective roll diameter and these must be controlled to ensure equal rate of output from all stands.

3.2.3 Rolling of plates

Plates are made mostly from steel and some amount from aluminum alloys. Following discussion will be confined to steel plate, unless stated otherwise.

(i) ***Definition***

Hot rolled steel plates generally have thickness from minimum 3 mm (1/8 in) and above and having width of several meters. Plates may be divided into three groups according to production process:

(a) Universal plates are made from rolling slabs between horizontal and vertical stands simultaneously and trimmed on ends only. These are usually 1525 mm wide.
(b) Sheared plates are rolled between a pair of horizontal stands and sheared to size on their edges and ends. These are up to 5000 mm wide at the present time and widely used for construction of ship body.
(c) Rolled floorplates which are flat hot rolled finished products having a pattern rolled on their surface in the finishing pass.

(ii) ***Method of rolling***

Slabs are sheared into lengths suitable for rolling into plate sizes desired. Due to the large width to thickness ratio of input stock, the spread is quite small and it is never considered for achieving the final plate width. Therefore, for rolling very wide plates, broadside rolling (i.e., the length of the input stock is used as width for rolling) is resorted to in one of the following two ways:

The slab may be rolled in length in normal way till its length is approximately equal to the required width of the plate, when it is turned broadside on and further rolled to finished thickness. This method is called the *length to width rolling,* as the original length becomes width of the plate. Some experts feel that this is detrimental to the properties of plate for certain uses. If the original slab length can be made equal to the width of the plate, *broadside rolling* can be commenced without any length rolling.

Alternatively, the input slab is turned broadside at the very beginning and rolled until its original width has become elongated to the desired width of the plate, when it is again turned back to its original lengthwise orientation and rolling is completed to the desired thickness. This method is called the *length to length rolling.*

After rolling in plain barreled horizontal roll, plate edges that get deformed due to spread are cold sheared to actual width. In universal mill, width of the plates is accurately controlled. Suitable grooves may be used to give round edged plates. The adjustments available in the vertical rolls limit the range of width of these universal plates.

(iii) ***Rolling techniques***

In rolling of thin flat products including wide plates, one major consideration is the uniform thickness along the width of the product. The three main factors that affect the thickness are:

(a) *Roll bending:* During rolling, the rolls bend in an arc due to roll separating force produced over the area of contact between the plate and the roll. Conforming to the shape of roll gap due to roll bending, a *crown* is produced on the plate. Crown is the difference of thickness between the middle and edges of the plate. Roll bending under a given load may be minimized by increasing the roll diameter. But when the roll diameter increases, the rolling load increases for the given reduction, and the roll bending also increases. Therefore, reduction of roll bending by increasing the roll diameter is rather limited in scope. In a 2-Hi mill this bending can be calculated. In 4-Hi mills, this roll bending effect is much less. Smaller work rolls used lead to less roll separating force, and the work rolls are backed up by much larger diameter rolls whose bending is quite small. Similar condition prevails for Lauth type 3-Hi plate mill where small work roll is supported by large roll on each pass.

(b) Roll shape: In order to offset the roll bending effect, the rolls are machined with a positive *camber, i.e.,* the center of the roll is made slightly more in diameter than that of the two edges. The amount of this camber follows from calculation of roll bending. If the draughting schedule is designed to result in almost equal rolling force in each pass including the finishing pass, and if the roll camber is based on this constant roll bending under nearly constant rolling load, it can be assumed that a good shape of the plate should result. If the roll load is set to the mill capacity, i.e., the maximum safe load, the mill will be working with maximum productivity. This, however, means giving maximum camber to the roll.

Unfortunately, under large cambering, rolling may not be stable (the plate develops tendency to run from center towards roll neck). Therefore, the safer practice is to roll with maximum permissible load until last three or four passes, when the load is gradually reduced and the roll cambering is based on the last pass roll load.

It can be easily seen that as rolling continues, the camber will get worn off the rolls gradually. So the common practice is to start rolling of the widest plate with new roll and then progressively work towards narrower plates as camber is reduced. When thickness across the plate goes beyond tolerance range, the roll must be changed and replaced with redressed rolls with original camber.

(c) Roll temperature: During hot rolling, the temperature of the roll also increases and after some time (a few hours of rolling) a state of equilibrium is reached. The equilibrium temperature at the middle of the roll barrel is higher than that at the two edges because the plate does not cover the entire barrel length. This is known as *thermal camber* which assists machined camber. This thermal camber should be taken into account while calculating the camber to be given in the roll.

(iv) ***Types of mills***

(a) 2-Hi mills: For limited production, often one reversing 2-Hi stand is used. However, the disadvantage of single stand mill is that the rolls are worn out fast. Therefore, the surface finish and gauge accuracy are difficult to be preserved for long.

Two stands in open train or in tandem overcome these difficulties to a great extent. The first stand is used as a roughing stand with heavy reduction passes, including any broadside rolling. The second stand preferably with independent drive is used for finishing passes.

(b) 3-Hi lauth mill: Single stand lauth mill is also used for plate production. The stand is fitted with tilting table on both sides. The backup rolls are not cambered and changed infrequently, e.g. weekly, while the smaller middle roll is changed more frequently, e.g. daily. As the rolling progress, the wear of the backup rolls increase progressively which cause crowned plates. To counteract this tendency, each day new middle rolls with progressively higher cambers are used. At the end of the week, the large rolls are changed and the cycle is repeated.

(c) 4-Hi mills: Single stand mill of this type suffers from the same disadvantages as the 2-Hi mill, but some are used. More common is to use 4-Hi stands in open train with a 2-Hi rougher or in tandem with 2-Hi, 3-Hi or 4-Hi stands, to give more output and improved quality. With the introduction of 4-Hi *hydraulic mill** with roll bending control during rolling, the quality of the plates could be improved to a large extent.

* Refer to chapter 1 and also chapter 4 for further information on hydraulic mills.

(d) Other layouts: Semi-continuous mills with multi-pass 2-Hi, 3-Hi or 4-Hi rougher followed by two or more single-pass stands usually 4-Hi, set in tandem have been used. Fully continuous mills with a 2-Hi roughting train and 4-Hi finishing train are sometimes used for high quality high production purposes.

A typical plate mill consists of a scale breaker and broadside 2-Hi mill preceding a 4-Hi rougher equipped with vertical edging rolls, and four high finishing stands. The roughing stand is a 915 mm and 1370 m×3050 mm 4-Hi mill with 0.2 mm work roll camber, while the finishing stands are 685 mm and 1370 mm×2540 mm 4-Hi stands having camber varying from 0.2 m to 0.1 mm on work rolls. The product range is down to 2.38 mm (3/32 in) thick.

Figure 3.17 shows a schematic layout of a plate mill set up in Urals heavy engineering works[1], the USSR.

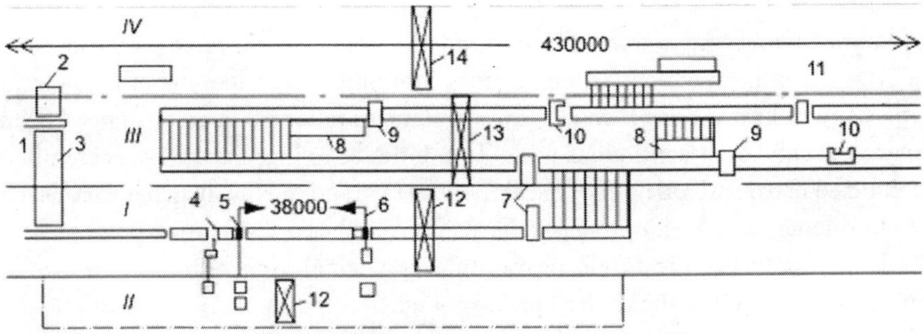

Fig. 3.17 Layout of a plate mill

2- pushers; 3- continuous furnace; 4- vertical stand; 5- roughing stand; 6-universal finishing mill; 7- straightening machine; 8- tilter; 9- marking out car; 10- inclined blade shears; 11- rotary shears; 12- 75/15 T crane; 13- 20/5T crane;14-15T crane.

The mill is for rolling plates from 4 to 50 mm thick and up to 2500 width and 18 m length. The input is steel slabs from 110 to 250 mm thick × 800 to 1550 mm wide and weighing up to 7.5 tons. It consists of a 2-Hi breakdown stand, one 2-Hi roughing stand and one universal 4-Hi finishing stand. The details of the rolling stands are as follows:

Stand no. reference Fig. 3.20	*Roll arrangement*	*Roll dimensions, mm*		*Drive details*	
		Diameter	*Barrel length*	*Power, kw*	*rpm*
4	Vertical	1000	700	750	0–70–210
5	Horizontal	1150	2800	2×1820	0–25–60
6	Vertical	600	–	2×200	0–200–500
	Horizontal	700/1350	2800	5000	0–50–120

3.2.4 Rolling of bars and rods

(i) ***Definitions***

Steel bar and rod mill may be part of the facilities of the integrated steel plants, but these products are often produced by re-rollers catering to market needs.

The starting material for any bar and rod mill is billet (rolled or continuous cast), small rolled slabs or ingot cast in pencil moulds (about 150×150 mm size).

Large sized squares and rounds from 40 to 80 mm size are rolled in medium section mills which will not be covered under present section. By the term ***bar***, it is generally meant rounds, hexagons, squares, etc. from 8 to 40 mm which are rolled in ***light section mills***, and finished as straight lengths. ***Rods*** are generally of circular cross-section and 5–10 mm in diameter, in coil form.

(ii) ***Types of mills***

The types of mill used for rolling bar and rod vary from hand operated (hand operated mills mean mills in which feeding of stock into mill rolls or looping of metal from one stand to next is done manually by operators using tongs) open or stepped open trains to high speed continuous mills. The duties of rolling bar and rods are often combined in bar and rod mills of the high speed type and rods are in general produced in continuous or semi-continuous mills mainly from heat conservation point of view.

Layout of bar and rod mills or rod mills vary widely depending on the size of input stock as well as the desired product–mix.

A number of rolling stands along with the ancillary equipment are so arranged that the layout offers suitable flexibility in producing different finished products starting from a range of sizes of input material. All the stands may not be used depending on the input and output. Layout of a light section continuous mill is shown in Fig. 3.18. The product from the mill are: rounds from 10 to 40 mm diameter, strip up to 100 mm wide and thin angles up to 50×50 mm and other similar products. The input is billet of size 75×75–100×100 mm and 9.0 m length.

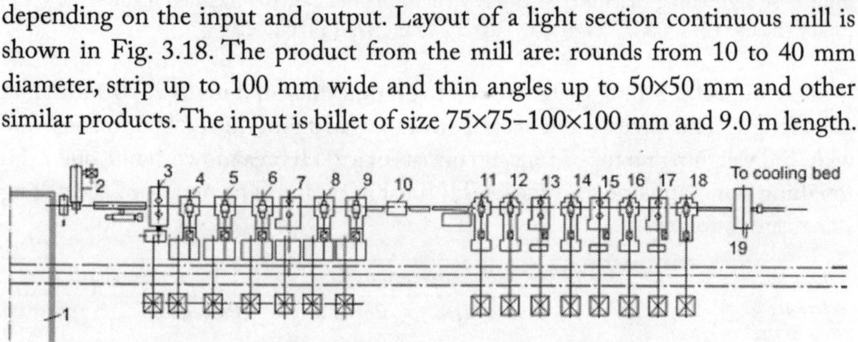

Fig. 3.18 Schematic layout of rolling equipment and shears in a 350 mm diameter continuous section mill

1- Reheating furnace; 2- shears; 3–9- stands of roughing train; 10- flying shear for cropping front ends; 11–18- stands of finishing train; 19- flying shear

In the Fig.3.19, a typical layout of a bar and wire rod mill is shown. The technical data of this plant is as follows:

Starting material	:	150×150 mm^2 ingot, 150 kg 100×100 mm^2 billet, 220 kg
Finished Products	:	Round bars, 10–32 mm Wire rod in coils, 6 –15.9 mm
Rolling speed, max	:	5.6 m/sec for bars 15.5 m/sec for wire rod

For large quantity production of wire rods, the high speed finishing stands in horizontal–vertical orientation are put in one block with a single drive. These are called *mono-blocks. Figure* 3.20 shows the layout of a mill with three mono-blocks for production of small diameter of steel wire rods and bars. The details of the mill are given below:

Technical data	
Starting material :	billet 120×120 mm
Finished products:	bars ∅ = 8–12 mm
	Wire rods in coils ∅ = 5.2–22 mm
Rolling speed:	bars 15 m/sec wire 32 m/ sec

(iii) ***Roll pass design***

Many of the principles of roll pass design for billet are also applicable to rolling of bars and rods. However, many more finished shapes and sizes are involved and that too at much higher speeds. Therefore, the pass design has to be more accurately done and executed. The following details are restricted to round products.

(a) General

The majority of round products are rolled by the "guide round" method in which a suitable oval is held by guides with major axis vertical (in a horizontal stand) and fed into the finishing round pass so as to form, in one pass, an accurate round section. The oval itself which is called the '*leader*' may be formed from a square, a round, or slug oval (an oval which has been edge worked) which in their turn are produced in the *strand pass* (the last pass but two). The method of getting to this strand pass are varied and many use box passes followed by diamond–square, oval–square, oval–round, or oval–edge–oval sequences.

In rod rolling it is necessary to have a speedy reduction form billet to coil in terms of time as the stock with small cross-section loses its heat very quickly. The

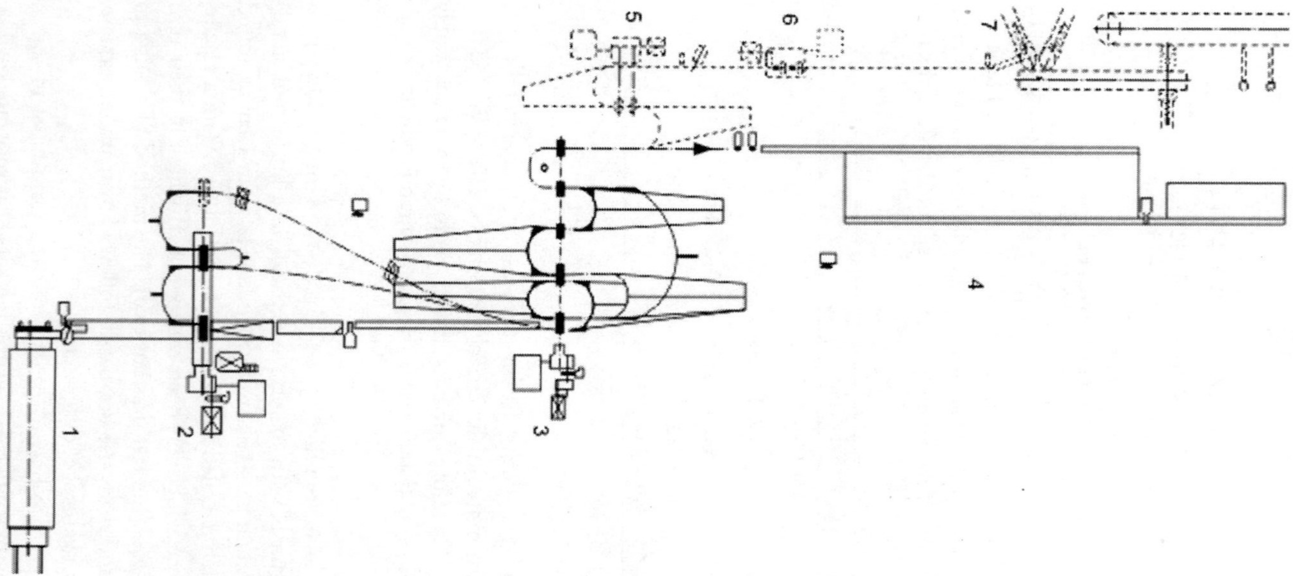

Fig. 3.19 Layout of a bar and wire rod mill

Legends
1. Reheating furnace, 12 ton / hr.
2. Break-down train with 2 number 3 Hi stands size 500×1600 mm, with fixed and tilting table, and automatic repeaters.
3. Intermediate finishing train with 5 stands size 280 to 340×1000–700 mm with automatic repeaters.
4. Cooling bed, 28m length.
5. First finishing train with 2 pre-stressed stands size 310×600 mm
6. Second finishing mono block train with 4-Hornizontal–vertical stands size 260×140 mm
7. Coiling unit with 2 pouring reels (Garrett type) with coil conveyor and hook transporter.

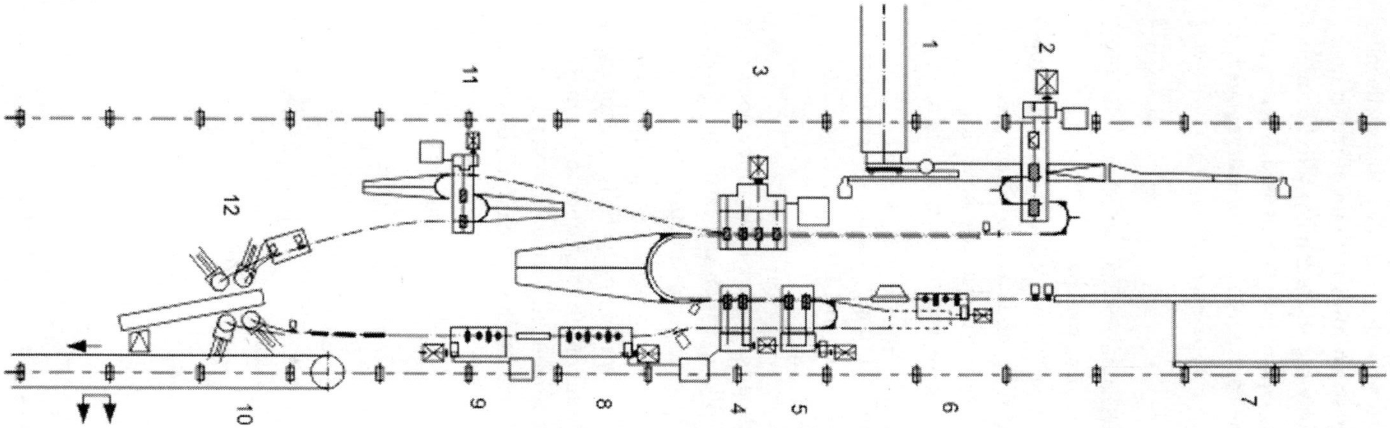

Fig. 3.20 Layout of a bar and wire–rod mill

Legends

1. Reheating furnace: 14 tons/ hour
2. Break down train with two stands ∅ = 430×1400–1200 mm mechanized with fixed roller table, tilting table, automatic repeaters and crop shears.
3. First continuous intermediate tandem train with four pre-stressed stands ∅ = 320×600 mm
4. Second continuous intermediate tandem train with two pre-stressed stands ∅ = 320×600 mm
5. Third continuous intermediate train with two pre-stressed stands ∅ = 320×600 mm
6. Monoblock finishing train with H–V (horizontal –verticals) stands ∅ = 260×140 mm
7. Cooling bed for bars
8. Monoblock finishing train with six H–V stands ∅ = 260×140 mm
9. Second monoblock finishing train with four H–V stands ∅ = 260×140 mm
10. Coils hook conveyor
11. Finishing train with two stands ∅ = 310×600 mm mechanized with automatic repeaters
12. Wire coiling unit: two pouring reels (Garrett) and two laying reels (Edenborn), with coil transporter

oval– square sequence is often favored but, for reasons of quality, other sequences are sometimes desirable and in the higher speed mills the rate of reduction by these other methods is satisfactory. Obviously the starting billet and the number of passes available will directly influence the type of pass design used, whereas this is flexible in an open train type of mill. The continuous or semi-continuous type of mill is usually designed around a given type of pass sequence.

(b) The finishing pass

The reduction taken in a finishing pass may vary from 12% to 16% and the actual figure used will depend on the form of the leading oval pass.

The final round pass groove is cut equally in the two rolls but any one of three methods may be specified by the designer in his endeavor to obtain greater life form the rolls before re-dressing. Each groove must of course be basically semi-circular but some side relief is necessary at the roll gap to prevent marking of the stock when the pass fills correctly.

The sharp corners are removed from the final round pass by using small radii, e.g. about 0.8 mm for up to 25 mm (1/32 in for up to 1 in) diameter and 1.6 mm for 25–40 mm (1/16 in for 1 to1½ in) diameter rods.

Tool relief or single plug method: By this method the groove is cut to the hot size of the finished bar making allowance for the roll gap which will exist during rolling (e.g. nominal roll gap and expected spring) so the groove depth is 1/2 × (hot bar diameter– roll gap). The sides of the groove are then relieved to a specified amount (e.g., 0.2 mm to 0.25 mm for a 25 mm round) with a substantially straight edged tool. The accuracy depends entirely on the skill of the roll turner for he must use templates or feeler gauges and extreme care is required particularly on the smaller sizes. The resulting groove is illustrated in Fig. 3.21 from which it will be seen that as the maximum wear normally takes places at the "shoulders" these will become high in dimension on the finished bar as no extra allowance for wear is made at that point on a profile cut in this way.

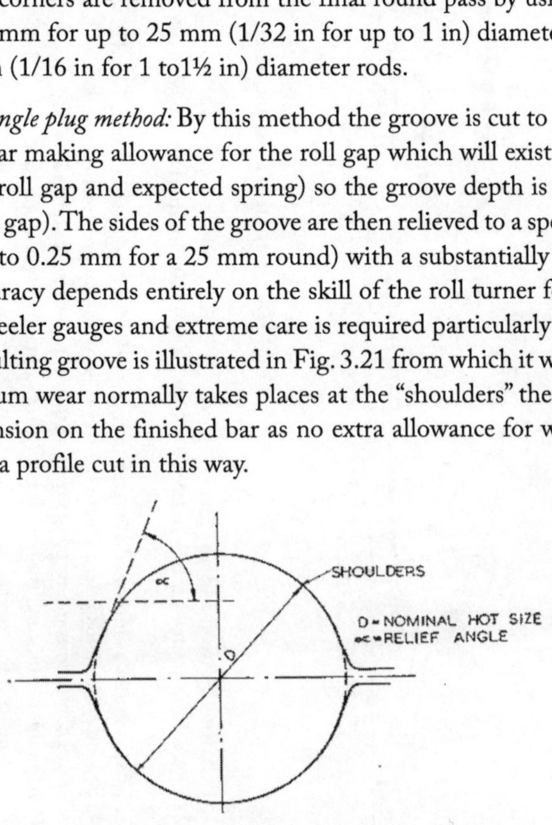

Fig. 3.21 Single plug method of cutting finishing pass for rounds

Cutting of the groove is achieved by the use of a plug tool which is a cylindrical tool ground to the round size required (in this case the hot size) and as the point of cutting is below the centerline of the roll the resulting groove is slightly oversize across the shoulders (even when new) (e.g., 0.025–0.05 mm on a 25 mm round). The hot size of the bar is found by using an expansion constant together with the expected finishing temperature.

For mild steel:
Hot size = cold size (1+.000012×drop in temp. °C)
and for high chrome or nickel steels:
Hot size= cold size (1+.000017×drop in temp. °C)

where the drop in temperature referred to is the difference between finishing temperature and "room" temperature. In practice, if the drop in temperature is 900°C, the cold size is multiplied by 1.011 for mild steel and 1.015 for the high alloyed steels, the mean being 1.013, i.e., .013 mm per mm of dimension.

There are two more methods of machining the roll grooves by the use of two and three plug tools which are known as two plug method and three plug method, respectively, to increase the groove life.

(c) The leading oval pass

For producing a good round it is advisable to use a plump oval which results in a low reduction in the finishing pass, less wear on the pass, better surface quality on the finished bar and greater accuracy. Theoretically, the more nearly the oval shape approximates to the round shape, the better is the finished round, as less draught gives less spread and hence less chance of incorrect filling of the round pass. However, this type of oval is difficult to hold up in *fixed guides* unless they are tight and this introduces a risk of marking the bar surface, hence *roller entry guides* are advantageous. As a result of these difficulties of guiding it was the normal practice to use plump ovals for larger products and slim ovals for the smallest rods, but the introduction of roller guides has made it possible to reduce this tendency for slimness on small ovals. Still, the width to thickness ratio of the ovals tends to increase as the product becomes smaller.

Winton[2] has constructed graphs of the area of the finished oval for given rounds and also of the major and minor axes of an oval for a given round. The latter are based on the following equations for the width, W, of the oval stock on the major axis and thickness, T, of the oval:

$$W = 1.26\,D_c + 0.1$$

$$T = 1.00\,D_c - 0.1$$

where, D_c is the nominal cold diameter of the finished round. All dimensions in above equations are in inches.

Two straight lines representing these equations are given in Fig. 3.22, and from these a curve for the area of the oval is derived, where single radius ovals' area appoximates to $\frac{\pi}{4}$ W.T. = 0.786 W.T. This is the geometrical area of a true ellipse and in practice many people take the actual area as 0.75 W.T.

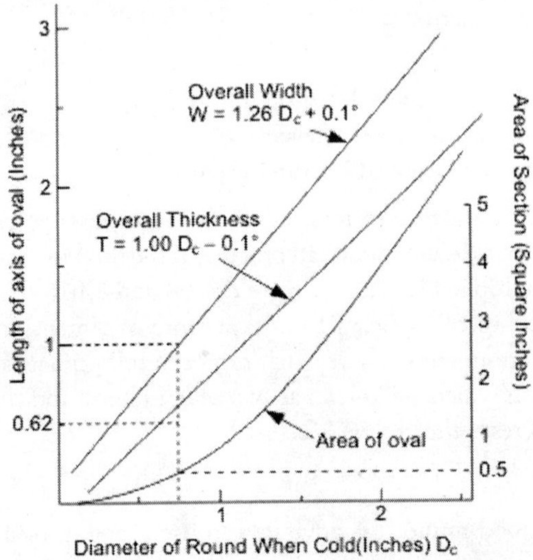

Fig. 3.22 Determination of leading ovals

The reduction from oval to round recommended by Kirchberg is about 12 percent, but Trinks[3] points out that this should be variable according to the size of round, tolerance, etc. and suggests 5–15 % with the larger reduction applied to smaller rounds.

In the case of rods less than 12 mm diameter a high reduction is more difficult to achieve due to the relatively large roll diameter giving an increased degree of spread. A flatter oval is used to offset the greater spread. In the interests of quality the reduction in the final pass is in some cases as low as 8%. However, figures which may be considered suitable for rolling small rods are 12–14% in the finishing pass and 15–17% in the leader with an overall reduction for the two passes of about 27%.

(d) The strand pass:

The pass preceding the leader pass is often called the strand pass and it may take several forms depending on the type of breakdown sequence used. If a diamond–square sequence is used the stock entering the leading oval will be square; with an oval–square sequence it will be square; with an open square sequence it will again be square; with an oval–edged oval sequence it will be an edged oval, and with an oval–round sequence it will be a round.

Strand square

It is important that the strand square should give a correctly filled leading oval so as to avoid the possibility of surface defects on the final round. Most rules of thumb and formulae which have been put forward, relate the size of the square to that of the finished round and once this relationship has been established, the same between the square and the oval is automatic. The side of the strand square is given by multiplying the hot size of the round by 1.10, resulting in an overall reduction from square to final round of about 35%. This represents reductions of about 14% in the final pass and 24.5% in the leader which usually works satisfactorily in practice on bar rolling.

Hoff and Dahl gave the following formulae for determining the strand square of bars above 1.2 in. dia.

Round size D_c, in.	*Side of square, in.*
1.2–2.7	$1.03D_c+0.16$
2.75–3.75	$1.04D_c+0.16$
3.85–5.1	$1.05D_c+0.16$

These squares seem to be a little on the high side for rounds of 2 in diameter and less, as they give overall reductions from square to round of 36% on a 2 in round and nearly 42 % on a 1.2 in round.

For smaller rounds typical square sizes are:

Round size D_c, in.	*Side of square, in.*
0.50	0.562
0.875	1.016 (or 1.031 gothic)
1.0	1.156 (or 1.172 gothic)
1.125	1.281

Strand edge oval

The square as a strand pass is frequently abandoned when a quality rod or bar is required and low reduction is to be taken. The alternative is an edge oval, and it is illustrated in Fig. 3.23 (a). An edge oval is produced by edging a wider oval in an edging groove with the sides relieved. This type of oval is sometimes called a slug oval and the groove may be flat in the bottom instead of concave. Another shape of pass which approximates to this shape is an 'off round', known as 'faux-round' in Europe. This is formed by two half round grooves and a large roll gap so giving a vertical ellipse, Fig. 3.23 (b). There is more danger of side marking with this type of groove. Bottom part of figure 3.23 shows the improved draughting from the quality point of view of using slug oval or faux-round as compared with the use of a strand square. In strand square there is a possibility whereby small quantities of scale are rolled into laps on the sides of the oval and subsequently give hairline seams on the finished product.

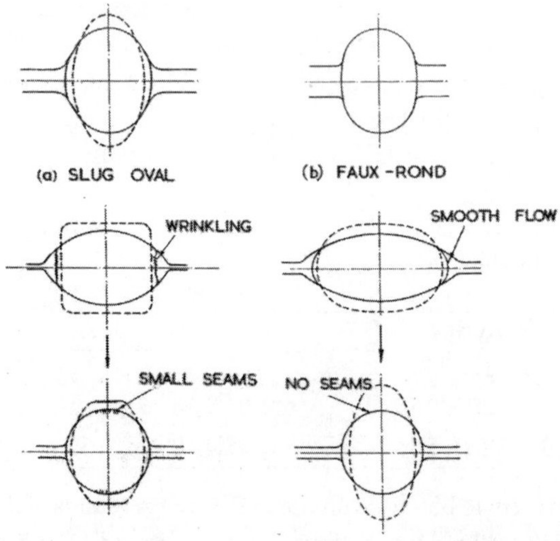

Fig. 3.23 Comparison of square–oval and slug–oval method

Strand round

A round is used as the strand pass when a round–oval–round finishing sequence is applied. This implies a light reduction from strand round to finished round, a typical figure for this might be 23–30%, the lower figure being used for the smallest rod while with bar sizes the higher figure will apply. The strand round is similar to the edge oval in the sense that it does not induce defects.

Sequence leading to the strand pass

There are several sequences which may be used to reach the strand pass and the one which is most satisfactory in any particular case depends on such considerations as the average reduction per pass required to get from the billet to the finished product, the quality required, and the type of steel being rolled.

Oval–square sequence

This sequence is primarily a high reduction one. It is achieved by placing a square on the flat into an oval and then turning up the oval 90° into a square pass set on the diagonal. These two stages are shown in Fig. 3.24a and b, the first one yielding up to about 42% reduction and the second one 30% reduction so giving a maximum overall square to square reduction in area of 59% on normal qualities of steel. Unfortunately the formation of wrinkles in the oval pass is quite likely so it must be used with discretion. The uneven draughting from center to edge in the oval pass also gives differential elongation which can result in surface defects from the shear forces induced.

The general trend in bar and rod mills is to use this sequence quite extensively at some stage of the rolling. In hand looping mills it is still common to use this as finishing sequence with hand looping of the oval and automatic repeating of the square which lend themselves to this method.

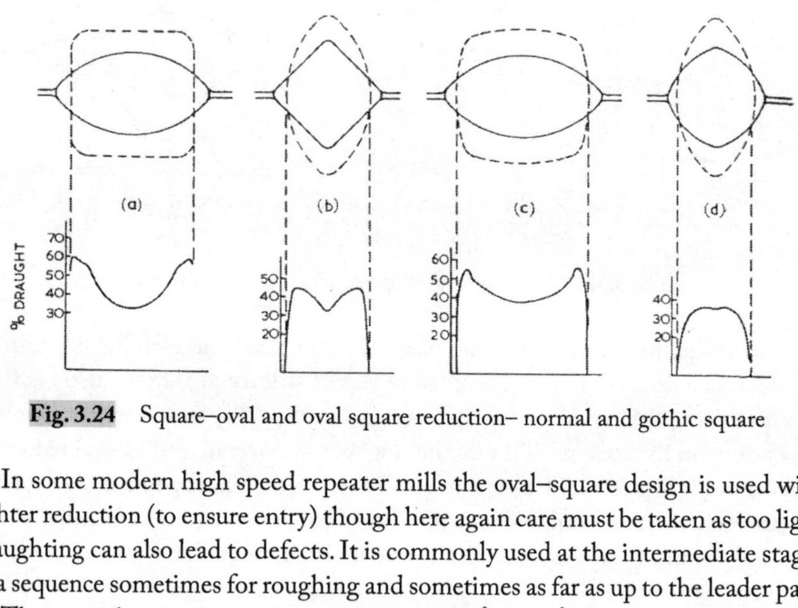

Fig. 3.24 Square–oval and oval square reduction– normal and gothic square

In some modern high speed repeater mills the oval–square design is used with lighter reduction (to ensure entry) though here again care must be taken as too light draughting can also lead to defects. It is commonly used at the intermediate stages of a sequence sometimes for roughing and sometimes as far as up to the leader pass.

The square has an apex angle usually between 90° and 92° as a greater angle will give a 'square' which rides badly in the oval and gives a poor shape at the break of the oval.

Oval–slug (edge oval) sequence

It was mentioned earlier that the slug or edged oval is formed by placing an oval vertical and edging. This slug may then be turned through 90° and passes through an oval groove so giving a sequence of relatively light reduction. An example of the last part of such a sequence is shown in Fig. 3.25. The reductions in the slug passes are approximately 25.5%, 11%, and 15%, respectively, and it is obvious that considerable flexibility of reduction is possible.

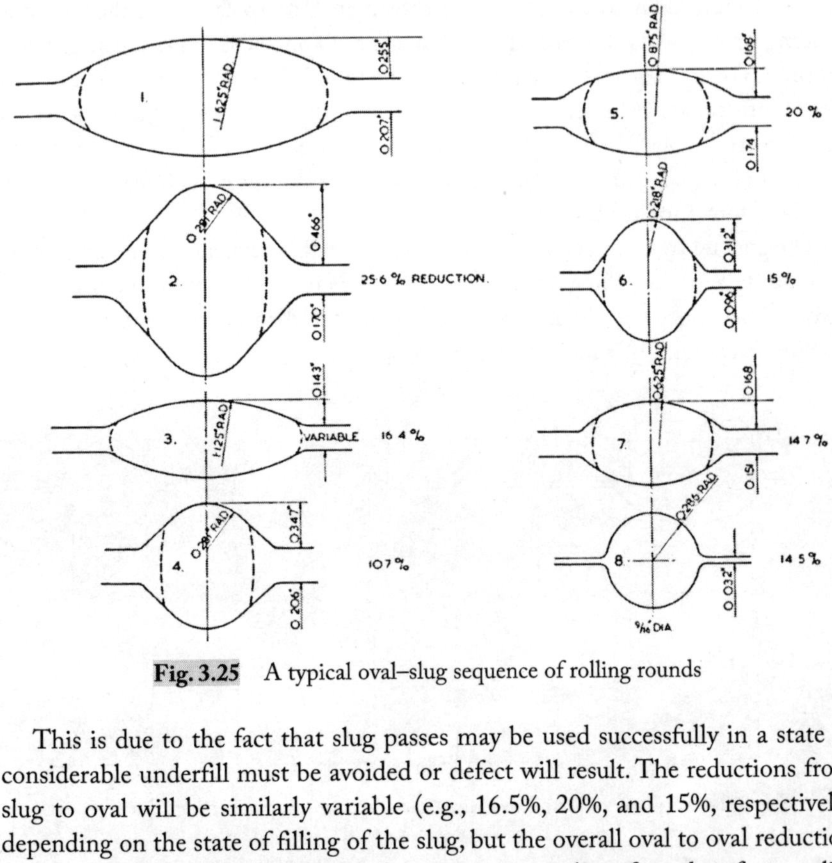

Fig. 3.25 A typical oval–slug sequence of rolling rounds

This is due to the fact that slug passes may be used successfully in a state of considerable underfill must be avoided or defect will result. The reductions from slug to oval will be similarly variable (e.g., 16.5%, 20%, and 15%, respectively) depending on the state of filling of the slug, but the overall oval to oval reduction may be as high as 38%. This type of sequence gives a bar of good surface quality but less reduction is possible than with oval–square sequence. It is frequently used in the finishing train but may also be used throughout the sequence after the production of an oval.

Oval–round sequence

This sequence achieves an oval to oval reduction via a round pass and it is similar to oval–slug in that it is a light reduction sequence requiring considerably more passes to achieve a given bar size form a given billet. It is frequently used in the finishing train of high speed rod mills and in some cases the round passes which are reducing passes in one sequence may also be used as finishing passes for a larger size of bar. The general order of reductions is 20% in the oval–round pass. The design is flexible in that the oval grooves may be used for rolling more than one size by lifting the rolls.

If this design is used throughout the sequence some aid to de-scaling is required particularly at the roughing stage as the method of reduction does not de-scale the stock efficiently. High pressure water is frequently used and if diamonds and squares are used in the roughing stage the early de-scaling is achieved satisfactorily. An oval round pass design is shown in Fig. 3.26[4].

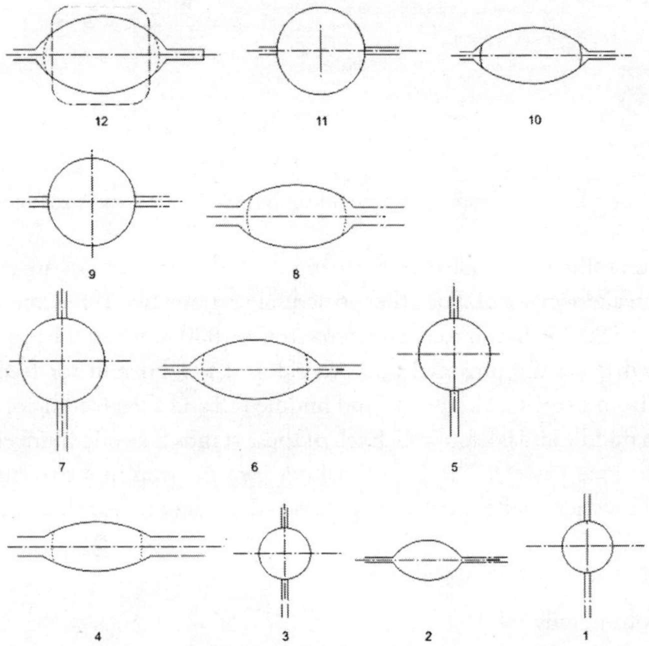

Fig. 3.26 Rolling of rounds by the oval–round method: 3/4 in. rounds from 4 in. billets in 12 passes

3.2.5 Rolling heavy sections

(i) ***Rolling mills***

The rolling stands and equipment used in manufacturing of different heavy steel sections like rails, beams or joists, large channels, and angles is called *rail and structural mill.* This facility generally forms a part of an integrated steel plant.

A typical layout is shown in Fig. 3.27. The starting material for such a mill is reheated bloom. The first stand of the mill is a 2-Hi reversing breakdown stand having roll dimension of about 900 mm diameter × 2300–2500 mm barrel length. It rolls the heated bloom to required smaller size for starting profile passes. Rough pre-forms may be rolled in this mill. The subsequent stands are 2-Hi reversing stands or 3-Hi stands depending on the final product.

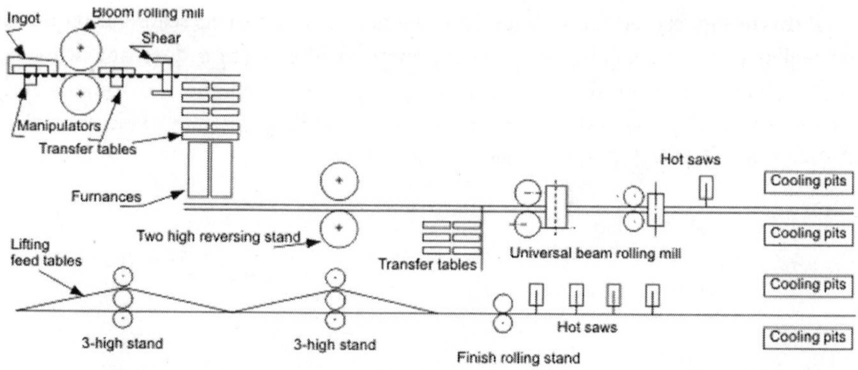

Fig. 3.27 Layout of rolling stands and equipment in a rail and structural mill

For beam rolling, generally two reversing 2-Hi universal stands are used in line with the breakdown stand. For other structural sections two/three numbers 3-Hi stands of size 800–650 mm roll diameter × 1600–1800 mm barrel length are used. Tilting feed tables are provided on both sides of a 3-Hi mill for feeding to or accepting from a stock from the top and middle roll and then feeding or accepting stock form middle and bottom roll. Each of these stands take odd number of passes (minimum three passes). The final finishing pass is given in a two high rolling stand. Hot saws are used for cutting the finished product to length before cooling.

(ii) ***Rolling methods and pass sequence***

(a) Flat bottom rails

A few basic types of passes are used in rail rolling. Different pass sequences consisting of different combinations of these types of passes are followed.

The types of passes are shown in Fig. 3.28. The rolling starts with a bloom whose size and number of passes is interrelated and also depends on the size of the rail to be produced.

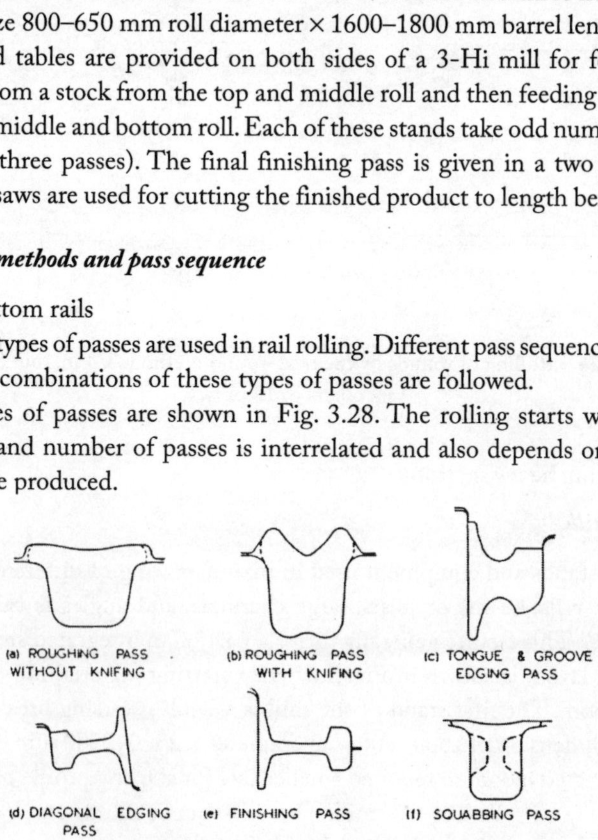

Fig. 3.28 Types of passes for rolling rails

The pass number (a) is often referred as trapezoidal passes, the majority of which are cut in the bottom roll. In the knifing pass (b), two protrusions are created at the top which will be opened up later to form the flanges. In the third type of pass called the edging pass (c), the rail is turned to 90° and rolled on edge to divide the mass into two areas corresponding to the head and the flanges, connected by a neck which forms the web of the rail. The finishing pass (e) controls the head size and flange length and delivers required radii. For a flat bottom rail, the entire flange length should be in one roll as shown. It is essential to prevent formation of fin at the joining line of two rolls at the head. Squabbing pass (f) is a pass which is generally applied after rail shape is formed, to reduce the overall height and flange thickness by compressing the rail vertically. This pass needs an additional turning of the stock, but is beneficial form the point of view of mechanical properties by giving heavy reduction on the head and the base. Experimental research has shown that the knifing passes generally result into improved physical properties and less rejection due to hairline cracks of the base, and other defects like laps and bend test failures.

Figure 3.29 shows two typical rail rolling sequence with 9 and 10 passes, respectively.

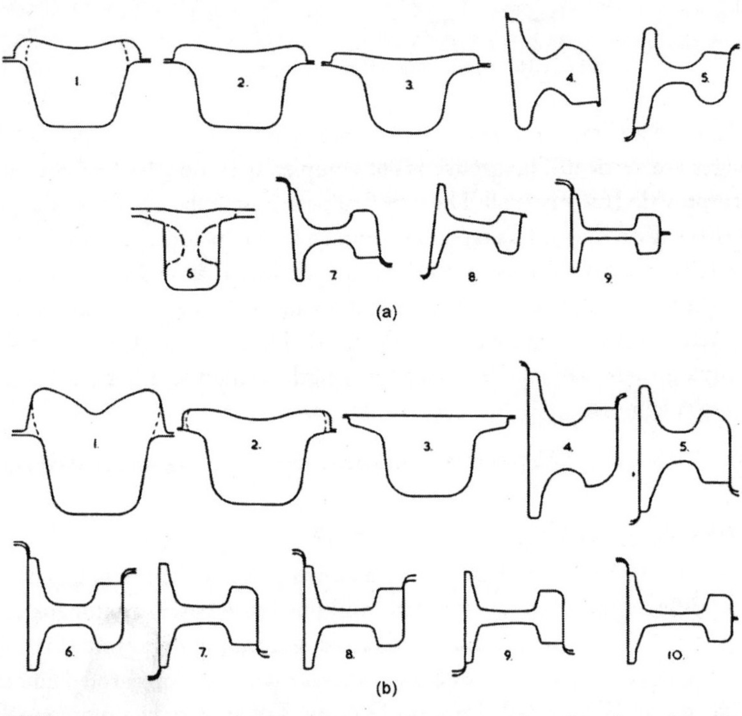

Fig. 3.29 Two types of rail rolling sequence (passes numbered in the order of rolling)

(b) Beams

Types of Passes: The finishing pass for producing a beam is always set with the web horizontal and flanges vertical or nearly so. The basic type of passes employed in joist or beam rolling is shown in Fig. 3.30.

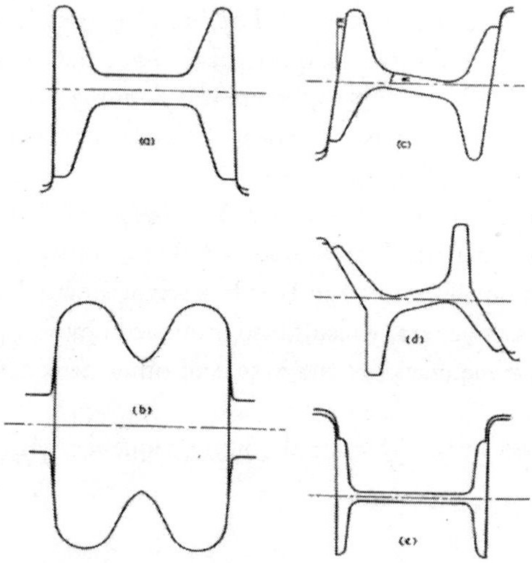

Fig. 3.30 Basic type of passes for joist rolling

The tongue and groove pass (a) is a common pass where web is horizontal and the flanges are vertical. The groove is cut completely in one roll and the closing tongue is provided by other roll. The very first passes could be a multipass sequence of passes as shown in (b). The taper on flanges and the included angle of the knife is progressively adjusted to permit entry into the first diagonal pass. In diagonal pass (c), the web is inclined to the horizontal and the "break" is at opposite sides of the pitch line, so that groove is equally divided into each roll. In this pass two diagonally opposite flanges are reduced in length, while the other pair is mainly reduced in thickness.

Pass sequences: Starting with rectangular blooms generally two types of pass sequences are used:

(a) Tongue and groove pass sequence: A typical ten pass sequence is shown in Fig. 3.31 (a). The first two breakdown passes have the break at the middle i.e., equal grooves in each roll. The last 6 passes are tongue and groove passes where breaks are alternating from top to bottom. The 10-pass sequence is divided into three stands each of 860 mm roll diameter for production of 250×150 mm H beam. The pass reductions from 1st to 10th pass are: 48%, 33.8%, 19.4%, 26.5%, 21.8%, 20%, 19.4%, 15.9%, 9.9%, and 8.3% respectively.

(b) Diagonal pass sequence, which is shown in Fig. 3.31(b). The 8-pass sequence distributed in two stands is for production of 5 in×4½ in joist from 9¼ in ×5½ in blooms. In this sequence diagonal angle is progressively reduced form as high as 15° to about 2.5° in the final pass. A diagonal sequence allows greater reduction than the tongue and groove sequence.

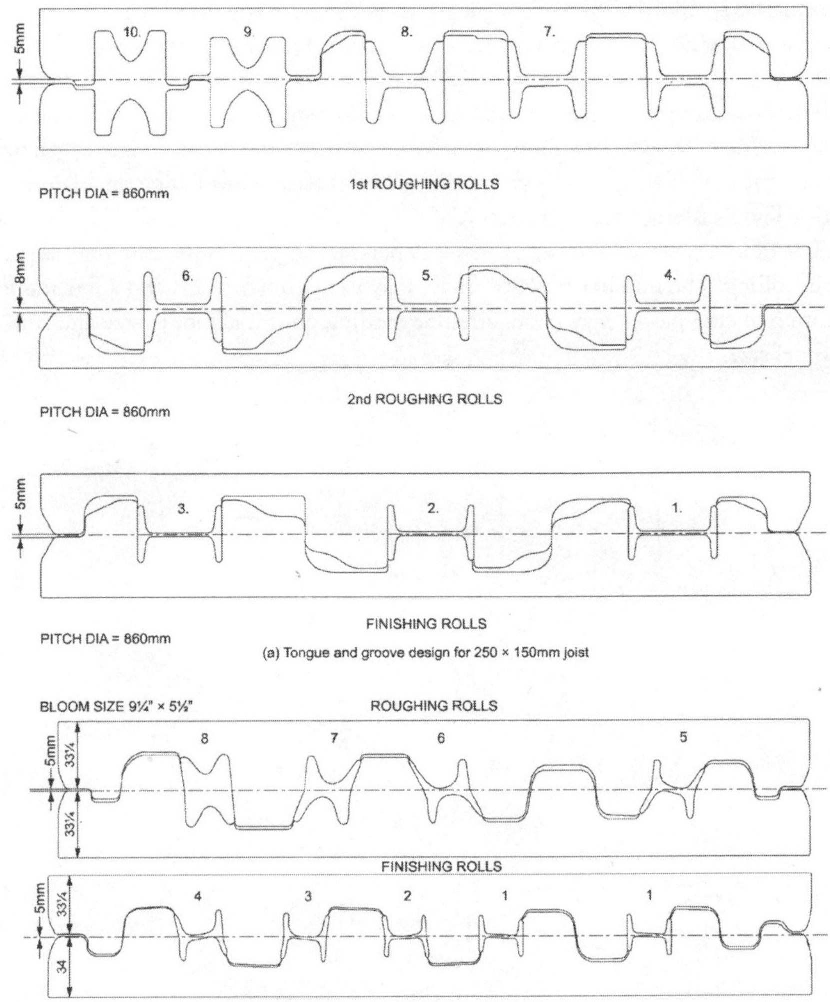

(a) Tongue and groove design for 250 × 150mm joist

(b) Diagonal design for 5in. × 4½ in. joist

Fig. 3.31 Pass sequences for beam rolling

The bloom size and number of passes depend on many factors like mill layout, mill size (roll diameter and barrel length), screw down limit in first-mill, mill power, etc.

(c) Channels

Types of Passes: To roll a channel, the finishing pass should always be with web horizontal and the flanges vertical up or down. Different types of passes may be used to arrive at the final pass, starting with a rectangular bloom.

Different types of passes used are shown in Fig. 3.32. The type (a) is a first forming pass. Another type of forming pass is the butterfly pass (c) to produce a shallow and wide W shape from a slab like bloom. This butterfly shape is further rolled to produce shapes as per pass (d). A similar method makes use of inverted butterfly resulting to the pass (e) with bent web, known as convex or gothic shaped flanges. One or more controlling pass may be used to control the flange length by closing the end of flange and putting the roll joint part below flange tip as shown in (b). The finishing pass is the pass (f).

The bloom size and number of passes depend upon the factors, as in the case of beam rolling. The number of shaped pass may vary from 5 to 11, and a few more bloom reducing passes may be required depending on initial bloom size and final channel size.

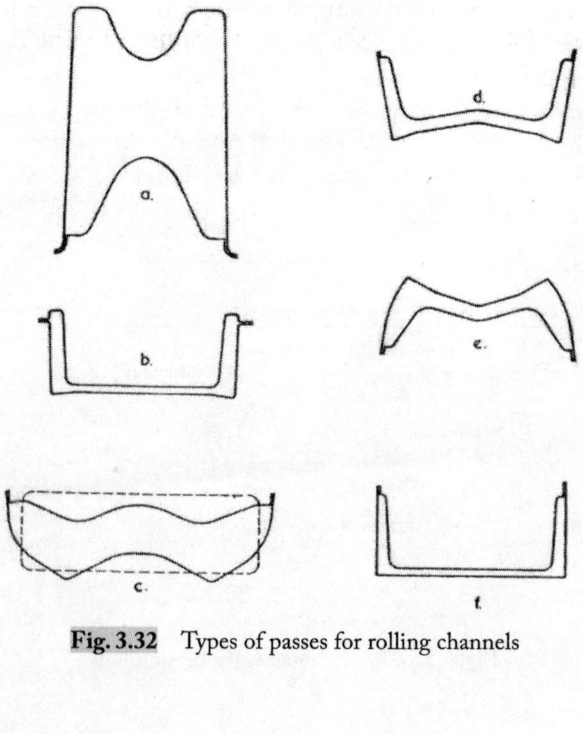

Fig. 3.32 Types of passes for rolling channels

3.2.6 Rolling of non-ferrous metals and alloys

This discussion will be limited to copper and copper based alloys, and aluminum and its alloys which together form the bulk of non-ferrous metals and alloys being rolled. For both the groups, rolling process is mainly used for production of flat finished product like plate, sheet, strip and foil. Various sections and tubes are produced by extrusion process. Therefore, in case of copper, aluminum and their alloys, only flat rolling is discussed here.

(A) ***Copper and its alloys***

Normal copper alloys are brass of different copper and zinc proportion, leaded brass, bronze, nickel-silver, cupro nickel, etc.

(i) *Production method:* The conventional production method is to hot roll cast slabs down to about 10 mm thickness, then surface milling of the hot rolled coils or sheets and then cold roll to desired thickness with appropriate inter-annealing and cleaning.

Where large production is required, direct chilled (DC) cast slabs of size from 125 to 200 mm thick and 650–1050 mm wide weighting from 3 to 6 tons are used. However, for smaller plants, statically cast slabs in book moulds of thickness in the range of 50 mm are used for alloys whose demand is small.

The high production hot mills are normally 2-Hi reversing type fitted with rolls between 900 to 1000 mm in diameter and designed for a maximum rolling load of 1500 tonnes, which is capable of reducing the maximum thick hot slab down to 8–12 mm thickness within maximum 11 passes. The mill speed may be up to maximum 200 mpm driven by a pair of DC motor having total power of 2500 kw.

The 2-Hi hot mill is fitted with a separately driven edger mill for controlling slab width and also for consolidating the edges of the slab to minimize edge cracking during cold rolling operations.

Runout tables of sufficient lengths to accommodate rolled strip length are provided on both sides of the mill. The finished strip is either cut to length by a shear installed at end of the output side table, or coiled in a roll type hot coiler installed at the table end.

The hot rolled coil or cut sheets are pickled to remove oxides and oils before cold rolling. However, to improve quality of copper and alloy strip products, particularly in case of thin strip, it is necessary to remove the oxides and impurities from hot rolled strip surface by milling operation.

In many of the modern plants, continuous strip casting machines have been installed, which can produce up to 1300 mm wide strips in thickness range of 17–20 mm. Narrow strips can be cast in even lesser thickness range in 2–3 strands simultaneously. The continuous cast strips are milled on line and then coiled into loose coils by up-coilers.

Hot rolled or continuous cast strips are cold rolled for producing the finished rolled material suitable for further processing for making final products. Cold rolling, in general, has been discussed in details in section 3.2.7. Here certain special practices applicable for copper and copper alloys will be discussed.

There are certain copper alloys which are hot short and are not hot rolled. These alloys are generally cast into thin slabs of thickness between 25–50 mm. These materials and also the continuous cast strips of thickness range 12–20 mm are cold rolled in 4Hi cold breakdown mills. A typical breakdown mill for rolling continuous cast strip of 1050 mm width would be:

Work roll diameter	: 420 mm
backup roll diameter	: 1120 mm
Roll barrel length	: 1200 mm
Max. rolling load	: 1400 tons
Rolling speed	: 250 mpm

The finishing cold mills are generally 4-Hi hydraulic mills with automatic gauge control* to produce thin strip with close gauge tolerance. Cluster mills have also been used successfully in some high production units.

During high speed rolling copper and some of its alloys, having rather low crystallization temperature, if the mill is reversing type, the temperature of the rolled strip increases continuously. This problem is reduced in a non-reversing mill where batch rolling is employed to allow ample time for the rolled coils to cool between consecutive passes. At the same time idle time of the mill between coils is avoided by incorporating efficient coil handling and re-circulating system. Typical maximum finish rolling mill speed varies between 450–1000 mpm.

The size of finish rolling mills for copper and copper alloy strip is dependent on the strip width and production requirements. Basic specifications for three typical mills would be as given in table below:

Specification Item	*Mill 1*	*Mill 2*	*Mill 3*
1. Strip width, mm	200	600	1000
2. Work roll diameter, mm	130	250	350
3. Backup roll diameter, mm	350	650	1000
4. Roll barrel length, mm	300	700	1150
5. Rolling load, ton	125	600	1000
6. Rolling speed, mpm	400	750	1000

* See section 4.3.6 for detialed discussion on automatic gauge control system.

To roll very thin strips between 0.1–0.05 mm, it is necessary to use smaller work rolls than rolls used for thicker gauge. If the quality of production does not justify a separate finishing stand, use is made of a *combination 4-Hi mill* with backup roll drive, where two different size of work rolls may be accommodated as required.

Mineral rolling oil is used for lubricating and cooling the mill rolls during both breakdown and finish rolling.

(B) ***Aluminum and its alloys***

Major uses of rolled aluminum and their alloys are in aviation or aerospace, automobile, utensils, and flexible packaging industries. Very high strength to weight ratio for some of the aluminum alloys make them obvious choice in airplane body building.

(i) *Production method*: The production methods for hot and cold rolling of aluminum and their alloys are very similar to those for copper and their alloys discussed under section 3.2.6(A).

Direct chilled cast slabs are hot rolled in reversing 2-Hi mill and finished as hot coil at around 6/5.5 mm thickness. The starting slabs sizes vary from 340 mm up to 890 mm thick and 1320 mm to 1830 mm (72 in.) wide. Present day cast slabs weigh up to 20 tons. Cast slabs are surface milled on both sides before hot rolling.

Specification of a 2-Hi reversing hot mill for rolling 350 mm×1320 mm slab down to 6 mm in 9 to 11 passes depending on soft or hard alloy of aluminum would be 650 mm diameter roll × 1525 mm barrel, each roll driven by 800 kw DC motor. Maximum mill speed about 40 mpm. Imulsified oil in water is used as roll coolant cum lubricant.

Continuous cast aluminum strips are generally made 6 to 8 mm thick, ready for cold reduction. Continuous casters are now available which can cast up to 2000 mm wide strip down to 2.5 mm thickness. However, these thin continuous cast strips are suitable for producing thin gauge strip or foil but are not suitable for production of deep drawn quality strip.

Modern cold strip or foil mills are all 4-Hi high speed hydraulic automatic gauge control mill, fitted with shape control arrangement. Features of a hydraulic gauge control and shape control system have been described later, in section 4.3.6 of the book.

The work roll diameter of these high speed strip or foil mills are made comparatively large for better thermal stability and control of local thermal camber of the rolls during rolling. The mills are essentially non-reversing type and a number of coils are employed for batch rolling with the help of coil handling arrangement to achieve minimum idle time of the mill between coils.

Typical duty specifications of aluminum strip mills and foils mills would be as under:

Strip Mills including Light Gauge Strip Mills	*Foil Mills*
1. Input thickness : 1 –6 mm	1. Input thickness: 0.5–0.7 mm
2. Output thickness: 0.5–2×0.02 mm	2. Output thickness: 0.2 to 2×0.005 mm
3. Strip width: 800–1800 mm	3. Width: 800–1800 mm
4. Coil weight : 1–8 tons	4. Coil weight: 1–8 tons
5. Rolling speed: 500–1500 mpm	5. Rolling speed: 600–2200 mpm

Light gauge strip mills are in between proper strip mills, which normally rolls from hot band strip of thickness around 6 mm down to foil mill input gauge of around 0.6 mm, and foil mills. These mills generally take an input at around 1 mm and rolls down to mid foil range of 2×0.02 mm.

Some of the special techniques used for rolling thin gauge strips and foils have been discussed in more details under next section 3.2.7.

3.2.7 Cold rolling of sheet, strip and foil

(i) ***Definition***

The input material for cold rolling is hot rolled sheets and strips, or continuously cast strips in coils. The thickness of the input material can vary between 20 mm of continuous cast strip on the higher side down to approximately 2 mm hot rolled strip. For a few non-ferrous alloys whose production requirement is small, cast slabs up to a thickness of 50 mm may also have to be cold rolled. Cold rolling of thick gauges to comparatively thinner gauges is called the *break-down* cold rolling.

Cold rolling is resorted to primarily for obtaining flat rolled products like sheets, strips and foils. These flat products are mostly used for producing pressed parts and components in power presses and hence require appropriate metallurgical properties and accurate thickness with close *gauge tolerance.* Aluminum strips and foils are used for production of laminated and printed materials largely used in packaging industry. These thin gauge strips and foils not only need gauge tolerance but also very good *flatness.*

(ii) ***Cold rolling mills***

4-Hi mills are commonly used for cold rolling flat products. Finishing mills, particularly the high speed mills to produce thin strips and foils are essentially hydraulic mills with automatic gauge control system. The foil mills are further provided with automatic flatness control system.

2-Hi reversing breakdown mills are still used in some plants to cold roll thick sheets, slabs, or continuous cast metals or alloys.

When requirement of production is high, three or more 4-Hi stands may be placed in one line in tandem and the input material can be rolled down to the desired thickness by passing the material through all the stands one after another, getting reductions in each stand. Such a combination of a few stands is called a ***tandem*** cold mill. Figure 3.33 shows layout of a 4-Hi three stand tandem cold strip mill along with the associated mill equipment.

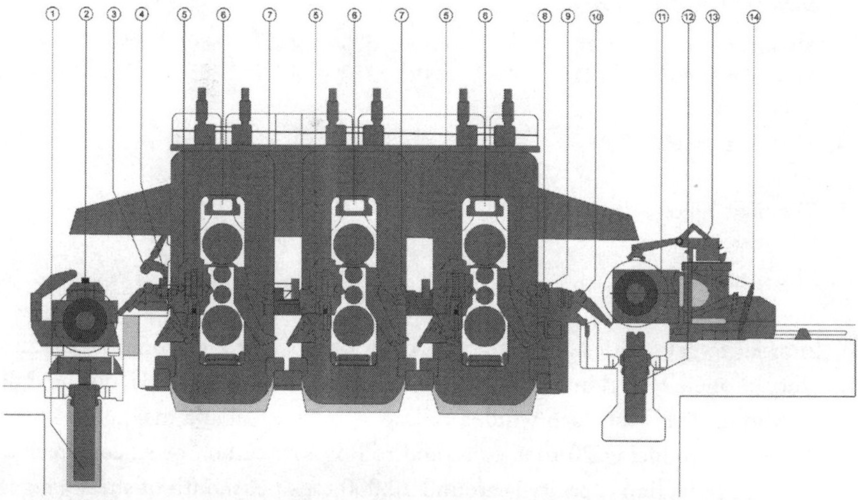

1. Coil transfer car
2. Uncoiler
3. Holdown roller
4. Side guards
5. Entry press
6. Roll load cylinders
7. Deflector rolls
8. X-ray gauge
9. Side trimming shear, if necessary
10. Deflector roll
11. Coiler
12. Coil removal car
13. Holdown roller
14. Belt wrapper

Fig. 3.33 Elevation of a three stand tandem cold strip mill

Each 4-Hi stand is 550 and 1300 × 2040 mm size driven by 4500 HP motor. Input is carbon steel 2–8 mm thick strip and output is 0.4–3.5 mm, maximum speed of last stand 1080 mpm.

12-Hi or 20-Hi Sendzimir mills have been widely used particularly for cold rolling of wide strip of copper alloys, alloy steel, silicon steel, stainless steel, high strength super-alloys, beryllium, copper, molybdenum, tantalum, titanium, etc. to an ultra-thin foil thickness. Dimensional specification of different models of Sendzimir 20-Hi (type 1-2-3-4) cold rolling mills is given in the Table 3.2.

Complete radial support of the rolls allows these mills to roll very wide strip with very small work rolls without the problem of roll bending. Constructional features of Sendzimir mills have been discussed in greater details in section 4.1.7 of the book.

Table 3.2 Specifications of Sendzimir 20-Hi cold rolling mills

Model/Item	*ZR 32*	*ZR 34*	*ZR 24*	*ZR 33*	*ZR 23*	*ZR 22*	*ZR 21B*	*ZR 20*
1. Nominal work roll diameter, mm	6.35	10	21.5	28.5	40	54	89	80
2. Backing bearing diameter, mm	47.6	76.2	120	160	225	300	406.4	456
3. Minimum / Maximum strip width, mm	110/ 225	190/ 445	215/ 500	330/ 1220	480/ 1575	660/ 3050	840/ 4200	840/ 4600
4. Maximum input thickness, mm			3.8	6.0				
5. Minimum rolled thickness, mm	0.003	0.01	0.02	0.025/ 0.05		0.05		
6. Mill speed, mpm				550		450	750	

In order to get an idea about the capacity of the larger mills, one can discuss the duty of one ZR 21B installation which is powered with 5000 HP on the mill motor and 4300 HP on each winder, rolling stainless steel at a maximum speed of 600 mpm producing 20 tons coils, and rolling low carbon or silicon steel at 750 mpm. Production capacity is around 20,000 tons per month of sheet gauges of carbon steel.

Skin pass rolling: Skin pass rolling involves a small amount of reduction between 0.5–1.5 %, to the annealed cold rolled strip by a large diameter 2-Hi mill in a single pass. The purpose of this pass is multifold: to produce a smooth and shiny surface, to obtain a uniform strip thickness, to impart a degree of surface hardening and to reduce the yield point elongation phenomena (Luders band) which is of vital importance in getting wrinkle free products through subsequent cold working (bending, deep drawing, etc.).

When reduction is raised above 1.5% up to 4%, it is called *temper rolling.* It gives added strength to the finished strip and eliminates the tendency of bending in non-uniform way in subsequent forming of the strip.

(iii) *Method of rolling*

Cold rolling is different from hot rolling of flat products in many ways:

(a) Cold rolled products are essentially rolled between flat cylindrical rolls. The width of the rolled product being much more than 12 times the thickness of the material, spread is very small and neglected in cold rolling.

(b) In hot rolling the major concern is heat loss from the rolling stock and the rolling schedule is completed before the temperature of the stock falls below its recrystallization temperature.

In cold rolling, with each pass the stock becomes work hardened and its yield strength increases. Therefore, the extent of total reduction given to the stock, in a series of passes, is generally limited by stock reaching high yield strength when reduction of subsequent passes is not economic. At this stage, if desired thickness of the product has not been reached, the rolled product is sent for annealing to bring down the strength to the material's annealed strength, which is then put to further rolling. These annealing operations performed between two sequences of rolling is called *inter-anneal.*

(c) Cold rolling is generally done from coil to coil. This means, the input material (hot rolled strip, continuous cast strip or cold rolled and annealed strip) comes in form of coils. The coil is unwound from an unwinder and the material passes through the roll bite, gets reduced in thickness and then rewound in a coiler. After the pass is completed, the rolls of the mill stand are reversed and the roll gap is adjusted so that the material is rolled again in the opposite direction. In this manner, a number of rolling passes can be given by using the same mill stand by reversing the mill stand in each pass.

(d) While material is unwound (or decoiled) and rewound (coiled) from and by the two winders (or coilers), the material may be given desired amount of back and front tension, and the resulting roll separating force may be reduced. Therefore, adjustment and control of strip tension during cold rolling is an important parameter. This is achieved through proper control of the winder drive motor.

(e) In case of small production of cold rolled sheets, *stack rolling* may be used. Stack rolling starts with one stack of annealed sheets of small lengths that are manually or mechanically fed from entry side and the rolled sheets are similarly collected from the exit side and stacked. After rolling of all the sheets, the roll gap is adjusted for the desired reduction for second pass. Either the mill is reversed or the rolled stack is brought to the entry side and second pass rolling is started. If the rolled product length becomes too long for handling, the sheets may be sheared at certain length of the product.

(f) During cold rolling, particularly during rolling of long strips at high speed, the rolls are heated up. In order to keep the temperature of the rolls within limit to avoid undesired thermal camber in the rolls and also to keep the temperature of the rolled material safely below recrystallization temperature, cold rolling mill rolls are cooled by spraying *roll coolant**. A lubricant is also mixed with the roll coolant which act as a lubricant between the rolls and stock during rolling and reduces the coefficient of friction between rolls and stock, which in turn reduces the roll separating force and rolling power.

(g) In case of certain metals and alloys, like in aluminum, the recrystallization temperature is low. Aluminum being a soft metal, it is possible to give a large reduction so that after very first pass, the temperature rise in the rolled material is high. The rolled material being coiled immediately, does not release heat quickly to become cool. Under such a case, if the material is subjected to a second pass immediately after the first pass, the temperature of strip may cross the recrystallization temperature. For this reason, the material after one pass is taken down from the coiler and allowed to wait till its temperature comes down sufficiently when it can be put to the next pass. For this reason, all the cold mills for aluminum rolling are non-reversing type.

However, in case of non-reversing rolling operation, *batch rolling* with a certain number of coils is employed in conjunction with suitable coil handling equipment to minimize down time between coils.

(h) While rolling very thin gauge at the foil stage, dark mode rolling (refer section 2.2.13 for details) condition is faced. In this condition the thickness of the stock cannot be reduced by trying to reduce the roll gap through operation of the screw-down mechanism. Actually the two ends of the rolls beyond the width of the stock touch each other and further screwing down is not possible.

To improve this difficult situation during foil rolling two different rolling techniques are applied, these are:

Pack rolling: In this process two strands of foils are placed one over the other and the pair is simultaneously introduced into the roll bite. By doing this doubling of stocks, the entry thickness becomes double and the dark mode rolling condition is either removed or at least improved. The two stocks are reduced more or less equally (the yield stress value of the two strips should be nearly equal i.e., both stocks should have

* See section 4.3.7 for discussion on roll coolant system.

same composition and same amount of previous reduction). The two strands of strips are separated after rolling into two separate coils in a separating machine. A very thin layer of oil is given between the two layers at the entry side to make the separation easy. Pack rolled foil can easily be recognized by the fact that sandwiched face becomes matt while the other face which is in contact with roll becomes polished and glossy.

Tension and speed control: During dark mode rolling of a single strip or doubled strip, the output gauge cannot be changed by screw down control. The output thickness of the foil can be controlled by changing the rolling speed and changing the input tension in the foil. When speed is increased the thickness reduces and vice-versa. Similarly increasing the back tension in foil reduces the foil thickness and vice-versa. During foil rolling the thickness of the foil is kept within close tolerance by controlling speed and back tension separately or jointly by varying one parameter after the limit is reached by the other parameter.

(i) Cold strip rolling is generally done from coil to coil in a reversing mill. The strip is coiled under tension on a winder mandrel. These winder mandrels are collapsible and made of three segments with provision of strip end gripper on one of the segments. When rolling thin strips or foils, the inner layers of strips may get impressions of the strip grippers and small gaps between drum segments. It becomes particularly problematic for softer materials and wound under large tension. To avoid such markings on the strip, cylindrical machined barrels, called *core* are used. These cores are fixed over mandrels. The end of the thin strip or foil is fixed on the core by adhesive tape or just by a belt wrapper manipulation, and then the coil is wound on the cores. All subsequent movements of coils take place with cores.

(iv) *Pass schedule for cold rolling*

Pass schedule for cold rolling (of flat products) basically means to determine the total number of passes and the respective reduction (draught) in each of the passes along with the number of inter-anneals to be given at what thicknesses, in order to roll down a strip or sheet of given thickness down to the desired finished thickness.

Obviously the productivity increases if the number of passes can be reduced, and if each pass can be rolled at the top speed of the rolling stand/s.

In order to reduce the number of passes, the reduction of each pass is planned to be kept at a high level keeping in view that the rolling load of each pass should be within the maximum roll load capacity of the mill stand, and also the maximum reduction of the pass should be within recommended maximum reduction for the material under consideration. Similarly while choosing the rolling speed of a pass, the power of the motor driving the mill should be kept in mind. While preparing the pass schedule, therefore, the expected rolling load and rolling torque /power for each pass is calculated to check whether these are within limit.

Once the rolling speed is determined, and knowing the length of the stock being rolled, the time for rolling each pass can be calculated. Total time taken by all the passes plus total handling time will determine the production rate for rolling a stock of certain weight from a given input thickness to some other finished thickness.

For obtaining a sound and defect free cold rolled product generally the maximum reduction on a single pass and total cumulative reduction between two anneals for different metals or alloys are kept within limits. Table 3.3 shows the recommended values of these limits for a few materials.

Table 3.3 Recommended maximum pass reduction and maximum cumulative reduction in cold rolling of different metals and alloys

Sl. No.	*Material*	*Maximum pass reduction on annealed material*	*Maximum cumulative reduction between anneals*	*Approximate no. of passes required for cumulative reduction*
1.	0.2% Carbon steel	25%	65%	5 passes
2.	0.36% Carbon steel	20%	50%	4 passes
3.	1.3% Carbon steel	18%	40%	6–8 passes
4.	0.08% carbon Rimming Steel	15%	65%	6 passes with equal % reduction
5.	18/8 stainless steel	15%	40%	7–8 passes
6.	Copper / brass	30%	60%	3 passes
7.	Nickel–silver (continuously cast)	25%	50%	3 passes
8.	Aluminum, 1100 series	50–55%	93–95%	4 passes

(v) ***Roll load and torque calculation in cold rolling***

There exist well proven methods of estimating roll load, torque and rolling power, as summarized below:

(a) The method based upon the cold rolling theory of Ford and Bland[6] in which the calculations are done using a computer. The method is an iterative one and it is very time consuming when done manually; but it does take into account strip tensions and further can be applied to any material provided its yield stresses at different degree of cumulative reductions are known. A series of experimental data in graphical form known as 'Kp', 'Kg' curves are available for different materials or alloys, which are used in this method.

(b) The alternative method is based upon to use of 'fc' curves, which have been derived experimentally, taking readings from experimental mills. Provided that the mill for which load and torque to be calculated is very similar in mechanical dimensions and power to the experimental mill on which the fc curve has been obtained, this method gives a very quick and easy way of determining mill power and load. This method, however, does not consider effect of strip tensions.

The two methods have been discussed in some detail below:

*A. Ford and Bland method**

The formulae for roll load and roll torque at each pass are given by:

$$P = Kp\sqrt{R'\delta}\left(1-\frac{t_b}{Kp}\right)\times f_3$$

and

$$T = 2RKg\delta\left(1-\frac{t_b}{Kg}\right)\times f_5 + R\left(t_b h_1 - t_f h_2\right)$$

where:

P = Roll load in tons per inch width of strip

T = torque (for both rolls together) in tons - ins per inch width of strip

K_p, K_g = mean yield stress in tons per sq in. at the mean reduction condition of the specific material, to be obtained form the respective Kp, Kg curve for the material.

R = work roll radius, inch

R' = flattened work roll radius, inch (to be calculated from Hitchcock's formula, and also given as a curve.)

δ = h_1-h_2, i.e., reduction in inches

t_b = back tension expressed as a stress in tons/sq.in

f_3 is a non dimensional roll load function obtained form series of different curves for different values of parameters 'a'

$a = \mu\sqrt{\frac{R'}{h_2}}$ and

μ = coefficient of friction

t_f = front tension in tons /sq in

f_5 is a non dimensional roll torque function, whose value also depends on factor "a" and expressed in form of series of curves for different value of 'a'.

* It is to be noted that all the input data and output results were worked out in old British unit by the authors of the formulae.

The following initial data have to be estimated by the pass designer for calculating the rolling load, torque and HP for a given pass, which are:

t_f, t_b, h_1, h_2, R, and rolling speed.

The different curves associated with Ford and Bland formulae and a sample calculation has been included at the end of this section as Appendix.

B. fc curve method

This method is particularly applicable to thin gauge strip mills, as the Ford and Bland method becomes suspect and less accurate for thinner gauges.

It states that the roll load is proportional to:

(a) the area of contact = $w\sqrt{R\delta}$
where, w = width of material
and δ =pass reduction = $h_1 - h_2$
However, flattening of the rolls results in an increase in the area of contact and an allowance must therefore be made for this. Call this factor f_F.
So, modified area of contact becomes $w\sqrt{R\delta}\, f_F$

(b) the average yield stress
If Ys_1 = yield stress of ingoing stock, and
Ys_2 = yield stress of outgoing stock, then
If it is assumed that yield stress increases proportionally with distance across the length of arc of contact,
Then average yield stress Y_{av} would be = $\dfrac{Ys_1 + Ys_2}{2}$
But two other factors must be taken into account
  (i) the friction hill factor, call this f_H
  (ii) the lateral spread, which is constrained and therefore gives the effect of increasing the yield stress. Call this f_L
Modifying for (i) and (ii)

$$Yav = f_L\, f_H\, \frac{(Ys_1 + Ys_2)}{2}$$

$$P\text{ (total)} = \left[f_L\, f_H\, \frac{(Ys_1 + Ys_2)}{2}\right] \times \left[f_F w\sqrt{R\delta}\right]$$

$$= \left[f_L f_H f_F \frac{(Ys_1 + Ys_2)}{2}\right] \times w\sqrt{R\delta}$$

If we define $fc = f_L f_H f_F \dfrac{(Ys_1 + Ys_2)}{2}$, and if yield stress is expressed in p.s.i. and w, R and δ are expressed in inches, then $P = \dfrac{fc\, w\sqrt{R\delta}}{2240}$ tons.

fc is not a constant, but will vary with the percentage reduction and the ratio $\dfrac{h_1}{R}$

A series of fc curves can be drawn for any material plotting fc against percent mean total reduction for different values of $\dfrac{h_1}{R}$ using experimental readings obtained on an existing mill.

This graph can then be used for the design of a new mill by plotting points on the appropriate $\dfrac{h_1}{R}$ curve for the percent mean total reduction at each pass and reading fc in each case.

Consider for example two consecutive passes of an annealed strip through the mill as follows:

h_1	h_2	δ	*Pass reduction*	*Total reduction*	*Mean total reduction*
0.100	0.080	0.020	20%	20%	10%
0.080	0.064	0.016	20%	36%	28%

One can continue along the same f_c curve for a specific $\dfrac{h_1}{R}$ value provided annealing does not take place between passes. With annealing between passes, continue the pass schedule from that annealed thickness as h_1 as though it is the first pass, using the new $\dfrac{h_1}{R}$ curve.

The rolling loads for each pass can then be computed knowing the values of fc.

The H.P. per pass can be calculated as follows:

Total torque (both rolls) T = 2Pa tons-in.

where a = length of lever arm in inch.

Now ratio of length of lever arm to length of projected arc of contact is a constant, λ, for any mill. As per equation (2.74),

$$\lambda = \frac{a}{\sqrt{R\delta}}$$

$$\therefore \qquad T = 2\,P\lambda\sqrt{R\delta}$$

and HP= 0.0355NT, where N is rpm of roll.

$$= 0.071\,P\lambda N\sqrt{R\delta}$$

λ has been found experimentally to be 0.42 (average of many experimental results).

$$\therefore \qquad H.P = 0.03\,PN\sqrt{R\delta}$$

Alternatively, this can be expressed in terms of line speed (f.p.m.) rather than roll r.p.m

As
$$\frac{2\pi RN}{12} = V \text{ or } N = \frac{6V}{\pi R}$$

$$H.P = 0.03P \times \frac{6V}{\pi R} \times \sqrt{R\delta}$$

or
$$H.P. = 0.057\,PV\sqrt{\frac{\delta}{R}}$$

Again replacing P by $\frac{f_c w\sqrt{R\delta}}{2240}$

$$H.P. = 25.6\times10^{-6}\times fc\, w\delta\, V$$

Series of fc curves for different metals and alloys published by various research organizations as well as mill builders are available for use. Different cold rolling mill organizations and mill builders use data of these curves to design the pass schedules of their need. Computer is used for doing this calculation. The maximum rolling load depending on the size of the mill, maximum pass reduction and maximum cumulative pass reduction depending on the materials to be rolled are all fed into the computer as limits and the computer calculates within these limits to produce a pass schedule with minimum number of passes for the entire rolling from input thickness to output thickness or thicknesses.

A few typical pass schedules are given below to get an idea of such pass schedules for cold rolling of metals and alloys.

Example 1:

Material: 1.3% carbons steel
Mill: 4-Hi mill with work roll diameter = 215 mm
Strip width: 350 mm
Input thickness: 2.94 mm annealed
Desired output: 0.50 mm

Pass no.	*ING, mm*	*Out mm*	*Reduction %*	*FRT*	*BCK*	*Speed, mpm*	*Fr.*	*Load, ton*	*Torque, t–m*	*Power kw*
				(kgf/mm²)						
1	2.940	2.520	14.28	4.42	0.97	100	.089	299	1.198	182
2	2.520	2.340	7.14	4.40	4.22	200	.078	257	0.830	252
3	2.340	2.200	5.98	3.81	3.66	250	.073	258	0.690	262
4	2.200	2.080	5.45	4.03	3.87	250	.073	256	0.611	232
5	2.080	1.980	4.80	3.52	3.41	300	.067	247	0.517	235
6	1.980	1.900	4.04	3.67	3.56	300	.067	235	0.416	189
7	1.900	1.840	3.16	3.79	3.71	300	.067	219	0.313	142
8	1.840	1.800	2.17	3.87	3.79	300	.067	199	0.207	94
				ANNEAL						
9	1.800	1.500	16.67	7.43	1.58	100	.089	289	0.867	131
10	1.500	1.340	10.67	6.26	6.00	250	.073	277	0.812	308
11	1.340	1.220	8.95	6.87	6.60	250	.073	274	0.649	246
12	1.220	1.120	8.20	6.22	6.04	300	.067	266	0.552	251
13	1.120	1.040	7.14	6.70	6.50	300	.067	253	0.445	202
14	1.040	0.980	5.76	7.11	6.97	300	.067	236	0.336	153
15	0.980	0.940	4.08	7.41	7.27	300	.067	216	0.223	101
				ANNEAL						
16	0.940	0.780	17.02	14.28	3.04	100	.089	257	0.381	57
17	0.780	0.680	12.82	12.33	11.84	250	.073	269	0.565	214
18	0.680	0.600	11.76	11.62	11.27	300	.067	280	0.485	221
19	0.600	0.540	10.00	12.91	12.52	300	.067	271	0.374	170
20	0.540	0.500	7.41	13.94	13.66	300	.067	252	0.252	115

Legend:

ING	=	Input gauge	F_r	=	Coefficient of friction
Out	=	Output gauge	Load	=	Rolling load
FRT	=	Front tension	Torque	=	Rolling torque
BCK	=	Back tension	Power	=	Rolling power
Speed	=	Rolling speed			
Reduction	=	% reduction of pass			

Example 2:

This example shows the effect of work roll diameter for rolling of same material under identical rolling conditions.
Material: 0.36% carbon steel
Strip width: 525 mm

Input thickness: 2.00 mm
Output thickness: 0.50 mm

Work roll diameter: 200 mm for first schedule and 265 mm for the second schedule. The two sets of result clearly illustrates effect of work roll diameter on rolling load, torque and power.

Pass no.	*ING, mm*	*Out mm*	*Reduction %*	*FRT* (*kgf/mm²*)	*BCK*	*Speed, mpm*	*Fr.*	*Load, ton*	*Torque, t-m*	*Power kw*
				Work roll diameter = 200 mm						
1	2.000	1.500	25.00	4.90	3.80	200	.078	372	2.083	680
2	1.500	1.100	26.67	5.46	4.90	350	.062	417	2.220	1269
3	1.100	0.790	28.18	5.88	5.44	475	.050	394	1.806	1401
4	0.790	0.590	25.32	6.44	5.88	500	.050	385	1.297	1060
5	0.590	0.500	15.25	6.72	6.44	550	.050	329	0.640	575
				Work roll diameter = 265 mm						
1	2.000	1.500	25.00	4.90	3.80	200	.078	457	2.864	706
2	1.500	1.100	26.67	5.46	4.90	350	.062	518	3.057	1319
3	1.100	0.790	28.18	5.88	5.46	475	.050	496	2.492	1459
4	0.790	0.590	25.32	6.44	5.88	500	.050	498	1.812	1117
5	0.590	0.500	15.25	6.72	6.44	550	.050	445	0.901	611

Production Calculation:

Let the coil weight of the material in above example 2 be 1700kg, then the initial length of the strip at 2.0 mm thickness would have been give by 'L' where,

$$\frac{52.5 \times 0.2 \times L \times 7.8}{1000} = 1700$$

or $$L = 161905 \text{ cm} = 1619, \approx 1620 \text{ m.}$$

One can tabulate for each pass the strip lengths, and also the time taken for rolling the pass, as below:

Pass no.	*Out, mm*	*Strip length, m*	*Speed, mpm*	*Pass Time, min*
1	1.5	2160	200	10.8
2	1.1	2945	350	8.41
3	0.79	4100	475	8.63
4	0.59	5491	500	10.98
5	0.50	6480	550	11.78
			Total	50.60

Assuming one minute time is required for screw down adjustment, reversing and rethreading of the mill between the passes, and total 5 minutes is required for unloading of rolled coil and reloading of a new coil, the total time required for rolling the 1700 kg coil form 2.0 mm thickness to 0.5 mm thickness in total 5 passes is = 50.6+4+5 = 59.6 minutes say 60mts. Therefore, the mill productivity, assuming 85% mill availability = 1.7×0.85 =1.445 tons per hour.

Appendix

Calculation of roll load and power requirements for cold rolling using Ford and Bland formula

1. General: The method used is based on the cold rolling theory of Ford and Bland which is given in "The Journal of the Iron and Steel Institute" May 1951 pp 57–72, July 1952 pp 239–249. The calculations being iterative are best carried out in a tabular form so that results are easily comparable or can be done in a computer. An example is shown at the end of this note.
2. Initial data: The following initial data are required:
   (a) Pass schedule – This should give a complete list of ingoing and outgoing gauges together with any anneals that may be considered essential.
   (b) Curves of kp, kg and indentation stress against percentage reduction for the referred material. These are used to determine the yield stress. See Figs. 3.34 and 3.35 for Al. alloys.
   (c) Curves of factors f_3 and f_5. These are independent of the material used. See. Figs. 3.36 and 3.37.
   (d) Graphical solution of Hitchcock's formula for flattened roll radius. See Fig. 3.38
   (e) Values of back and front tension considered for a pass, either as stresses or forces.
   (f) Roll radius and strip width.
   (g) Friction Factor
3. Method: The results are laid out in the example given. The calculation is carried out in the following steps:-
   (a) Draught, $\delta = h_1 - h_2$
   (b) Pass % reduction = $\delta / h_1 \times 100$
   (c) Total % reduction since anneal. This is given by

   $$\frac{\text{Gauge at anneal} - h_a}{\text{Gauge at anneal}} \times 100,$$

   where h_a = thickness of strip under consideration

(d) % reduction used to find kp
= 0.4 × Total % reduction before pass + 0.6 ×Total % reduction after pass (as considered by Ford and Bland)

(e) % reduction used to find kg
= 0.5 (Total % reduction before pass + Total % reduction after pass) (ref. as above)
N.B. When the pass is the first one after an anneal, items d and e have the same value as the pass reduction, item b.

(f) Assume a roll load, P. This will vary with the material being rolled and the draught, but first estimate should be taken from the following ranges:
- Steel 20–30 tons/in. strip width
- Aluminum strong alloy 10–13 tons/in. strip width
- Aluminum common alloy 5–7 tons/in. strip width

(g) Reduction stress= P/δ

(h) Roll flattening factor, $\frac{R'}{R}$. This is obtained from the graphical solution of Hitchcock's formula. Figure 3.44

(i) Flattened roll radius, $R' = R \times \frac{R'}{R}$

(j) Roll slip factor, $a = \mu\sqrt{\frac{R'}{h_2}}$

Values of μ are

a. Artificially roughened rolls	0.15
b. Smooth dry rolls	0.09 – 0.11
c. Smooth lubricated rolls	0. 07 (steel)
	0.062 (copper)

(k) Roll pressure factor f_3. Taken from graph Fig. 3.36 using appropriate values of a, (item j), and pass % reduction

(l) Torque factor f_5, similarly determined from Fig. 3.37

(m) Allowable roll load stress, kp. Taken from graph in Fig. 3.34 using % reduction value as given in (d).

(n) Allowable load stress, kg. Taken from graph in Fig. 3.34 using % reduction value as given in (e).
N.B. When the pass is the first one after an anneal, kp and kg are read off the kp and kg curves. For subsequent passes, they are to be read off the indentation curve.

(o) Back tension factor, $1 - \frac{t_b}{kp}$. If the back tension is given in tons, it must be converted to tons/in^2. by dividing by (strip width x h_1)

(p) Roll load $P' = kp \times \sqrt{R' \times \delta} \times f_3 \times \left(1 - \frac{t_b}{kp}\right)$

This must be compared with the roll load assumed in item (f). If the two values differ by more than 10%, a new value, of P must be assumed and items g to p repeated, according to the following rule:

If P>P′, assume a new P slightly less than P′
If P<P′, assume a new P slightly greater than P′

(q) Total roll load = P′× strip width

(r) Rolling torque per roll = $R \times kg \times \delta \times f_5 \times \left(1 - \frac{t_b}{kp}\right)$

(s) Tension correction torque = $\frac{R}{2}\left(t_b h_1 - t_f h_2\right)$

If tensions are given in tons, this can be replaced by

$$\frac{R}{2}\frac{(\text{Back tension , tons} - \text{Front tension , tons})}{\text{strip width}}$$

(t) Corrected torque T′= Rolling torque + Correction torque

(u) Total torque T = T′ × strip width ×2

(v) Roll horsepower= 0.0355× roll R.P.M. × total torque. The total torque must be in tons-ins, and the factor, 0.0355 includes all the necessary constants.

(w) Motor horsepower= Roll horsepower / drive efficiency

If it is required to limit the roll or motor horsepower, the corresponding roll rpm can easily be found.

One calculation based on Ford and Bland method is shown below:

Calculation of Roll Load and Power Requirements for Rolling Soft Aluminum Alloy

Legend

R = Roll radius = 8.96 in
R′ = Flattened roll radius
t_b = Back tension
t_f = Font tension

h_1 = Ingoing gauge
h_2 = Exit gauge
$\delta, \Delta h$ = Draught
f_3, f_5 = Constants
μ = Coefficient of friction = 0.08

Proposed pass schedule with tension and speed data:

Pass	*Ingoing Gauge (in)*	*Exit Gauge (in)*	*Draught (in)*	*Width (in)*	*Back Tension (ton/in²)*	*Front Tension (ton/in²)*	*Roll rpm/ speed fpm*
1A	0.2756	0.1181	0.1575	57	1.3	1.95	210/985
2A	0.1181	0.0630	0.0551	57	1.3	1.95	350/1640
3A	0.0630	0.0275	0.0355	57	1.3	2.6	400/1875
1B	0.2756	0.1102	0.1654	57	1.3	1.95	190/890
2B	0.1102	0.0520	0.0582	57	1.3	1.95	350/1640
3B	0.0520	0.0197	0.0323	57	1.3	2.6	400/1875

Sample calculation for pass 1A

Pass % reduction $= \dfrac{\Delta h}{h_1} = \dfrac{0.1575}{0.2756} = 57.2\%$

Total % reduction since anneal = 57.2%
% reduction to find kp = 57.2%
kp = 8 (from graph)
% reduction to find kg = 57.2%
kg = 7.5 (from graph)
Assume a roll load (P) of say 8.1 tons/ inch width

Reduction stress $= \dfrac{P}{\Delta h} = \dfrac{8.1}{0.1575} = 51.43 \text{ tons/in}^2$

Roll flattening factor $= \dfrac{R'}{R} = 1.017$ (from Hichcock's formula)

Flattened roll radius $= R' = R \times \dfrac{R'}{R} = 8.96 \times 1.017 = 9.11 \text{ ins}$

Roll slip factor, $= a = \mu\sqrt{\dfrac{R'}{h_2}} = 0.08\sqrt{\dfrac{9.11}{0.1575}} = 0.703$

Roll pressure factor f_3 = 0.97 (from Fig. 3.36)
Torque factor f_5 = 0.478 (from Fig. 3.37)

Back tension factor $= 1 - \frac{t_b}{k_p} = 1 - \frac{1.3}{8} = 0.8375$

Roll Load P′ $= kp\sqrt{R' \times \Delta h} \times f_3 \times \left(1 - \frac{t_b}{k_p}\right)$

$= 7.8$ tons / inch width

This is within 5% of assumed load P,

Therefore let us take P′ = 8 tons/ inch width

Therefore total roll load = 8×57= 456tons

Torque due to rolling $= R \times kg \times \Delta h \times f_5 \times \left(1 - \frac{t_b}{kp}\right)$

= 4.237 Tons ins.

Tension correction torque $= \frac{R}{2}\left(t_b h_1 - t_f h_2\right)$

= 4.48(1.3×0.2756 – 1.95×0.1181)

= 0.573 Tons ins.

Corrected torque, T′ = Rolling torque + Correction torque

= 4.237+0.573

= 4.81 Tons ins

Total torque, T = T′×strip width×2

= 4.81×57×2

= 548.3 Tons. ins

Roll Horsepower = 0.0355×Roll rpm×total torque

= 0.0355×210×548.3

= 4087 H.P.

= 3050 kw

Sample calculation for pass 2A

Pass % reduction = 46.7%

Total reduction before pass = 57.2%

Total % reduction after pass = 77.2%

% reduction to find kp = 0.4 total % reduction before pass + 0.6 × total % reduction after pass = 69.2%

Kp (from graph) = 8.65

% reduction to find kg = 0.5 (total reduction before pass + total reduction after pass)

= 67.2%

Kg (from graph)	= 8.6
Assumed roll load P	= 7.5 tons / inch width
Reduction stress $\frac{P}{\Delta h}$	= $\frac{7.5}{0.0551}$ = 136.1 tons/in^2
Roll flattening factor	= $\frac{R'}{R}$ = 1.045 (from Hitchcock's formula)
Flattened Roll Radius	= $R' = R \times \frac{R'}{R} = 9.36$ in
Roll slip factor, a	= $\mu\sqrt{\frac{R'}{h_2}} = 0.08\ \sqrt{\frac{9.36}{0.063}} = 0.975$
Roll pressure function,	= $f_3 = 1.17$ (from graph)
Torque function,	= $f_5 = 0.565$ (from graph)
Back tension factor	= $1 - \frac{t_b}{kp} = 1 - \frac{1.3}{8.65} = 0.85$
Roll load P′	= $kp\ \sqrt{R' \times \Delta h}\ \ f_3 \times \left(1 - \frac{t_b}{kp}\right)$
	= 6.28 tons./ inch width
Say roll load, P′	= 6.3 tons / inch width*
Total roll load	= 57×6.3=360 tons
Torque due to rolling	= $R \times kg \times \Delta h \times f_5 \times \left(1 - \frac{d_1}{kp}\right)$
	= 2.038 Tons. ins.
Tension correction torque	= $\frac{R}{2}\left(t_b h_1 - t_f h_2\right) = 0.133$ T/in
Corrected torque, T′	= 2.171 tons. ins.
Total torque, T	= T′×strip width×2
	= 247.5 tons. ins.
Roll Horsepower	= 0.0355×Roll RPM×total torque
	= 0.0355×350×247.5
	= 3075 HP
	= 2300 kw

* This value has been assumed arbitrarily. Actually the exact value should be found out by necessary number of iterations

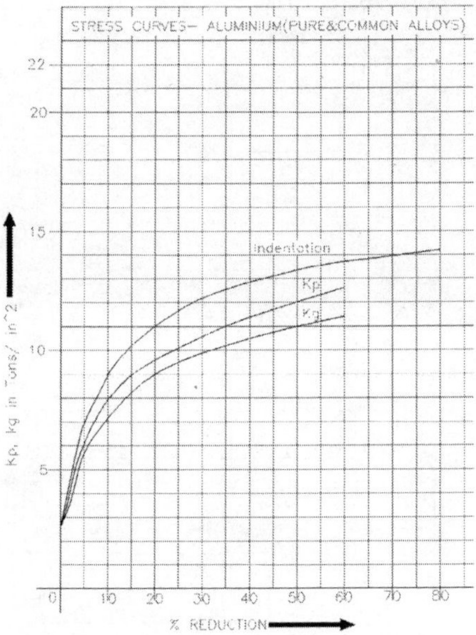

Fig. 3.34 Stress curve- aluminum (pure and common alloys)

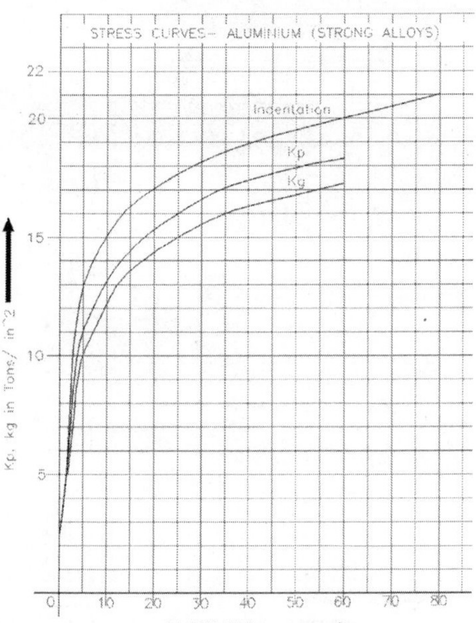

Fig. 3.35 Stress curve- aluminum (strong alloys)

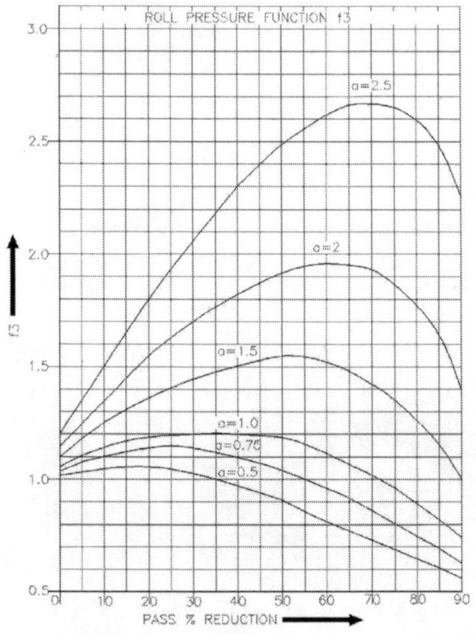

Fig. 3.36 Roll pressure function – f3

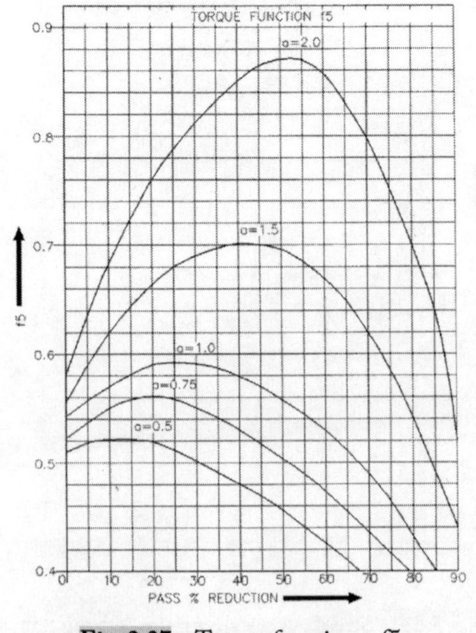

Fig. 3.37 Torque function – f5

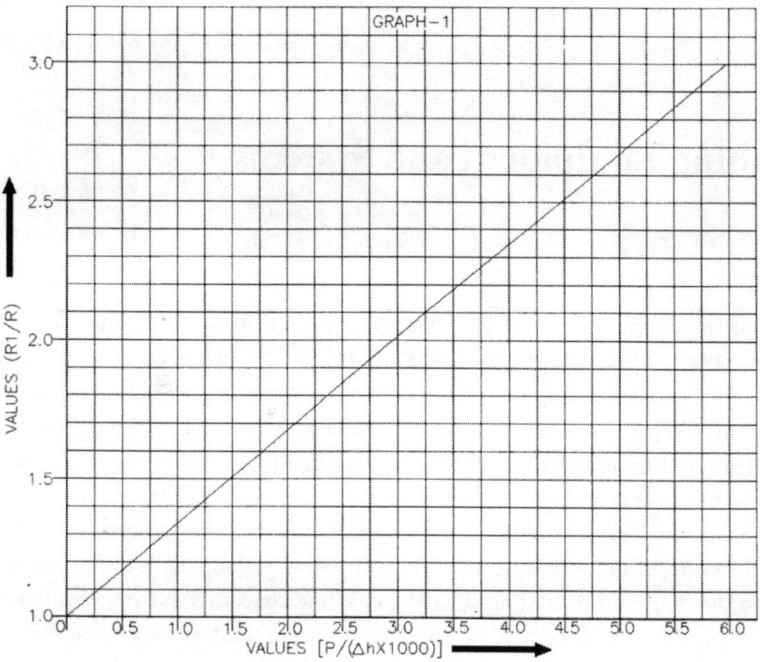

Fig. 3.38 Values of R1/R for reduction stress P/Δh as per Hitchcock's formula

References

1. Tselikov and Smirnov, "Rolling Mills", Pergamon Press, 1965.
2. Winton, D.A., "Roll Pass Design for Round Sections", B.I.S.R.A Publication, 1949.
3. Trinks, W., "Roll Pass Design", Penton Publishing Co., 1933.
4. Based on data provided by "Parkgate Iron and Steel Co. Ltd."
5. "Sendzimir Cold Rolling Mills", Catalog of T.Sendzimir Inc., Waterbury, USA.
6. Ford, H., Ellis, F. and Bland, D.R., "Cold Rolling with Strip Tension, Part I, J.Iron and Steel Inst., 1951 and Part II & III, J.Iron and Steel Inst., 1952.

CHAPTER 4

Rolling Equipment and Systems

Various types of equipment and systems are used in a rolling mill. The major equipment is obviously the *mill stand* in which the material is actually rolled between two or three rotating rolls. Apart from the mill stand along with its own different mechanisms and systems for rotating the rolls, positioning the rolls to effect amount of reduction on the stock, to guide the stock in perfect position and orientation with respect to the rolls, etc., various other equipment and systems are necessary based on the rolling process (hot or cold), size and shape of input stock, specifications of product mix, desired quality of the products, and handling of the rolling stock up to finished product stage. These are called *auxiliary equipment.*

In this section, for each of these equipment and systems the functions, constructional features and specifications and in some cases the basic design features and calculations have been discussed. The design aspects will be limited to mechanical design. Electrical drive, controls and instrumentation plays a very important role in modern rolling mills. A brief section has been devoted to these but their design aspects are beyond the scope of the present treatise.

4.1 Mill Stand Components and Mechanisms

Most type of rolling mill stands have certain common basic components, mechanisms and systems. These are:

(i) Work rolls between which rolling takes place.

(ii) Backup rolls to support work rolls, excepting in case of 2-Hi, 3-Hi mills or cluster mills.

(iii) Roll supporting bearing and bearing blocks (chocks).

(iv) Roll positioning (or screw-down) mechanism which alters the gap between the work rolls.

(v) Roll balancing arrangement which keeps the set of top work and backup rolls in position and does not allow the top work and back-up rolls to come down and touch bottom work roll when there is no stock in roll bite.

(vi) Mill housings (generally a pair for normal 2-Hi, 3-Hi, or 4-Hi mills) within whose windows the roll bearing chocks are supported and allowed to be slided up and down during screw down adjustment.

(vii) Base Plates (also called shoes) in the form of two girders onto which the housings are fixed. The base plates are mounted on the foundation.

(viii) Roll change device by which the used work rolls and / or back-up rolls are removed from the mill stand and new or reground rolls are fitted back in the stand.

(ix) Drive Arrangement which consists of :

(a) Gear box that is coupled with drive motor, reduces the motor speed to desired level.

(b) Pinion stand which is a gearing arrangement with one input shaft connected with gear box, and two or more output shafts to drive the rolls.

(c) Spindles and couplings which connect the pinion stand outputs to the driven roll ends.

(x) Guides that direct the stock in entering and leaving the work rolls.

These components and mechanisms are discussed in greater details in the following sections.

4.1.1 Rolling mill rolls

A. ***Types of rolls***

Mill rolls are the most important part of a mill stand, because the rolling stock gets rolled while passing through the gap created between the rotating rolls under compressive stress exerted by the rolls, the rolled stock being continuously in contact with surface of the two rolls (three rolls in very special cases of rolling like tube rolling). The rolls act as the tool for a rolling mill.

A typical roll (Fig. 4.1) consists of a middle part which touches the rolled metal, called barrel. At both ends of the barrel are the roll necks which are supported in bearings. The part extending beyond the necks, in one or both sides, is for connecting the roll to the coupling of the driving spindle and coupling.

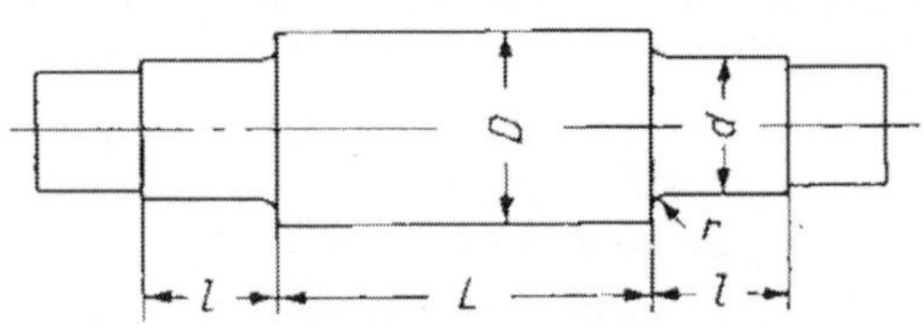

Fig. 4.1 Plain roll

For rolling flat products, the roll barrels are made plain cylindrical or with slight convexity (crowning of roll) to compensate for roll bending during rolling. Grooved barrels are used for hot rolling of different sections (Fig. 4.2).

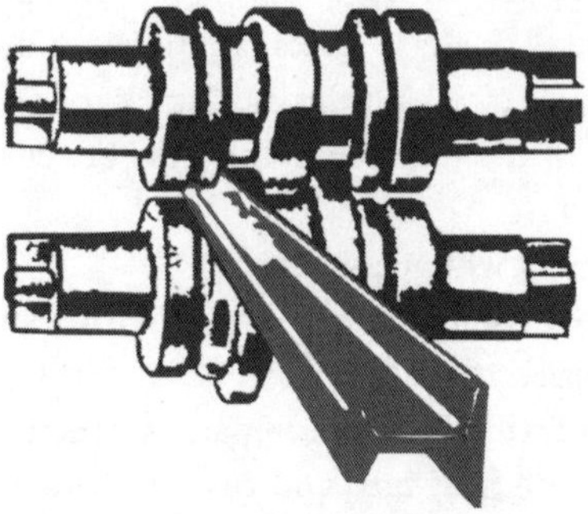

Fig. 4.2 Grooved roll

As rolls are in continuous contact under large pressure and under sliding friction with the rolling stock, one of the main properties of a roll is its barrel hardness apart from its strengths under shear, bending and torsion. It is customary to classify rolls into following categories based on barrel hardness.

Soft rolls

Hardness 150–250 Brinell (25°–35° shore).

Use: in primary mills, in roughing stands of heavy section and intermediate mills, in piercing mills, etc.

Materials: cast steel, forged carbon steel and grey cast iron when subjected to small loads.

Semi-hard rolls

Hardness 250–400 Brinell (35°–60° shore).

Use: in the roughing stands of rail and section mills and all section and flat rolling mills, in the finishing stands of heavy section and intermediate mills.

Material: semi-chilled iron, cast and forged steel.

Hard rolls

Hardness about 400–600 Brinell (60°–85° shore).

Use: in finishing stands of strip, sheet, rail and section, heavy section, medium section and light section mills, and also for back-up rolls in 4-High cold mills.

Material: chilled iron, alloy steel and rolls built up with various hard alloys.

Very hard rolls (rolls for cold rolling)

Hardness 600–800 Brinell (85°–100° shore).

Use: work rolls for cold rolling mills and for small rolls (in Sendzimir mill)

Material: forged alloy steel with chromium, tungsten carbide and other alloys with hardness more than 100° Shore.

B. ***Size of rolls***

Size of a roll is generally designated by the nominal diameter of its barrel and the barrel length.

With grooved rolls, the nominal diameter of the barrel is usually considered to be the distance between the axes of the two rolls at the rolling condition (in case of blooming mills that during the last pass). Thus nominal diameter is always greater than the actual or working diameter. The maximum ratio of nominal diameter to the working diameter is generally kept within 1.4 to avoid too deep grooves. In flat rolling, the actual diameter of the roll barrel is the actual or working diameter which remains fixed during rolling.

The working diameter of the rolls is chosen so that it allows for the permissible angle of bite and for adequate strength of the rolls against bending. The frequently used rolling angles are given in Table 4.1. The corresponding maximum Δh/D ratio is also indicated in the table.

Table 4.1 Maximum permitted angles of bite and Δh/D

Name of rolling	*Angle of bite (α) in degrees*	$\frac{\Delta h}{D}$
Cold rolling with lubrication on well polished rolls	3–4	$\frac{1}{700} - \frac{1}{400}$
The same, when rolling with roll pressure applied after the stock has been fed into the rolls	6–8	$\frac{1}{170} - \frac{1}{100}$
Cold rolling using coarser rolls	5–8	$\frac{1}{250} - \frac{1}{100}$
Hot rolling tubes on a short mandrel (automatic mills)	12–14	$\frac{1}{60} - \frac{1}{40}$
Hot rolling flat metal	15–22	$\frac{1}{30} - \frac{1}{15}$
Rolling sections	22–24	$\frac{1}{15} - \frac{1}{12}$
Rolling with ragged or built up roll surfaces	27–34	$\frac{1}{9} - \frac{1}{6}$

Depending on these rolling angles, the working diameter of the roll should satisfy the condition.

$$D \geq \frac{\Delta h}{(1 - \cos \alpha)}$$

In case of hot flat rolling, the diameter of the working roll is often fixed on the basis of width of the rolled material and the requirement for minimum deflection, which is particularly important for strip rolling. The length of the barrel 'L' (see Fig. 4.1) is chosen in proportion of roll diameter within following limits:

(i) 2-Hi mills for rolling tinplate : (1.3–1.5) D
(ii) 2-Hi mills for rolling strip : (1.5–2.2) D
(iii) Sheet and plate mills : (2.2–2.8) D
(iv) 4-Hi mill :
  (a) work roll : (2.5–4.0) D
  (b) back-up roll : (1.3–2.5) D

The barrel length of grooved rolls depends mostly on nature of passes, and the proportions generally are:

In blooming mill : (2.2–2.7) D
In roughing stands : (2.2–3.0) D
In finishing stands : (1.5–2.0) D

In case of cold rolling of flat products particularly for metals and alloys having high yield stress (like steel, alloy steel, etc.), generally 4-Hi or cluster mills are chosen whose work roll diameters should be based on the minimum thickness of the rolled material. In large diameter rolls, the roll flattening effect is large and rolling of thin gauge may become impossible. On such consideration, the diameter of work roll (hard rolls) for rolling ferrous material should satisfy the condition:

$D < (1500 \text{ to } 2000)h_{min,}$ when rolling under tension, and
$D < 1000\ h_{min,}$ when rolling without tension,

where h_{min} is the minimum thickness of the rolled strip.

If the rolls are made from tungsten carbide (very hard rolls) as used in cluster mills, the minimum strip thickness may be 2–4 times thinner.

For four-high cold mill, the back-up roll diameter is generally taken 2.4–4.5 times more than that of the work roll, major consideration being resistance to bending under the rolling force.

In each individual mill design, the selection of roll diameter and barrel length should be checked for strength and deflection under rolling condition.

Rolls wear with use and are re-ground or re-polished several times. With each re-grinding a metal layer of 0.5–5.0 mm thick is removed from the barrel surface, while in re-polishing it is from 0.01–0.5 mm approximately. With each such operation the roll becomes smaller and for hardened rolls its surface hardness also reduces. When the roll diameter is reduced to a certain limit, the roll is said to have lost is serviceable life and are discarded. The total re-grinding generally permissible, expressed as a percentage of the diameter of a new roll is:

In blooming / billet mills	10–12%
In section mills	8–10%
In sheet and plate mills	5–7%
In cold mills work rolls	5–7%
In cold mills backup rolls	10–17%

Roll neck diameter 'd' for rolls mounted on plain bearings are made in proportion to new roll diameter 'D' as shown in Table 4.2.

Table 4.2 Roll neck dimensions

Mills	$\frac{d}{D}$
3-high section	0.55
2-high section	0.6–0.7
Blooming	0.55–0.6
Sheet and plate	0.67–0.75
2-Hi strip	0.75–0.8

However, modern mill rolls are fitted with roller bearings. The OD of such bearings being proportionately large, the neck diameters have to be kept smaller as per following relation:

$$d = (0.5 - 0.63)D$$

This will be discussed further in roll neck bearing section.

The roll drive ends are sized to accommodate different types of drive spindle and coupling used.

C. Roll materials

Choice of Material of the roll is very important as it has to withstand heavy load without failing or getting deformed and also to undergo continuous friction from the stock. As far as materials are concerned, rolls are classified into (a) *iron rolls* and (b) *steel rolls*. Normally distinction between the two is made at a carbon content of 1.8%. In case of rolls the demarcation line is about 2.4% carbon. It is common practice to refer to rolls as steel base below this figure, and iron base above this figure. However, it is to be noted that in steel rolls, no free graphitic carbon exists.

Steel rolls may be further subdivided into *cast* or *forged* category. *Duplex rolls* is another subclass which are cast with a hard metal outer surface and a tougher and stronger metal forming the center. Brief details of the different types of rolls and their frequent uses are given below:

(a) *Iron rolls*

(i) Grain rolls: These are grey cast iron rolls cast in sand or loam, and containing flakes of free carbon. These have uniform structure throughout and resistant to fire cracking. They are to an extent self lubricating and particularly used as grooved rolls with collars. A typical material composition is as follows:

2.5–3% C, 0.5 –1% Si, 0.4–0.8%Mn, 0.5% max P and 0.1% max S.

To improve quality, various alloying elements like Ni, Cr and Mo may be added to form so-called *special grain rolls* with increased hardness. Uses of grain rolls are :

Sl. No.	*Type*	*Hardness, shore*	*Uses*
1.	Pearlitic grain	30°–38°	Plate mill: Roughing roll; Structural and billet mill: Roughing, Intermediate and Finish roll; Merchant mill: Strand rolls.
2.	Special grain	35°–40°	-Do-
3.	Alloy grain	38°–50°	Strip, structural and billet mills: Intermediate and finish roll; Rod and wire rod mill: Roughing rolls.

(ii) Clear chill or definite chill rolls: These rolls have a surface layer of white iron produced by inducing rapid cooling of the barrel surface (by using chill in the mould) which restrict formation of free graphite. The core remains grey iron and the intermediate zone consists of mottled iron (mixture of white and grey iron.)

The surface hardness may be between 55°–65° shore but the rolls have good resistance to temperature change and fire cracking. Alloying elements may be added to obtain higher surface hardness up to 65°–90° shore. The uses are as follows:

Sl. No.	*Type*	*Hardness, shore*	*Uses*
1.	Straight carbon chill roll	55°–65°	Plate mills: Roughing and finishing rolls; Universal plate mill: Horizontal and Vertical rolls. 2-Hi sheet mill: Roughing and hot and cold finishing rolls. Strip and bar mill: Hot finishing rolls. Merchant mill: small section rolls.
2.	Molybdenum alloy chill rolls	55°–65°	4-Hi plate mills: Work rolls. 2-Hi sheet mill: Roughing and hot and cold finishing rolls. 4-Hi continuous strip mill: work rolls. Rod and small merchant mill: Hot finishing roll.
3.	High nickel alloy chill rolls (Nichillite)	65°–90°	Plate mill: Finishing rolls. Univ. mill: Edging rolls. 4 Hi cont. strip mill: Work roll Strip mill: cold finishing rolls.

(iii) Composite or duplex rolls: These rolls having a very hard surface with a tough core is produced by double pouring. The first pour gives a shell of highly alloyed white iron which is rapidly cooled by chills, after which a second pour of grey iron displaces the molten center alloy iron establishing a tough core of grey iron. These rolls are primarily used for cold rolling.

Shell hardness	: 75°–90° shore
Uses: 4-High Plate Mill	: Work rolls
Continuous strip mill	: Finishing rolls
4-Hi skin pass mill	: Work rolls

(iv) Indefinite chill rolls: In these rolls there is a very thin layer of white graphite free chill which is machined during roll preparation, and no intermediate mottled zone. The surface consists of very small particles of graphite which gradually changes to grey iron core. These rolls may be of straight carbon or alloyed. After removal of the chilled skin, the hardness of straight carbon rolls may have a roll hardness of 38° to 45° shore. Alloyed indefinite chill roll containing nickel, chromium, and molybdenum has a hardness of 55°–75° shore. These alloyed rolls may be heat treated to toughen them against shock loadings, an example being the Adamite indefinite chill roll. Uses of these rolls are as follows:

Straight carbon Indefinite chill rolls, Hardness: 38°–40° shore	Bar and Small section mill intermediate stand rolls.
Alloy indefinite chill rolls, Hardness :55°–70° shore	Medium structural and billet mill, continuous billet mill, rod and wire rod mill, merchant mill: intermediate and finishing rolls. Strip mill, continuous narrow strip mill: Intermediate and finishing rolls.
Fully hard alloy indefinite chill roll (such as *Niromite, Manganite*); Hardness:65–85° shore.	4-Hi plate mill, continuous wide strip mill: work rolls. 2-Hi, 3-Hi plate mill, continuous billet and sheet bar mills: Finishing roll. Strip and hoop mill: Hot finishing and planishing rolls. 4-Hi skin pass mill: Work rolls.

(v) Spheroidal graphite (SG) rolls: Advantages of these rolls are that they can be easily cast, readily machineable, take good finish, they are rigid and have resistance to heat and corrosion. Little information is available on wear properties of SG rolls. These rolls are generally suitable for use where grey iron rolls are not strong enough and where steel rolls give poor life due to excessive wear. However, cost of SG is more than both iron and steel rolls, hence to be used judiciously. Hardness up to 80° shore is achievable. Uses are:

Reversing roughing mill	: Mill rolls	(45° shore)
billet, sheet, bar mill	: Roughing rolls	(55° shore)
	: Finishing rolls	(60°–65°shore)
Strip mill	: Roughing rolls	(60°–65°shore)
	: Work rolls	(65°–70°shore)
	: Back–up rolls	(50°–55°shore)
Small section and bar mill	: Most stand rolls	(60°–65°shore)

(b) *Steel rolls*

Steel rolls may be cast or forged. These are much stronger and tougher than iron rolls, and are used where iron rolls are not considered strong enough. They are specially used where deep grooves are required. However, steel rolls have less hardness, hence wear faster than iron rolls.

(i) Cast steel rolls: These may vary considerably in chemical composition. The straight carbon roll has from 0.40–0.90% carbon and the hardness is from 28° to 36° shore. Heavy mills (cogging, slabbing, and heavy roughing) use the lower grades (up to 0.60% C) while billet roughing stands use the higher grades. The addition of about 0.5% molybdenum to this type of roll together with small amounts of nickel and chromium (or higher manganese) gives increased strength and reduces the severity of any fire cracks which may occur. The hardness is 30°–42° shore. More highly alloyed rolls usually lie within the following ranges of analyses:

Carbon	*Manganese*	*Nickel*	*Chromium*	*Molybdenum*
0.80/1.0	0.60/0.90	1.0/2.5	0.50/1.10	0.20/0.40

A carbon–chrome roll (1% C, 1.5–1.75% Cr) is also made.

These rolls are usually heat treated, the hardness range is 35°/55° shore and they are commonly used as back-up rolls in 4-Hi mills. An alloy steel containing tungsten and with a hardness of 40°–50° is very resistant to fire cracking and is sometimes used for roughing rolls in wide strip mills.

Cast alloy steel base rolls are made also in the range of analysis of:

Carbon	*Silicon*	*Manganese*	*Nickel*	*Chromium*
0.9/2.5	0.5/1.0	0.4/0.6	0.25/1.0	0.5/1.5

It is seen that the carbon content is in a higher range than in the cast steel roll, but the carbon is in combined form and not in free graphitic form. The hardness range is 30°–55° shore, according to carbon content, and the rolls are wear resistant and are strong. The life is in line with the cost–about twice that of a grain roll. Good water cooling during rolling is required. An example of this type is the adamite range of steel base rolls.

(ii) Forged steel rolls: Forged steel rolls are forged from cast steel ingot and the necessary mechanical working results in an improved tougher structure. In the carbon steel form (0.35/0.75% carbon) they are used for cogging, slabbing, and heavy roughing mills in the lower end of the carbon and for smaller intermediate mills in the higher end of the range. This is somewhat arbitrary and depends on the particular mill conditions. They are normalized before use and the hardness range is 24°/30° shore.

In the alloy steel form they may be heat treated to give a wide range of hardness. In the range 50°–55° they are used for large back-up rolls, around 80° for small back-up rolls in cold rolling, and 90–100° (fully hardened) for work rolls in cold rolling. A typical analysis is 1% carbon, 1.5–1.75% chromium and 0.5% nickel.

Forged steel rolls in the hot rolling hardness range are highly resistant to shock loading.

Uses of forged steel rolls are:

Forged carbon steel roll Hardness: 24°–30° shore.	Cogging and slabbing mill and heavy section mill: Roughing rolls.
Forged alloy steel roll, Hardness: up to 100° shore	Hot sheet and wide strip mill: Back-up rolls (50°–55°) Cold mill : Back-up roll (70°–80°) : Work roll (90°–100°)

D. Roll strength

The strength of a roll is dependent on the type of material it is made of. Table 4.3 shows the equivalent tensile strengths and hardness figures for the material constituting the body of the roll (which is the part controlling the general strength of the roll though the surface of some rolls is much harder).

Table 4.3 is applicable to steel rolls only.

Table 4.3 Hardness figures, tensile strengths and approx. shear strength for steel rolls

Sclerescope hardness number	*Brinell hardness number*	*Tensile Strength*		*Approx. shear strength*	
		tons/sq in.	*kg/cm²*	*tons/sq in.*	*kg/cm²*
42	269	59	9260	39	6120
39	255	56	8790	37	5800
37	241	53	8320	35	5490
35	229	50	7850	33	5180
33	217	48	7530	32	5020
31	207	45	7060	30	4710
28	187	41	6440	27	4240
24	170	36	5650	24	3770
-	156	34	5340	23	3610
-	143	31	4870	21	3300
-	126	29	4550	19	2980

Few data are available with regard to shear strengths for cast iron rolls and most of what is published is not available in precise form.

A Shore hardness is not quoted below 24° shore as this measurement is not reliable for such soft material. In any specific case it is recommended that the roll user asks for the tensile and shear strength figures from the roll maker. It will be noticed that the shear strengths quoted are approx. two-third of the tensile strength. This is a general assumption which is not always valid, the shear strength often falling below this figure.

During rolling, a roll is stressed by compression, tension, and shear resulting from the roll separating force, roll bending, and the driving torque required to turn the rolls. An analysis containing all these is not simple and it is sufficient here to consider bending and torsion together and shear separately (direct compression being ignored in hot rolling except in so far as the deformation of the roll radius affects the rolling load).

(a) *Direct shear stress*

The shear stress resulting from a roll load P being used to roll the stock in a given groove is P divided by the cross sectional area of the roll at this point and if the diameter is D this result in:

$$\text{Shear stress, } f_s = \frac{4P}{\pi D^2}$$

The strength in shear must exceed this.

$$\therefore \qquad \text{As the shear strength } S_s > \frac{4P}{\pi D^2}$$

it follows that D must be greater than $1.13\sqrt{\dfrac{P}{S_s}}$

(b) *Bending stress:*

The stress, f_b, at any point in the roll due to bending is given by:

$$\text{Stress, } f_b = \frac{\text{Bending moment M} \times \text{Distance from centre of roll}}{\text{Moment of inertia of cross section}}$$

The maximum stress is at the roll surface and is given by:

$$\hat{f}_b = \frac{M.\dfrac{D}{2}}{I} \text{ where, } I = \frac{\pi D^4}{64} \quad \therefore \ \hat{f}_b \cong \frac{10M}{D^3}$$

Considering Fig. 4.3 it may be shown that

$$M = \frac{Px}{2} - \frac{P}{2b}\left(x - \frac{L-b}{2}\right)^2,$$ the roll being considered as a simply supported beam.

where, x is the distance of the point being considered from the assumed point of support at the roll end and L is the distance between the two reactions $\frac{P}{2}$ at the roll necks. By Macauley's notation the second term is neglected when

$$x < \frac{1}{2}(L - b).$$

This stress at any point should not exceed the tensile strength to avoid failure. The expression for bending and torsional stress is modified by several factors:

(i) Special properties of cast iron: The structure of cast iron in the material at the center of the roll taking a higher proportion of the load than suggested by the linear distribution of stress, i.e., stress proportional to distance from center. The material at the periphery of the roll takes less stress than expected and to compensate for this the bending stress at the roll surface may be taken as $\frac{6M}{D^3}$* instead of $\frac{10M}{D^3}$ which is used for steel, though for alloy iron the figure approaches nearer to that for steel.

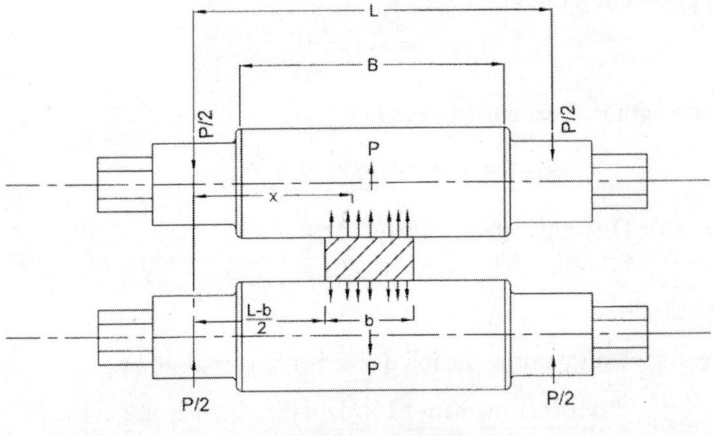

Fig. 4.3 Evenly distributed load centrally placed in the rolls

(ii) Stress concentrations: It is caused by sharp corners, in roll grooves or by narrow deep grooves, magnifying the stress to which the material is subjected. For instance if the collar diameter of the roll is D and the diameter at the bottom of the close flange is d, a stress concentration is set up at the bottom of the flange hole which depends for its intensity on the radius at the tip. Let this be r. Hitchcock has determined curves for stress magnification factors plotted against the ratio $\frac{r}{d}$ at given values of the $\frac{d}{D}$ as shown in Fig. 4.4. These data are referred particularly to

* This is an empirical relation based on practical evidence.

the determination of the stress magnification in such cases as the radius between the roll neck and body. Trinks states that with alloy steel the above factors should be used but with low carbon steel only 60% of the factor should be taken and with cast iron only 35% need be applied.

(iii) Fatigue: The effects of fatigue should be taken into account, because the rotation of the rolls gives a continual alteration of stress throughout the roll and can cause failure when it would not otherwise be expected. A factor of safety is usually applied to allow for this.

If failure through fatigue is likely, it is usual to take the material strength to be half the normal strength. This is equivalent to applying a safety factor of 2.

(iv) Torsional stress: A torque or twist is required to turn the rolls under a given rolling load P. This twisting action sets up stresses in the roll and

$$\text{the stress, } f_t\text{, at any point} = \frac{\text{torque T} \times \text{distance from centre of roll}}{\text{polar moment of inertia}}$$

$$\therefore \quad \text{at the surface of the roll, maximum stress } \hat{f}_t = \frac{T.\frac{D}{2}}{\frac{\pi D^4}{32}} = \frac{16 \times T}{\pi D^3}$$

For the roll neck stress the factor D must be replaced by the neck diameter d.

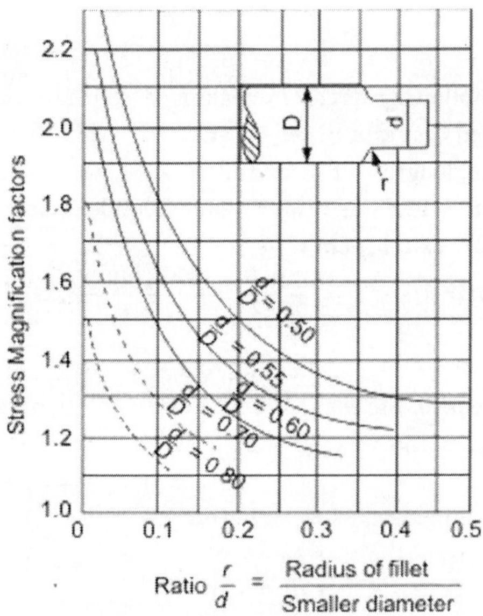

Fig. 4.4 Determination of stress magnification factors

(v) Combined stresses and overall safety factor: When a roll is operating under both bending and torsion stresses it is not sufficient to add the stresses at any point to obtain the total stress. Alternative theories exist to asses the working stress but the one by Rankine gives the greatest factor of safety. By Rankine's hypothesis an equivalent torque is found which accounts for the effect of the bending. This is given by $T_e = M + \sqrt{M^2 + T^2}$ where M and T are the bending moment and torque, respectively at the point in question. The stress under this equivalent torque is the total stress at that point and if failure is not to take place it must be less than the shear strength of the material.

There are always quantities which cannot be accurately estimated but which affect the strength of a roll. Fire cracking caused by alternate heating and cooling of parts of the roll, overloading due to bad heating of the stock, hard spots in annealed material, or turning down and cobbling, shock loading and faults in the roll material (e.g. blowholes, etc.), make the use of a safety factor essential in the design stage. The magnitude of the factor will depend on how closely these variable influences are controlled by such measures as quality roll manufacturing, efficient roll cooling, accurate and consistent heating or annealing of the stock, precision guiding, etc. A minimum safety factor of 3 is suggested where close control on all factors is not possible.

(c) *Example*

Let us consider the following data for checking strength of rolls in a rolling mill:

Flat rolls of chilled cast iron of tensile strength = 2350 kg/cm², Roll diameter = 45 cm, roll barrel length = 100 cm, roll neck diameter = 30 cm; neck length = 25 cm; stock width = 12.5 cm; rolling load = 100000kg; roll neck radius = 5 cm; rolling torque = 277,000 kg-cm

(i) The *direct shear* stress, 'f_s' $= \dfrac{4 \times P}{\pi D^2} = \dfrac{4 \times 100000}{\pi (45)^2}$

$$= 62.9 \text{ kg/cm}^2$$

This is well within the shear strength of the roll material even allowing for a large safety factor.

(ii) *Bending stress*

The bending moment, M, at roll center

$$= \frac{100000x}{2} - \frac{100000}{2 \times 12.5}\left(x - \frac{100 - 12.5}{2}\right)^2$$

At center of roll from roll neck center, x = 62.5 cm. Taking this value of x,

$$M = 31250000\text{-}1406250 = 1718750 \text{ kg-cm at roll center.}$$

$$\therefore \quad \text{Bending stress, 'f}_b\text{'} = \frac{6\,M}{D^3} = \frac{6 \times 1718750}{(45)^3} = 113 \text{ kg/cm}^2$$

There is no stress concentration at the center of the barrel. To consider fatigue, consider the fatigue strength to be half of normal tensile strength, which gives a fatigue strength of 1175 kg/cm². The bending stress is well within this allowing a large safety factor of 10.

(iii) *Torsional Stress* at roll barrel:

The torsional stress $f_t = \dfrac{277{,}000 \times 16}{\pi(45)^3} = 15.48\text{kg / cm}^2$

Considering combined stress, the equivalent torque is:

$$T_e = M + \sqrt{M^2 + T^2}$$

$$= 1718750 + \sqrt{(1718750)^2 + (277{,}000)^2}$$

$$= 3459677 \text{ kg-cm}$$

$$\therefore \quad \text{Equivalent torsional stress 'f}_{te}\text{'} = \frac{3459677 \times 16}{\pi(45)^3} = 193.4 \text{ kg/cm}^2$$

This value is much below the shear strength, even considering fatigue strength.

(iv) *Stress at the roll neck:*

For checking the strength of the roll neck, check directly the maximum shear stress considering combined bending and torsion.

The bending moment at the end of the neck (shoulder), i.e., at x = 12.5 is

$$M_n = \frac{100000 \times 12.5}{2} = 625000\text{kg.cm}$$

$$\therefore \quad \text{Bending stress, } f_b = \frac{6 \times 625000}{(30)^3} = 138.9 \text{ kg/cm}^2$$

A stress concentration will occur at the roll neck. The neck radius 'r' being 5 cm; $\dfrac{r}{d} = \dfrac{5}{30} = 0.167$ and $\dfrac{d}{D} = \dfrac{30}{45} = 0.67$

∴ From Fig. 4.4, the stress magnification factor is approximately 1.22. This value is applicable to alloy steel rolls. Applying Trinks' criterion, the magnification factor for cast steel roll is 35%. Thus the factor will be = 1 + 0.35 × 0.22 = 1.077 only.

Hence increased bending moment = 625000× 1.077 = 673125 kg.cm

$$\text{Equivalent torque, } T_e = 673125 + \sqrt{(673125)^2 + (277000)^2}$$

$$= 673125 + 727825 = 1400950 \text{ kg.cm}$$

∴ Equivalent torsional stress at roll neck

$$f_{te} = \frac{16 \times 1400950}{\pi(30)^3} = 264 \text{ kg/cm}^2$$

Considering a safety factor of 3, this max stress comes to 792 kg/cm^2 which is about 33.7% of tensile strength of the cast iron roll, hence much below the ultimate shear strength of the roll material. Thus neck is also strong enough for the rolling operation.

E. Bending of roll

Bending of rolls during rolling under rolling load plays an important part in the resulting shape of the rolled product. This becomes particularly critical while rolling thin gauge strips in cold rolling process. The effect of roll bending is attempted to be annulled by cambering the rolls and also by deliberate bending of the rolls in opposite direction during rolling (this is called 'as you roll bending'). However, in order to control the effect of this roll bending, it is essential to estimate the amount of roll deflection caused by roll bending.

Figure 4.3 shows a pair of rolls subjected to a load resulting from rolling rectangular stock of breadth b in a roll of barrel length B and diameter D. Roll deflection can be caused by bending and shear force and two separate expressions can be derived for these. The rolls are considered as simply supported beam. If the neck length is n and the distance between the reactions is L then the point of support is usually taken such that: L = B + n

The expression for the deflection due to bending is:

$$\delta_b = \iint \frac{M}{EI}\, dx\, dx$$

where M is the bending moment, E is Young's Modulus and I is the moment of inertia of the roll barrel.

In the rolling of plate, sheet and strip, the main concern with regard to bending is the difference in deflection of the roll at the edge and center of the stock, respectively, which leads to crown, i.e., difference in deflection at

$$x = \frac{L - b}{2} \text{ and } x = \frac{L}{2}$$

This is given by : $$\delta_b^t = \int_{\frac{L-b}{2}}^{\frac{L}{2}} \int \frac{M_x}{EI} dx\, dx$$

In the case of the top roll, $M_x = \frac{P}{2}x - \frac{P}{b}\left(x - \frac{L-b}{2}\right)\frac{1}{2}\left(x - \frac{L-b}{2}\right)$ and integrating twice

$$\delta_b^t = \frac{Pb}{384\,EI}\left(7b^2 - 12bL\right)$$

The expression for deflection due to shear is:

$$\delta_s = \int \frac{Sdx}{GA}$$

where S = shear force, G = modulus of rigidity, A = cross section of roll = $\frac{\pi D^2}{4}$.

(Assuming uniform distribution of shear stress over the cross sectional area.)

The change in deflection from $x = \frac{L-b}{2}$ to $x = \frac{L}{2}$ is:

$$\delta_s^t = \int_{\frac{L-b}{2}}^{\frac{L}{2}} \frac{S_x}{GA} dx$$

For the top roll $$S_x = \frac{P}{b}\left(x - \frac{L-b}{2}\right) - \frac{P}{2}$$

And integrating $$\delta_s^t = -\frac{Pb}{2\pi GD^2}$$

∴ Total differential deflection between the center and the edge of the stock for the top roll only is:

$$\delta = \delta_b^t + \delta_s^t = -\frac{Pb}{D^2}\left(\frac{12bL - 7b^2}{18.8ED^2} + \frac{1}{2\pi G}\right)$$

It will be appreciated that this quantity is negative due to the fact that the roll is deflected upwards and by the usual convention upward deflection is negative and downward positive. To indicate the magnitude of this deflection the following calculated data referred to a 2-high 10ft. plate mill with cast iron roll, are shown:

D = 42 in., L = 147 in. (assuming L = B + n)

E = 10,200 tons/sq in., G = $\frac{2}{5}$E = 4080 tons/sq in

Total roll deflection = 2δ and is given below:

Bearing load $\frac{P}{2}$ *tons*	*Roll deflection, in.* × 10^3				
	b = 40 in	*b = 60 in*	*b = 80 in*	*b = 100 in*	*b = 120 in*
1000	19.6	38.4	59.2	80.4	100
800	15.7	30.6	47.3	64.2	80
600	11.8	23.0	35.5	48.3	60
400	7.8	15.3	23.7	32.1	40
200	3.9	7.7	11.8	16.1	20

The deflection of any other point relative to the end of the roll barrel can be found substituting suitable limits in the equations above.

4.1.2 Roll neck bearings

The rolling mill rolls are mounted on bearings at their necks, and are called roll neck bearings. These bearings are subjected to the roll separating forces (in case of 2-Hi, 3-Hi and Back-up rolls of 4-Hi mills) and thus subjected to very high specific load which is generally several times more than permissible for normal journal bearings. The work rolls neck bearings of a 4-Hi mill is not subjected to the rolling force. They may be subjected to roll bending forces, if provided in the mill design, which is much less in comparison to rolling forces.

A. *Types of roll bearings*

Roll bearings are classified into following three major groups:

(i) Open journal bearings with plain metal or non-metallic liners (also called inserts), generally cooled by pouring water on roll necks.
(ii) Enclosed bearings with forced oil circulation e.g. Morgoil design oil–film bearings.
(iii) Antifriction roller bearings specially designed in multiple rows for withstanding heavy loads, to be used as roll neck bearings. These are grease or oil–mist lubricated.

B. *Open bearings*

Open bearings with metal inserts, like white metal bearing, have high coefficient of friction and short life. These are hardly used now.

Non-metallic inserts are used in roll bearings of primary mills like blooming, slabbing, intermediate, plate, rail and section mills, i.e., majority of hot mills excepting rod mills and cold mills where gauge precision is of paramount importance.

Most used materials for non-metallic linings for roll bearings are: *textolite* and its substitute *lignofol* for high specific pressure up to 250–300 kg/cm^2 and *lignoston* for lower specific pressure up to 100kg/cm^2.

These materials are basically composites made from polymer resin and suitable filler material. Textolite contains cotton cloth as a filler. Usual laminated texolite is made by hot pressing of layers of cloth which have been saturated with synthetic resin. Composite linings are made from a heavy coarsely woven cloth of the belting type. This composite textolite is stronger than the laminated kind, it does not flake and has greater resistance to splitting. Lignofol (also called laminated wood plastic) has birch plywood as a filler. Its strength and antifriction qualities are almost equal to textolite, but cheaper in price.

Lignoston is made from birch wood blocks, first impregnated with a 20% solution of glucose and then pressed across the grain with a reduction up to 60% at a temperature of 140–160°C and a pressure of 250–300 kg/cm^2. Its main advantage is its resistance to heat up to 300°C and low cost.

Comparative properties of these non-metallic inserts in comparison to bronze is shown in Table 4.4.

Table 4.4 Basic properties of textolite, lignofol, lignoston and bronze

Property	*Textolite*		*Lignofol*	*Lignoston*	*Bronze*
	layered	*composite*			
Specific weight	1.3–1.4	1.35–1.4	1.35–1.45	1.2–1.25	8.8
Maximum strength in compression, Kg/mm^2					
Along fibers	13.0–13.5	not less than 11	15–20	10–12	45
Across fibers	23.0–24.5	not less than 10–13	-	6–7	45
Modulus of elasticity, kg/mm^2	500–1100	-	-	-	10,000
Thermal conductivity, kcal/ m hr ^{0}C	0.2–0.3	–	-	-	55
Heat resistance, ^{0}C	140–150	125	-	300	700

Because of extremely low thermal conductivity of these inserts, these need to be cooled by liberal spraying of water on exposed roll necks. The cooling water entering the bearing also acts as a lubricant. When the surface speed of the roll neck is more than 1.0 m/sec, these liners can be lubricated by water alone. At surface speeds higher and 2.0m/sec and with water lubrication, the coefficient of friction of these bearings remains between 0.003–0.006 which is 2–10 times less than well lubricated metallic linings. The relation of coefficient of friction to sliding speed of textolite bearings working under different specific pressures is shown in Fig. 4.5.

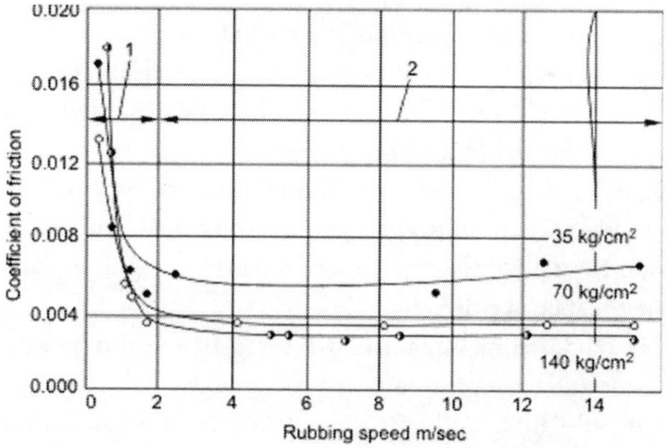

Fig. 4.5 Dependence of the coefficient of friction of textolite bearings, with water lubrication, on sliding speed at various specific pressures.

1. zone of semi-fluid friction; 2. zone of fluid friction

Besides saving power, use of non-metallic linings considerably increase life of bearings— up to 200 roll changes.

The arrangement of the position of the inserts are shown in Fig. 4.6.

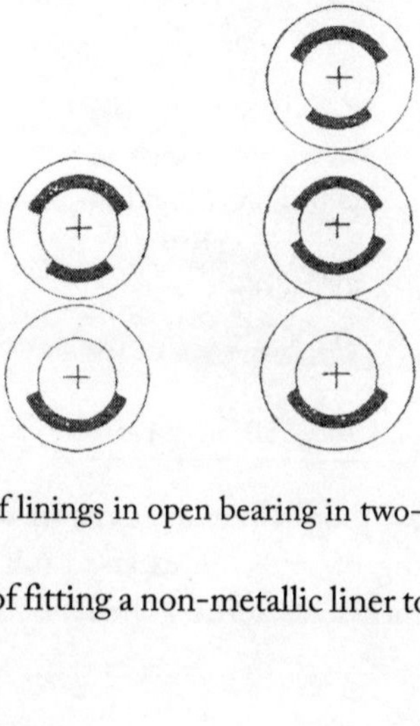

Fig. 4.6 Arrangement of linings in open bearing in two-high and three-high mills

A typical arrangement of fitting a non-metallic liner to the chock (bearing block) body is shown in Fig. 4.7.

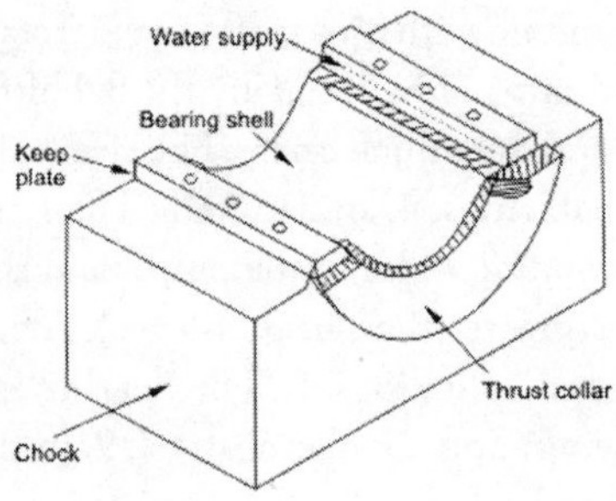

Fig. 4.7 Fabric bearing

C. ***Oil–film enclosed bearings***

An oil–film bearing, popularly known as Morgoil bearings which were first introduced by Morgan Construction Company, USA during early 1930s, is designed to maintain a thin film of oil between a steel sleeve mounted on the roll neck and the casing or bushing of the bearing assembly and thus achieving fluid friction with low coefficient of friction (0.001–0.008, i.e., the same as in antifriction roller bearing) and workability under high specific pressure of 250 kg/cm^2 and more.

Figure 4.8 shows an oil–film bearing (radial thrust) of a back-up roll of a four-high cold rolling mill.

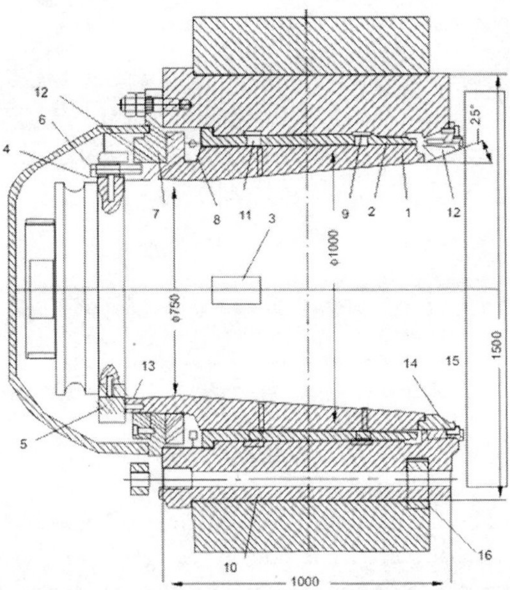

Fig. 4.8 Radial thrust oil–film bearing for back-up roll of 2500 mm wide continuous cold rolling mill

The conical roll neck is fitted with a journal sleeve; 1. rotating in a bush, 2. covered with a thin layer (about 5 mm) of high quality tin Babbitt (a thin layer of Babbitt withstands pressure better than a thick layer). The sleeve, 1, is fastened to the neck by a key, 3. split thread half rings, 4. and a nut, 5. The latter is firmly locked by a screw, 6. fitted in corresponding holes in the nut, 5. and the sleeve, 1 (the number of holes in the nut is one greater than in the sleeve). Axial thrust is taken by the flange, 7. and the babbitted split rings, 8. The bearing is lubricated through an inlet (not shown in drawing) and circular channel, 9, in the chock, 10, and holes, 11, in the bush, 2. The used oil is colleted in drain passages situated at the ends of the bearing and returns by gravity to the oil collecting tank of the lubrication system. The bearing pressure is maintained by seals, 12, and rubber rings, 13 and 14. In order to avoid wear of the sleeve 1 by rubbing against the cups of packing, 12 a changeable ring, 15 is fitted. The working surface of the bushing as shown in Fig. 4.9, is bored from two centers (each offset 0.3 mm). This is done to satisfy requirements or the size of the oil wedge (layer) in the load-carrying part of the bearing and at the same time it ensures the necessary clearance in the part of the bearing diametrically opposite, which is not carrying a load, for sufficient cooling oil to pas through it.

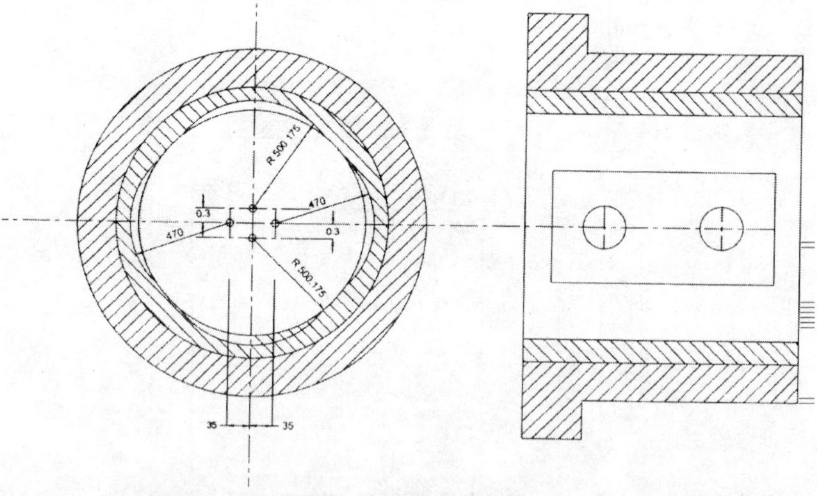

Fig. 4.9 Diagram of boring of sleeve-lining of bearing

This method of boring makes the load-bearing and non-load bearing surfaces of the bushing interchangeable. The side oil passages in the bushing are made by boring from centers offset 35 mm. This makes it easier for the oil to penetrate to the working area (due to the more gradual reduction in the clearance) and helps to carry away heat from the bearing by the intensive supply of oil to the journal in the area of these borings. The tapered shaped of the roll neck, although it is more difficult to make than cylindrical, has the advantage in strength and ease of removing the bearing. At roll changes, the bearing is not dismantled but is removed whole, together with the chock, from the tapered neck of the roll.

Both bearings of a roll are generally of the radial thrust kind. In an axial direction related to the housing only, the chocks are fixed on the non-drive side. In recent years oil–film bearings without thrust rings have begun to be widely used. Axial forces are taken-up by special thrust roller bearing.

D. ***Antifriction roller bearings***

SKF bearing company[1] in Sweden, one of the largest manufacturers of rolling bearings, in 1922 experimentally mounted spherical roller bearings on the roll necks of a 9 inch size wire rolling mill at their own steel rolling plant. Since then, roller bearings of different designs, manufactured by different rolling mill bearing manufacturers of the world, have been used in different mills. These bearings have been used starting from primary mills like blooming mills to structural mills, plate mills and almost exclusively for rod mills and cold mills (both for work and back-up rolls).

The major advantages of using the roller bearings are their long life, accuracy of rolling, ease of mounting and dismounting of the bearing form roll necks. For rolling at very high speed, a tendency to replace roller bearings by oil–film bearings exists. Another shortcoming of the rolling bearings was large thickness of these bearings and thus reducing the roll neck diameter corresponding to a specific roll barrel diameter. Manufacturers, therefore, manufactured specially designed bearings with small OD but with multiple rows of rolling elements to offer large designed load bearing capacity suitable for roll neck in rolling mills.

Different types of rolling mill rolls are best suited for different types of bearings. In the following some of the rolling mill bearings with their preferred applications along with some of the typical roll neck bearing arrangements are described:

(a) *Double row spherical roller bearing:* These bearings can carry both radial and axial loads and their use often simplify the design of mill stand. These may be used in (i) moderate speed wire rod mills and medium section mills; (ii) back-up roll for small and medium sized cold mills, (iii) rolls of a vertical edger mill, (iv) vertical rolls of a universal beam mill, (v) cold tube mill, etc.

Figure 4.10 shows the arrangement of a medium section mill roll neck fitted with double row spherical roller bearing. As the bearings can adjust themselves, they can operate well with roll bending. One of the bearings in one side is located on chock while the other side bearing is kept free to move axially in the chock to accommodate axial roll expansion during rolling. This arrangement of locating bearing is true for other types of rolling bearings also. Toleranced dimensions of different size of these bearings along with recommended chock and roll dimensions are available in bearing manufacturers' catalogues.

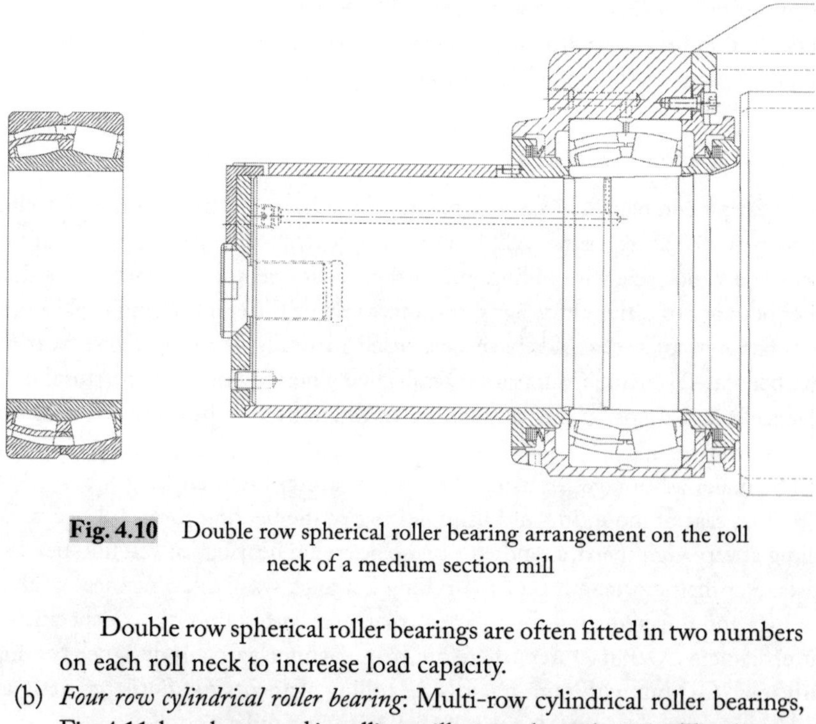

Fig. 4.10 Double row spherical roller bearing arrangement on the roll neck of a medium section mill

Double row spherical roller bearings are often fitted in two numbers on each roll neck to increase load capacity.

(b) *Four row cylindrical roller bearing*: Multi-row cylindrical roller bearings, Fig. 4.11, have been used in rolling mills since the early 1950s. The friction in these bearings is less than in any other type of roller bearing, and they therefore permit very high rolling speeds. Their radial section is small so that the roll neck diameter can be large relative to that of the roll body. However the design of the bearings allows incorporating a larger number of large rollers so that their carrying capacity is extremely high .Owing to their design, multi row cylindrical roller bearings can not accommodate thrust loads. However, this is not to be considered a disadvantage as the carrying capacity of the bearings can be fully utilized to take radial loads. When using roller bearings of other types, e.g. four-row taper roller bearings, the calculation must take the combined bearing load into accounts.

The inner and outer rings of multi-row cylindrical roller bearings are manufactured to very close tolerances, and where the rings are in two sections they are very closely matched and must not be separated.

Each pair of outer rings with its roller sets fits any pair of inner rings, and these permits roll changes without removing the inner rings, from the roll necks.

The inner rings are given a tight fit on the roll necks. Nevertheless mounting and dismounting are easily done as the chocks with outer rings and roller sets are pushed on and withdrawn independent of the inner rings.

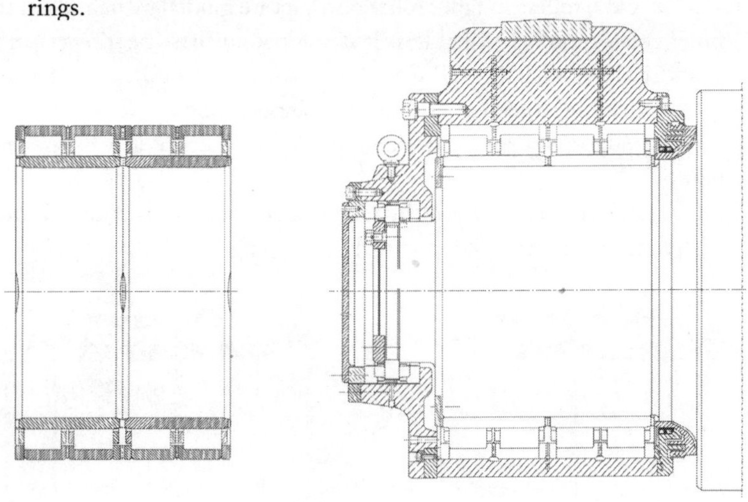

Fig. 4.11 Back-up roll neck of a four-high hot mill fitted with four row cylindrical roller bearing and double acting taper roller thrust bearing.

A thrust roller bearing is always used with cylindrical roller bearing to accommodate the axial thrust in a rolling operation.

(c) *Four row taper roller bearing*: These bearings can accommodate radial as well as axial loads. These bearings, pioneered by Timken Company, USA, have been successfully used as roll neck bearings since 1930s. Four-row taper roller bearings have been intensively used for 2-Hi mill roll, back-up roll for 4 Hi mills, work roll of 4-Hi mills, tube mills, etc.

To facilitate the mounting and dismounting work involved in frequent roll changes, taper roller bearings are mounted on the roll neck with a close clearance fit; however, this causes the inner rings to rotate or creep on the neck. The creep will result in smearing and wear of the mating surfaces. The damage can be avoided, or the risk reduced, by making the surface of the neck seating and the faces of shoulders and collars abutting the bearing inner rings sufficiently hard; in the first case 45° to 50° Shore C and in the latter case 60° shore C. The mating surfaces must also be lubricated with heavy cylinder oil or a grease with corresponding properties.

This problem has been investigated in detail by bearing manufacturers, and it has been found that helical grooves in the bores and radial grooves in the side faces of the inner rings effectively reduce these difficulties. Lubricant is stored in the grooves and distributed over the complete mating surface. Any wear debris collects in the grooves and cannot cause further damage.

Since the rollers in taper roller bearings are guided by flanges on the inner rings, more frictional heat is developed in these bearings than in cylindrical roller bearings. Taper roller bearings cannot therefore be used at the same high speed as cylindrical roller bearings.

These bearings are available with both cylindrical as well as taper bore.

Figure 4.12 below shows the roll neck of a bottom backup roll fitted with four-row taper roller tapered bore bearing.

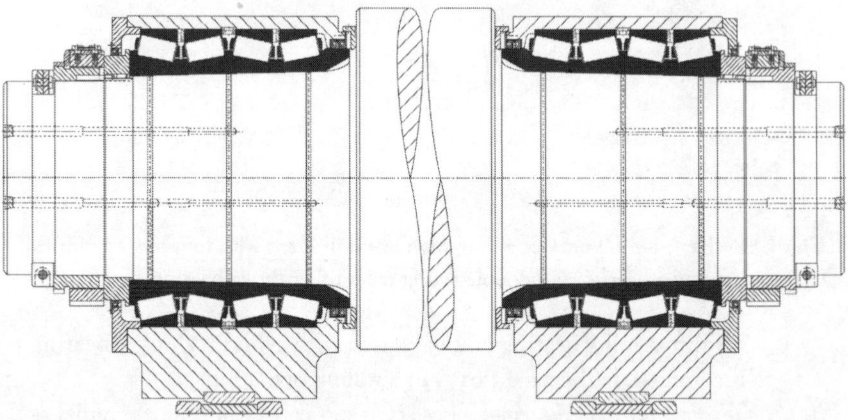

Fig. 4.12 Cut sectional view of a four-row taper roller tapered bore bearing assembly fitted on a back-up roll neck of a 4-Hi mill

Mounting and dismounting of bearing on and from roll neck is a critical operation and to do this, bearing manufacturers' instructions should be followed. For interference fit of inner races of cylindrical roller bearing on roll neck, the inner races are generally heated up to 70° to 80°C above atmospheric temperature in an oil bath or up to 200°C in an induction heater before pushing them into position. Same heating method may be employed for cylindrical bore spherical roller bearing mounting.

Mounting and dismounting of spherical roller bearings, or tapered bore taper roller bearing may be done by hydraulic means. Oil under high pressure is pumped by a hand oil pump through oil groove made in the roll, to the mating surfaces of the seating and bearing. Under this condition the bearing is pushed by a hydraulic nut, as shown in Fig. 4.13. After the bearing is pressed to position, the oil pressure in the seat is released and then after some time the hydraulic nut is taken off.

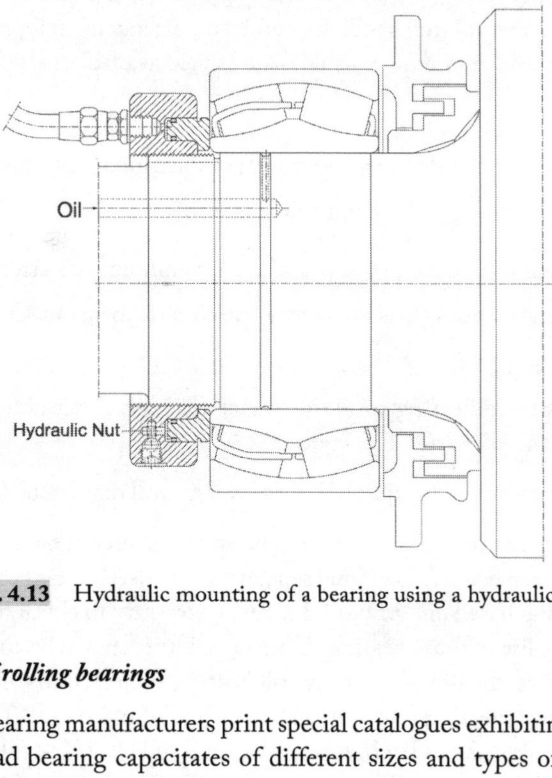

Fig. 4.13 Hydraulic mounting of a bearing using a hydraulic nut

E. ***Selection of rolling bearings***

Rolling mill bearing manufacturers print special catalogues exhibiting dimensional details and load bearing capacitates of different sizes and types of mill bearings manufactured by them.

Once the tentative minimum barrel diameter of the rolls and the maximum rolling load of a mill is estimated, the roll neck bearings are selected based on its O.D., I.D. and desirable working life in hours or revolutions under operation with rolling load.

The selection procedures are as follows:

(i) To determine *maximum bearing outside diameter* (O.D.) based on minimum (discard) roll diameter.

For 2-Hi mills:

Approximate bearing O.D.

$$= \frac{1}{1.08} \times (\text{min.roll dia} + \text{min.material thickness})$$

for thin gauge strip mill / foil mill, neglect minimum material thickness.

For 4-Hi mills back-up roll:

Approximate bearing O.D. = $\frac{1}{1.08}$(min.roll diameter)

For 4-Hi mills work roll:

Work roll bearings are not subjected to rolling load but may be subjected to crown control (roll bending) load, if this arrangement is provided. The bearing selection requires analysis of space available between the mill pass line and the bottom of the back-up roll chocks. The requirement for a large roll neck diameter is also to be considered.

Based on above data, the approximate bearing O.D. (for symmetric chocks*) = $\frac{1}{1.125}$(min imum roll diameter)

Alternate procedure to allow a smaller minimum roll size, using non-symmetric chocks*, is to take approximate bearing O.D. = $\frac{1}{1.10}$ (min. roll dia.)

(ii) Approximate bearing inside diameter (I.D.) is determined based on maximum roll barrel diameter:

Approximate bearing I.D. = 0.6 × max. roll diameter.

(iii) *Checking bearing life:* The type of bearing to be used is based on the mill type, size, speed and load (mill separating force). After tentative selection of bearing based on O.D. and I.D., it is necessary to check the expected working life of those bearings. The procedure is shown by an example of calculating the life of a backup roll bearing subjected to roll separating force (each bearing is assumed to carry half of rolling load).

Example: A 4-Hi High reversing mill of size 125 × 500×500 is to roll 400 mm wide mild steel strip from 2 mm down to 0.3 mm. The maximum designed roll separating force is 200 tons (= 448,000 lbs). The average rolling speed is 200 mpm.

A four row tapered bore taper roller bearing is selected (Timken series no. M252300)[2] having 271.462 mm I.D., 381.0 mm O.D., 269.875 mm wide (over cup) which allows a discard roll diameter of 411.2 mm (which means 17.76% grinding allowance), having 208,000 lbs. of basic dynamic load rating. (Dynamic load rating of a rolling bearing is statistically on an L_{10} life of 3000 hours at 500 rpm or 90 million revolutions.)

* See section of 4.1.3 on chocks.

Calculations:

The bearing manufacturer's catalogue[2] gives following data:

Life of a roller bearing = $(LF)^{10/3} \times 3000$ hours;

Where life factor 'LF' = $\dfrac{\text{dynamic load rating} \times \text{SF}}{\text{Rolling load}}$

and Speed factor 'SF'= $\left(\dfrac{500}{\text{roll rpm}}\right)^{3/10}$

Now, the average roll rpm = $\dfrac{200}{\pi\left(\dfrac{500 + 411.2}{2}\right)} = 143$

$$\therefore \qquad \text{SF} = \left(\frac{500}{143}\right)^{3/10} = 1.456, \text{ and}$$

$$\text{LF} = \frac{208{,}000}{\left(\dfrac{448{,}000}{2}\right)} \times 1.456 = 1.352$$

$\therefore$ Expected bearing life = $(1.352)^{10/3} \times 3000 = 8200$ hours or 70.35×10^6 revolution at 143 rpm.

For rolling mill bearings, a life of more than 5000 hours is acceptable. Hence selection of this bearing is ok.

N.B. In strip rolling, it is reasonable to assume that each neck bearing will be subjected to half of total rolling force. However, in grooved roll rolling, one of the roll neck bearings may be subjected to higher share of the rolling force for a part of total rolling campaign. This larger bearing force to be estimated from position of loading in the roll, and this higher force is to be used for calculation of bearing life.

4.1.3 Roll chocks

Roll chocks are the bearing blocks housing the roll neck bearings. The chock for each roll bearing is mounted with close clearance fit within the machined window gap of the mill housing. For gap adjustment* between rolls, the chocks along with the rolls are moved up and down between the window gaps of two housings. Design of chocks depend on the type of bearing used. Figure 4.14 shows arrangement of chocks in a 4-Hi mill.

* Roll gap adjustment screw-down mechanism is discussed later in a separate section.

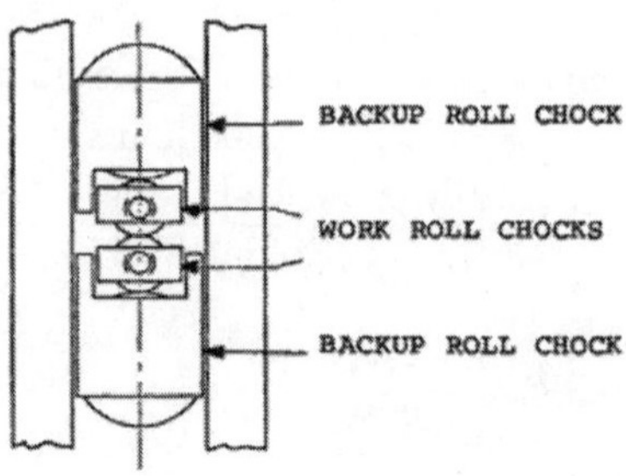

Fig. 4.14 Arrangement of work roll and backup roll chocks within window gap

A. ***Chocks for open bearings***

For accommodating the open bearings, the chocks are usually of U-shaped block made of cast steel. In small mill, the linings are usually mounted and fixed directly to the chock. In large mills where the roll is adjusted frequently, like in blooming, primary and plate mill, linings arranged in boxes are mounted in the chocks.

B. ***Chocks for enclosed bearings***

The chocks for enclosed bearings are parallelepiped blocks made from cast steel, suitably machined both inside and outside. Internal shape is governed by the type of bearing used.

A general requirement for bearing chocks is that they should be self aligning. This requirement is equally applicable for oil–film bearings as well as cylindrical or tapered roller bearings. This is generally achieved by providing a slight curved surface on which the chocks sit (bottom of bottom chocks and top of top chocks.)

C. ***Dimension of chocks with roller bearings***

(a) *Work roll chocks for 2-Hi mill:* Figure 4.15 shows sectional view of a bottom chock of a 2-Hi mill. The basic dimensions are dictated by the chock bore (i.e., bearing O.D.) dimension. The bottom thickness of chock 'k_b' below the bearing bore has to withstand the rolling force and hence is made thicker. The thickness of chock 'K_a' on the top side is not subjected to any load and hence kept smaller. The 'K_a' dimension determines the minimum (or discard) roll diameter, the relation being,

$2\ k_a \leq$ minimum roll diameter + minimum material thickness

Therefore, smaller is 'K_a' smaller can be the discard diameter of roll (i.e. maximum roll utilization).

The 'K_a' and 'K_b' dimensions are proportioned as follows:

Minimum K_a = 0.54 × bearing OD.

K_b = 0.75 × bearing OD., average

= 0.70 × bearing OD., minimum for small mills with low rolling force.

The minimum width of the chock 'K_c' = 1.2 × bearing OD

The width of the chock has to be closely fitted with the window gap of the mill housing, and therefore both the dimensions has to be machined with proper tolerances. Typical clearance between chock and window is shown in Table 4.5.

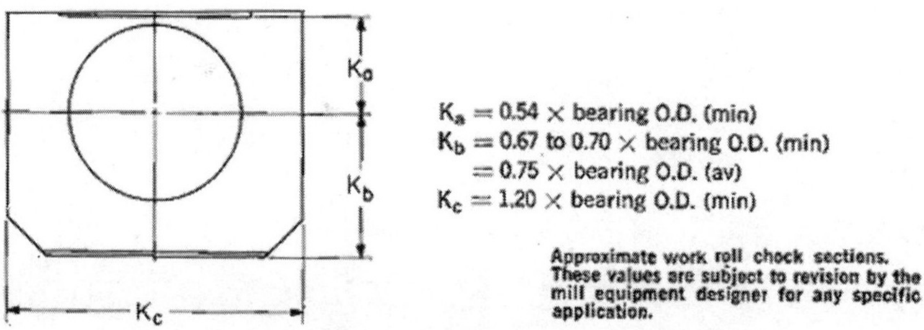

Fig. 4.15 Dimensions of a 2-Hi work roll chock

It is to be noted that the top roll chock of a 2-Hi mill will be upside down view of the bottom chock, in which the thickness 'K_b' will be on top side.

Table 4.5 Clearance between chock and window

Chock width range, mm	*Clearance, mm*	
	Minimum	*Maximum*
160–240	0.10	0.25
240–360	0.11	0.29
360–500	0.12	0.44
500–630	0.25	0.51
630–800	0.36	0.65

The length of the chock along the axis of the bearing is dependent on the bearing width as well as bearing mounting arrangement. The detailed dimensions of the chock bore also depend on the bearing O.D. dimension and bearing mounting arrangement. For these details, the recommendation by the bearing manufacturers are to be followed.

(b) Back-up chocks for 4-Hi mill: Typical sectional view of a bottom back-up chock of a 4-Hi mill is shown in Fig. 4.16. These chocks are similar to 2-Hi work roll chock dimensionally, with the exception that there are two projections called horns of the chock on the work-roll side. The inner faces of the horns provide the side guides for the work-roll chocks. The height of the horns are calculated from the height of the work roll chocks and maximum and minimum (discard) diameters of both work and back-up rolls.

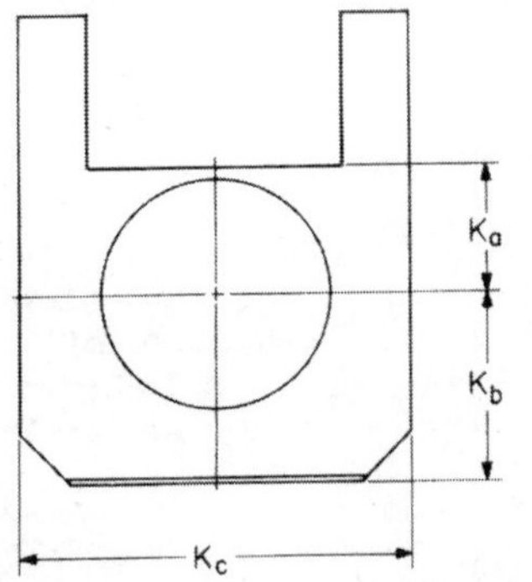

Fig. 4.16 Dimensions of a 2-Hi back-up roll chock

(c) Work-roll chocks for 4-Hi mill:
As the work roll necks are not subjected to rolling force, the thickness of chock section on top or bottom of work rolls can be kept small. Fig.4.17 below shows two design of work roll chocks, one called symmetrical and the other non-symmetrical.

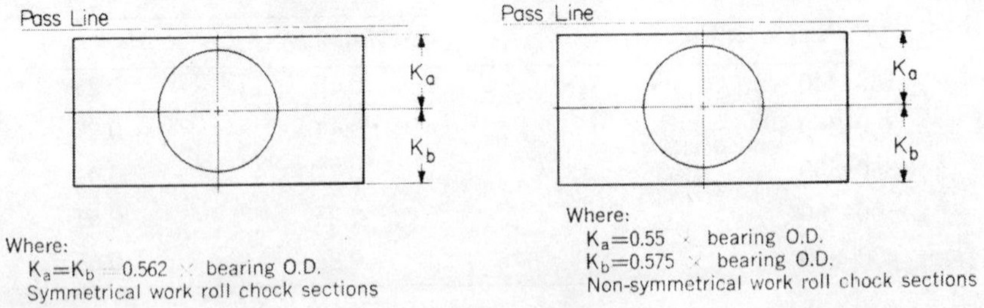

Fig. 4.17 Approximate minimum work roll chock sections of a four high mill. These values are subject to change by the mill equipment designer to suit specific mill

Width of the chocks are dimensioned to have close running fit in the gap between the horns of the back-up chocks, the clearance being similar to those indicated in Table 4.5.

4.1.4 Mill housings

The main body or frame in which the rolls with their bearings and bearing blocks of a rolling stand are accommodated is called *mill housing*. The shape and design details of the housing for special design mills like cluster mill, etc. are distinctly different, but conventional 2-Hi or 4-Hi mills have fairly conventional shape for the mill housing which will be discussed in this section.

A conventional mill consists of two housings which when assembled together form the structure of a mill stand. Each housing is a rectangular shaped cast steel (for small mill, it can be made from rolled thick steel plate also) structure. The central portion of the structure has a machined rectangular opening, called window, between the vertical edges of which the roll chocks are retained and slide. Two such housings are placed face to face, and fixed at two flat bottom ends called *feet*, on two base plates grouted on foundation. The housings are also connected at the top by tie beams. The rolls lie between the two housings as two roll chocks on two ends are fitted in the windows of two housings.

A. ***Types of housings***

The entire rolling force is taken by the two housings, which are under tension. Therefore, the housings need to be strong and rigid. The conventional housings can be of two types: the ***closed top*** in the form of a solid frame and the ***open top*** with a removal cap (Fig. 4.18).

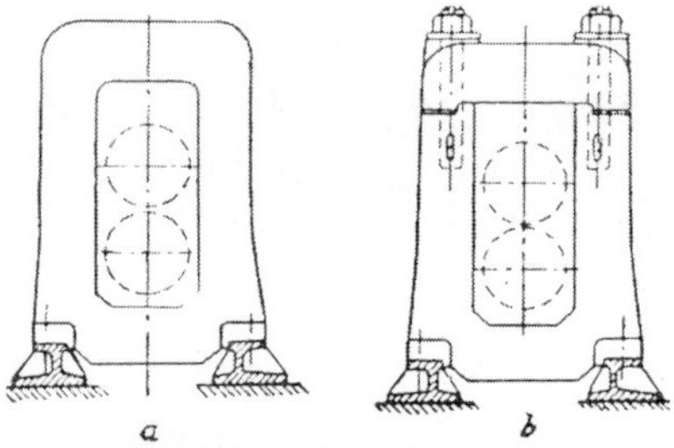

Fig. 4.18 Stand housings
a- closed–top; b- open–top

Closed–top housings are more rigid, cost less and their use is obligatory in all flat rolling mills and also in blooming, slabbing, tube, rounds and sometimes section mills.

Open–top housings are generally not used. They are used only in section mills to facilitate roll changing, where the stands are arranged in train. Roll changing* with closed–top housing would be more difficult.

* Roll changing is discussed in section 4.1.6.

B. ***Design of housings***

Mill housings are usually made of graded cast steel.

In determining the permissible stresses in the windows during rolling, it should be so that permanent deformation or breakage is not possible. Large increase of the housing cross section above the strength required for high rigidity is not justified since deformation of the housing is only a small part of the total roll gap deviation in the stand.

In order to protect the stand it should be strong enough against damage from fatigue and also against damage from overloading, which in practice is very often unavoidable owing to incorrect entry of the metal into the rolls or for other reasons.

In order to protect the housing from fatigue damage the safety factor should be equal to approximately two.

A detailed calculation for stresses can be made considering both roll separating force 'P' acting on each housing and the moment due to chock pressure on the housing. Neglecting the bending moment, the tensile stress on the housing post is given by

$$\sigma = \frac{P}{2F_2},$$

where, F_2 = area of each side of the window (post)

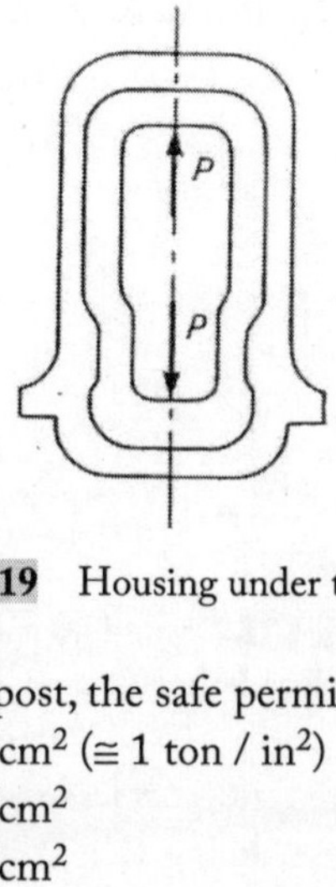

Fig. 4.19 Housing under tension

For calculating the area of the post, the safe permissible stresses are taken as:

In cold rolling mills : 155 kg/cm^2 ($\cong$ 1 ton / in^2)
In primary mills : 240 kg/cm^2
In other mills : 200 kg/cm^2

Example: If a 4-Hi cold strip mill is designed for maximum 400 tons of designed rolling force, calculate the post size of its housing.

Considering in a strip mill, both the housings are subjected to equal rolling force, each housing is under a tensile force of $P = \frac{400}{2} = 200$ tones

∴ The safe cross sectional are of each side of the housing post

$$F_2 = \frac{200,000}{2 \times 155} = 645 \text{ cm}^2$$

The dimension of the posts in the roll axis direction is made somewhat less than the chock length so that the edge of the posts does not interfere with the rolling of the stock. The housing axis should exactly coincide with the middle of the bearing. On the inner side, in places of contact with both top and bottom roll chocks, the posts are usually faced with wear plates to protect them from wear. Posts are made with I or rectangular, cross-section.

I-sections are more convenient for fixing the wear plate since bolts can be passed through the extremities of the Tee.

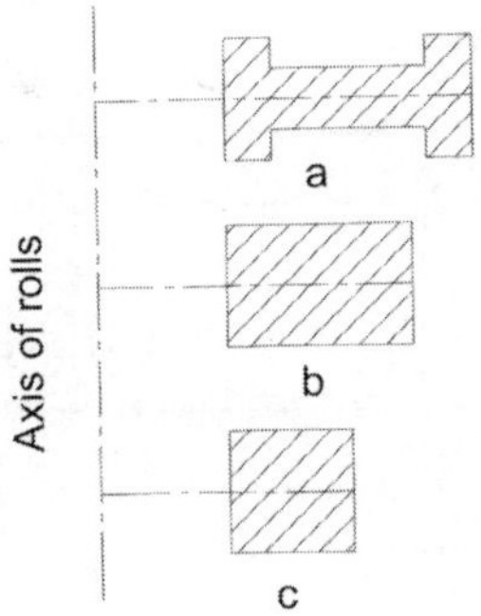

Fig. 4.20 Various cross-sections of housing posts

I and elongated rectangular post cross-sections owing to their greater rigidity, are advisable with comparatively wide housings (two-high mills), particularly when there are considerable horizontal forces. In addition, with comparatively broad closed-top housings, such cross sections can relieve the crossbeams from bending moments.

With high and comparatively narrow housings (e.g., four-high stands), and insignificant side forces, the strength and weight of the housing are more satisfactory with a less elongated cross-section, e.g., one which is nearly square. A typical housing of a four-high stand is shown in Fig. 4.22.

Both housings of a stand are connected at the top and the bottom by special cast /fabricated tie-beams and bolts with distance tubes or lugs on the housings themselves.

Lugs, called feet, are provided on the sides of the housings at the bottom. These feet are for fastening the housings by bolts to the shoes. Also in large primary mills, to allow for the considerable impact load, the feet are usually fitted in lateral grooves in the shoes and, besides bolts, are fastened with wedges.

The housings of a blooming mill (Fig. 4.21) have an I-shaped post cross-section. On top the housings are connected by a cast tie-beam. 1. at the bottom by the shoes with bots, 2. and wedges, 3. with an intermediate strip, 4. fitted together with the housing feet in lateral grooves in the shoes. The strips, 6. are for the axial adjustment of the bottom roll. The longitudinal beam, 7. acts as a skid along which set of rolls are moved during roll changes. The insides of the housing posts at the top are faced with removable wear plates, but in places of contact with the bottom roll chocks they are not faced with plates. This is a disadvantage since with shock loads the housings become worn in these places. The cramp (or guide) bars, 9. are set up on top in grooves in the housings. The shoes are connected together by lateral beams, 10. secured with foundation bolts and distance wedges. the strips, 12. protect the housing from blows by ingots. Overall dimensions of the housing are also indicated in the drawing.

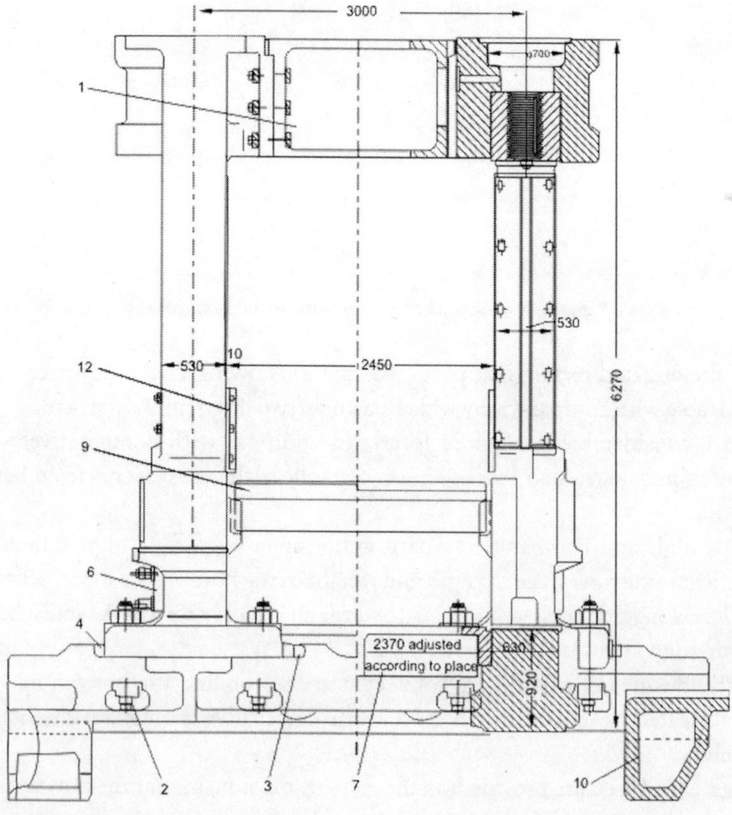

Fig. 4.21 General view of the housing assembly of a 1000 mm blooming mill. Weight of one housing 62 t, material-graded cast steel

Figure 4.22 shows dimensioned drawing of the two housings in assembled condition of a four high stand. The housings are connected to each other at the top and bottom by cast tie-beams. The drive side housing window is faced inside by protective wear plates; the operating side housing window is not faced (which is a disadvantage) and to facilitate roll changing the width of this housing window is 50 mm greater (twice the thickness of the facing plates). Comparatively small moment of inertia of the housing post cross-sections, as pointed out earlier, is characteristics of housings of four-high stands.

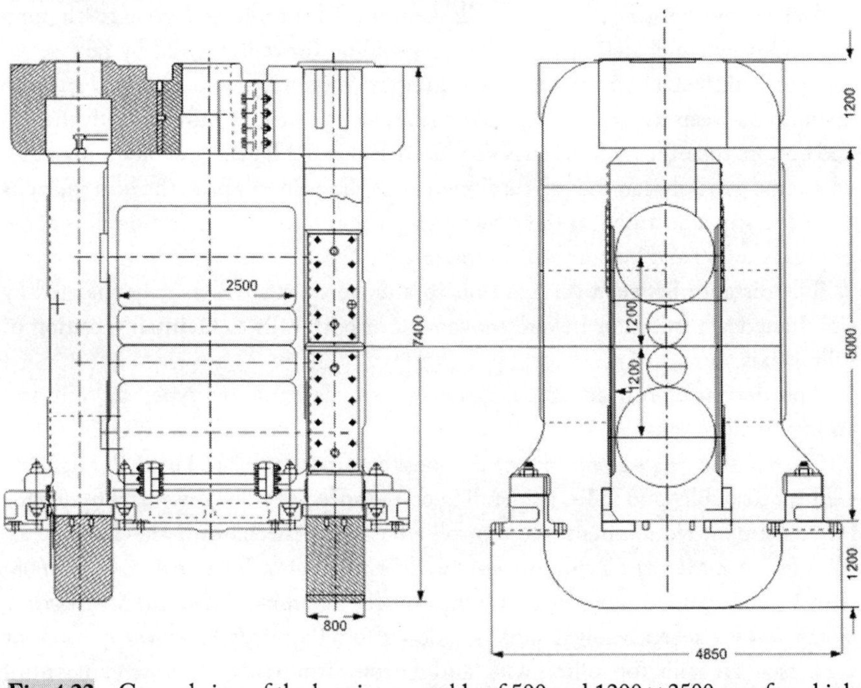

Fig. 4.22 General view of the housing assembly of 500 and 1200 × 2500 mm four-high stand in a tandem cold rolling mill. Weight of one housing 111 T

4.1.5 Mechanisms for roll adjustment and roll balance

Up and down movement of the roll and chock assembly in the mill housing is called *roll adjustment.* It has two basic purposes:

(i) To adjust the gap between the work rolls to roll the stock into given sizes or with desired draughts. In addition, in primary mill like blooming, billet, plate or universal mills, the gap between the rolls needs to be changed after almost every pass.

(ii) To adjust the position of the rolls in relation to the level of the roller tables, and for continuous mills, to adjust the rolls of one in relation to the rolls of the other stands, in order to keep the rolling axis in a straight line.

In mills with horizontal rolls, level with roller table is often achieved by adjusting height of bottom rolls. For continuous grooved roll mills, the relative position of rolls to each other is achieved by axial adjustment of rolls.

In 2-Hi, 4-Hi, and double–duo stands the distance between the rolls is achieved exclusively by adjusting the upper rolls. Lower rolls are adjusted when re-aligning the mill or when changing rolls after re-grinding (or replacement by new set of rolls) to maintain a constant level of rolling. In 3-Hi mills with fixed roll gap, the distance between the rolls is adjusted with new pair of rolls moving both the top and bottom rolls, the middle roll is kept fixed. But in 3-Hi plate and blooming mills, where the gap between the rolls is altered after almost every pass, the bottom rill is fixed and gap adjustment is done by moving the top roll. The middle roll is made to press the top and bottom roll alternately by a special mechanism.

The distance between vertical rolls in a universal stand is usually changed by simultaneous outward or inward movement of both rolls to maintain position of rolling axis.

The design of roll adjustment mechanism depends on types of mill and arrangement of rolls.

Top roll adjusting mechanisms are the most common type, used in all 2-Hi, 3-Hi, 4-Hi, and double–duo mills. Depending on the speed of adjustment and frequency for adjustment requirement, the top roll adjustment mechanism are classified as: (i) *Hand operated;* (ii) *Fast electro-mechanical* for lifting or lowering speed of more than 1 mm/sec up to more than 100 mm/sec for blooming mills; (iii) *Slow electro-mechanical* for speed below 1 mm/ sec down to 0.05 mm/sec, generally used for 2-Hi or 4-Hi mills for rolling wide and narrow strip where accurate gap control is necessary.

Bottom roll adjustment mechanism use both hand and electro mechanical operation. They are used in 3-Hi section mills (with fixed middle roll) and in 2-Hi section mills where constant rolling level is to be maintained. In almost all other mills, the bottom roll is adjusted by inserting packing strips under bottom roll chocks, in order to maintain constant roll pass line after each roll change.

Hydraulic roll adjustment system : Modern 2-Hi and 4-Hi mills for rolling strips and foils are provided with this system for very quick and accurate roll gap control even during rolling. Either top rolls or bottom rolls are adjusted by use of suitable hydraulic rams between chocks and mill windows at the top or bottom, respectively. Hydraulic roll gap adjustment and control will be discussed in details in section 4.3.6.

Adjustment arrangement of vertical rolls is also termed as *screw-down mechanism.*

Roll balancing arrangement is a mechanical or hydraulic system which does not allow the top roll chock assembly to fall under gravity on the bottom roll, when there is no stock in the roll gap. In other words, this mechanism pushes the top roll–chock assembly (top work and backup rolls in 4-High mill) up and maintains the gap between the rolls to set value during fresh entry of rolling stock in mill. Roll balancing is also called the *counterbalancing.*

A. ***Hand-operated screw-down mechanism:***

These mechanisms are used in most section mills with fixed passes and in small hot and cold mill for rolling narrow strips.

In hand operation the roll can be moved by:

(i) Moving tapered wedges between mill housings and top of top chocks, at both ends (Fig. 4.23 [a]).
(ii) Turning a screw moving though a nut fixed on top part of the housing. Two such screw forces the top chocks with roll to move down against roll balance system (Fig. 4.23 [b])
(iii) Turning a nut on a non-rotating screw (Fig. 4.23 [c]).

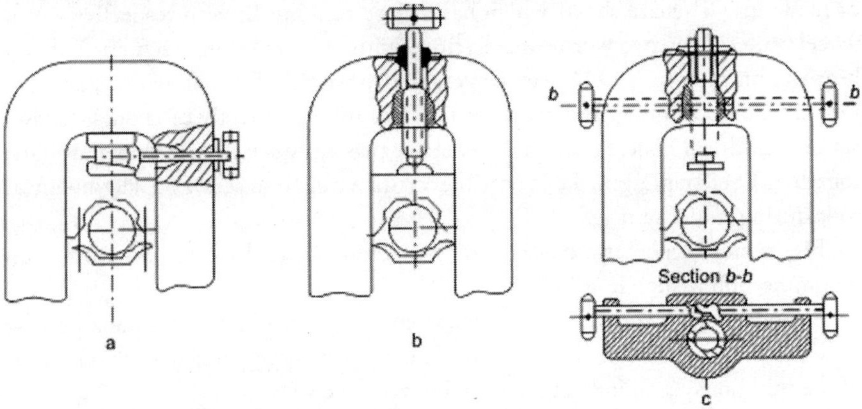

Fig. 4.23 Schemes of manual top roll adjusting mechanisms

a- with wedge; b- with rotating screw-down; c- with rotating nut.

The second type of manual screw-down is most common. The screw is turned directly by a lever or wheel or though worm–wheel mechanism. All these mills are provided with top roll balancing (also called counterbalancing) arrangement. Due to this, no gap exists between the chock and the screw down, and also between the screw and nut when there is no stock between passes. This overcomes the back-lash error in gap setting with hand operated mechanisms. The roll balancing arrangement essentially consists of four springs (two per chock) connected with the upper roll chocks. During screw down, the chocks move against the spring force. The force at any position of the chocks, should be 20–40% higher than the combined weight of the roll chock assembly.

B. ***Fast electro-mechanical screw-down mechanism***

This motorized mechanism ensures lifting speed of more than 1 mm/ sec. and is used mainly in blooming, slabbing, plate and universal mills. A particular feature of this mechanism, besides its screwing speed, is the large height to which the top roll can be lifted, for example 1600 mm or more in a 2-Hi blooming or slabbing mill. In this type, the roll is moved only by screw-down mechanism.

The screw-down mechanism of a 1000 mm roll diameter blooming mill with 1200 mm screw-down travel is shown in Fig. 4.24.

The mechanism consists of two long screws made of alloy steel of requisite diameter, with square / trapezoidal thread of proportionate pitch. The screw passes through a nut made of phosphor bronze of length about 2 times screw diameter, and permanently fixed on window. The top part of the screw is splined or made into a square shape, which has a close running fit with respective worm wheel centers. The two worm–wheels fitted with the screws, and securely fitted on bearings on their boss O.D.s, are driven by two worms. The two worm shafts are joined by one coupling, and are driven by two motors directly or though further set of cylindrical helical gearings. Generally two screws are moved up and down together. The coupling make it possible to drive only one screw for alignment of rolls during initial setting.

The typical kinematic diagram of the screw-down drive for the 1000 mm blooming mill is shown in Fig. 4.25.

A connection is taken from one of the screw-down drive, through suitable gearing to rotate the arm of a roll gap indicator. This circular indicator is mounted in front of the mill housing to indicate the gap between the rolls.

The roll balancing becomes difficult when the rolls are to be adjusted through a long distance as in a blooming or slabbing mill. The widely used methods of counterbalancing are:

(i) mechanical with counterbalance weights,
(ii) hydraulic with hydraulic cylinders, and
(iii) springs mounted on nuts which move equally with movement of the screws.

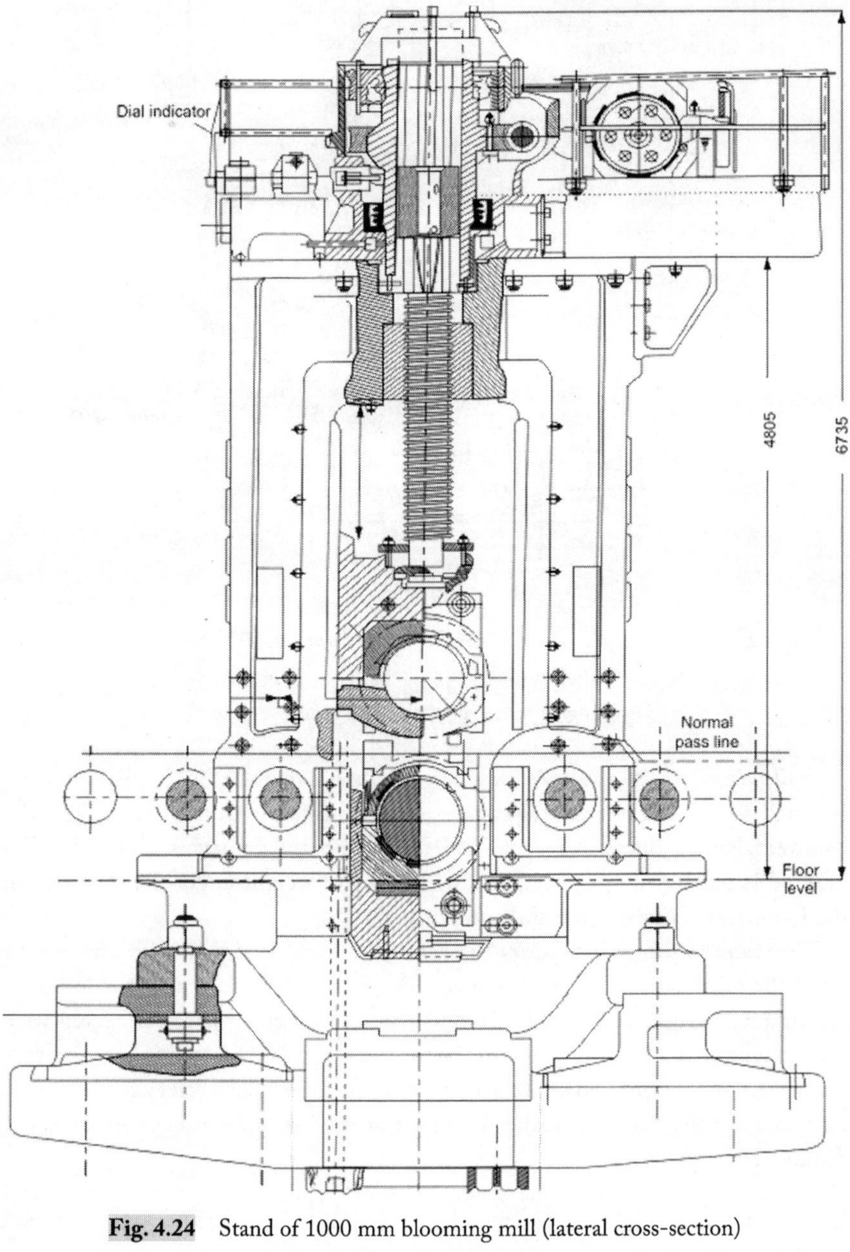

Fig. 4.24 Stand of 1000 mm blooming mill (lateral cross-section)

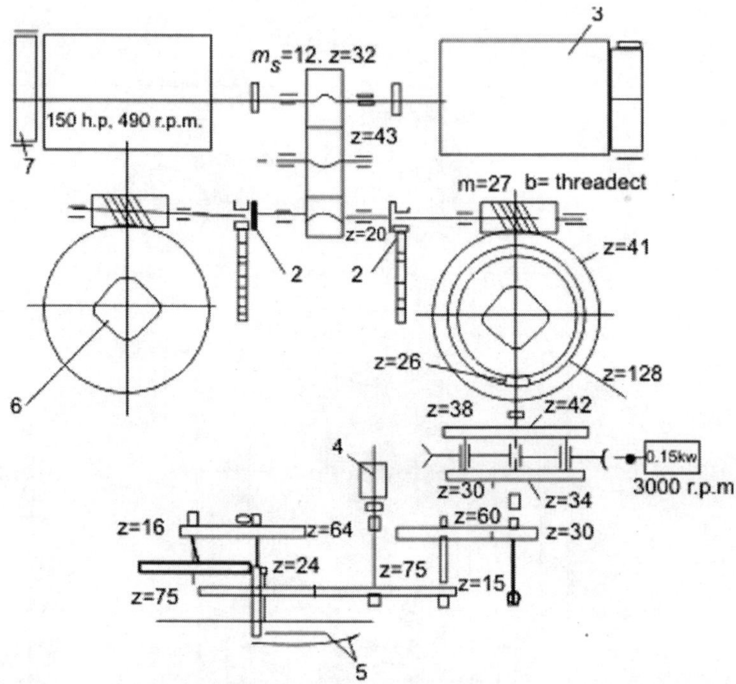

Fig. 4.25 Kinematic diagram of drive for screw-downs in 1000 mm blooming mill

1- brake; 2- connecting coupling; 3- shunt motor; 4- Selsyn; 5- rolling gap indicator arm; and 6- screw-down

Roll balancing with counterbalance weight is an old method, but is still employed for high roll lifts. Counterweights are usually located under the stand and the counterbalancing force is transmitted to the top roll by levers and rods. The method is reliable in operation, but the placing of weight under the stand make the foundation cumbersome and complicated.

Hydraulically balanced adjusting mechanism is usually designed with one or four hydraulic cylinders in the upper parts of the stand. Figure 4.26 shows a design of roll balancing with one central cylinder. The top roll chocks are connected with the piston rod by a longitudinal cross beam and tie-rods. The hydraulic cylinder is connected to a compressed air loaded oil accumulator for fixed roll balancing force. The smaller cylinder seen in the right side of the Fig. 4.26, is counterbalancing the top drive spindle.

Spring counterbalanced adjusting mechanism with springs placed on movable nuts may be of various designs. They have been used in the past in blooming, slabbing and plate mills. One such design is shown in Fig. 4.27, which is self-explanatory.

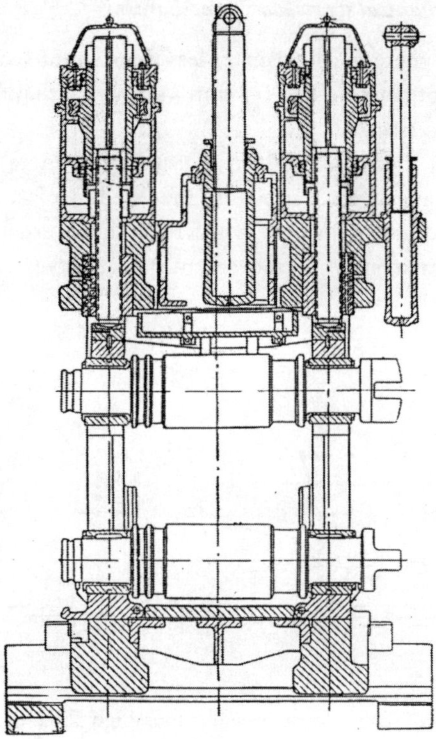

Fig. 4.26 Stand of a 1100 mm blooming mill with hydraulic top roll balancing

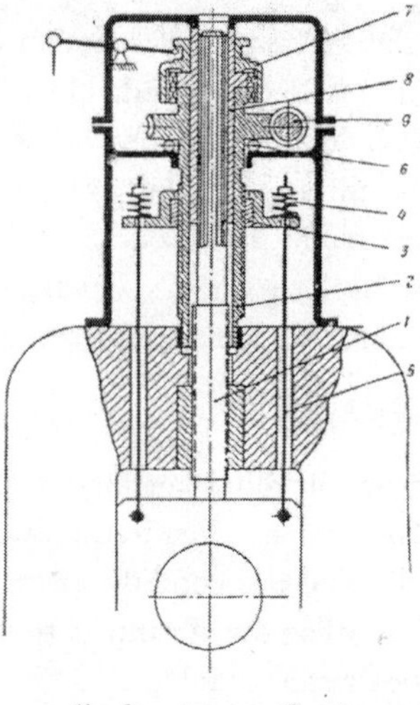

Fig. 4.27 Diagram of top roll adjusting mechanism with spring counterbalance arrangement in a three-high plate mill

C. ***Slow electro-mechanical screw-down mechanism***

The lifting speed of these mechanisms is less than 1 mm/sec. For 4-Hi hot strip mill, the speed is kept between 0.1–0.2 mm / sec, while that for cold strip mills it may be 0.05–0.1 mm /sec.

The roll lift with these mechanisms is usually not more than 100–200 mm. The screw-down arrangement is similar to that used in a fast screw-down system. However, usually two stages of worm–wheel reduction is used, and the pitch of the screw is smaller to attain slow roll movement speed. A typical kinematic diagram is shown in Fig. 4.28.

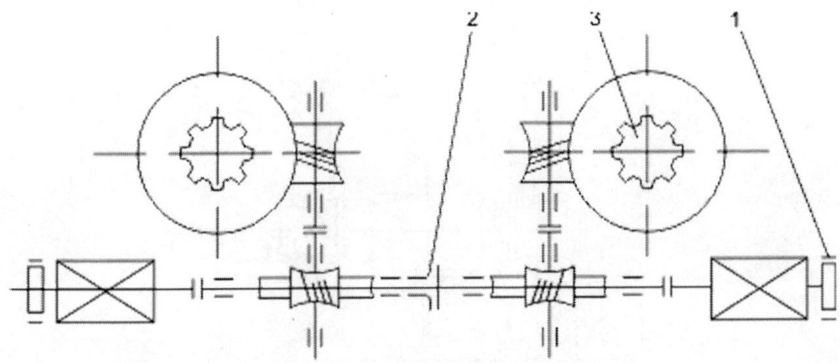

Fig. 4.28 Diagram of screw-down drive in a four-high 500and1250×1700 reversing cold rolling mill. Gear ratio from motor to screw-down is 1080
1- brake; 2- friction electromagnetic coupling; 3- screw-down

In small stands where the lift required is not more than 70 mm, ordinary spring counterbalancing of roll is employed. However, in modern mills, the counterbalancing is generally hydraulic. This design offer great convenience during roll change. Figure 4.29 shows the arrangement of top roll balancing in a 4-Hi mill, using total eight or twelve hydraulic rams. Four large diameter rams are placed on the horns of the two bottom back-up chocks which support the top back-up roll chock assembly. Two or four smaller rams are placed in the body of each bottom work roll chocks which support the top work roll chock assembly. Oil pressure used is in the range of 200 bar.

In modern mills, all the hydraulic roll balancing rams are housed in four separate steel blocks (called *Mae–West block*) attached to the window faces around pass line height and intruding into the housing window space. All hydraulic connections are made to these blocks. This offers the advantage that the hydraulic lines during roll change do not have to be disconnected.

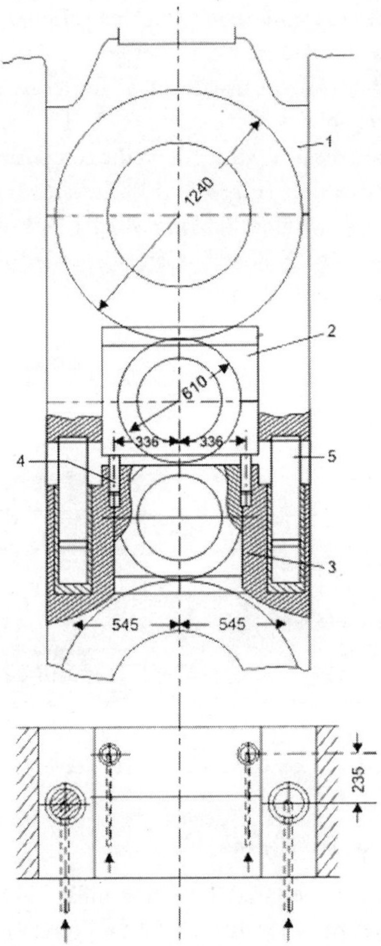

Fig. 4.29 Diagram of hydraulic roll balancing in four-high 610and1240×1680 stand
1- top back-up roll chock; 2- top work roll chock; 3- bottom work roll chock;
4- work roll balancing rams; 5- BU roll balancing rams

D. ***Safety devices***

Safety devices are put between the screws and chocks in some mills for protecting the rolls and housings from damage or breakdown in the event the rolling force becomes abnormally high due to some rolling defect. Presently they are chiefly employed in the stands of plate and sheet mills. Because of large width of rolled material, a small error in roll gap adjustment may impose a great influence on rolling load.

In practice safety devices are made in the form of ***sleeves*** or ***wedges,*** which gives-in when they are subjected to more than a designed force.

Safety sleeves (Fig. 4.30 (a)) are usually made from cast iron or cast steel, which is sheared at the periphery of the sleeve thickness 'S'. Disadvantage of safety sleeves are the dependence of the shearing strength on the composition of the cast material.

In wedge type safety devices (Fig. 4.30(b)), the shearing parts are bolts under tension. The bolts are made with necks along which they break. The length of these necks are kept small (about 0.8 bolt diameter) in order to reduce its elongation before failure, as this lowers rolling accuracy.

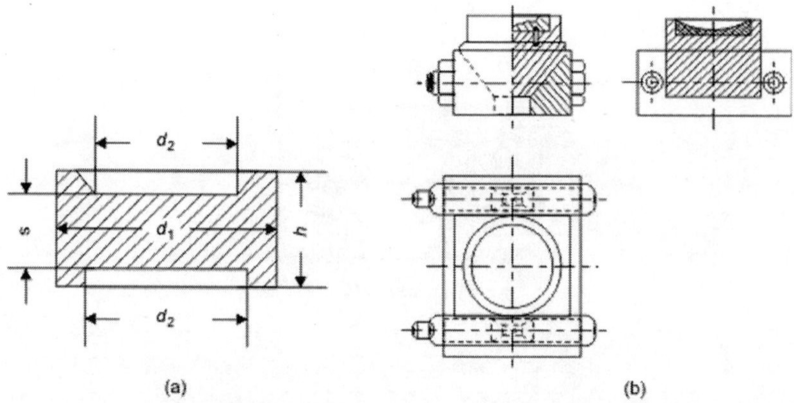

Fig. 4.30 Safety devices: (a) safety sleeve, (b) wedge.

E. *Axial adjustment and fixing of roll*

Axial adjustment of rolls is essential in section mills where correct alignment of pass grooves in top and bottom rolls are vital for correct product. Adjustment of top roll is more difficult in view of its vertical movement.

For those mills where the roll is raised and lowered rarely (excepting for setting up the stand), axial adjustment is effected by bolts passing through the housing, which act on the chock from the outer side either through flanges as in Fig. 4.31(a) or through strips, Fig. 4.31(b).

The second method is simpler and more convenient for roll changes. With roller or oil–film bearing when the axial movement of the roll is effected by moving only one chock (on the side opposite to the drive) a two-lever system is often employed Fig. 4.31(c). With the separation of threaded tie-rods the roll with chocks moves towards the housing, and when tie-rod length is reduced, roll moves away from the housing. This method is simple, convenient and avoids angular deflection of the chock owing to the self-aligning properties of the lever.

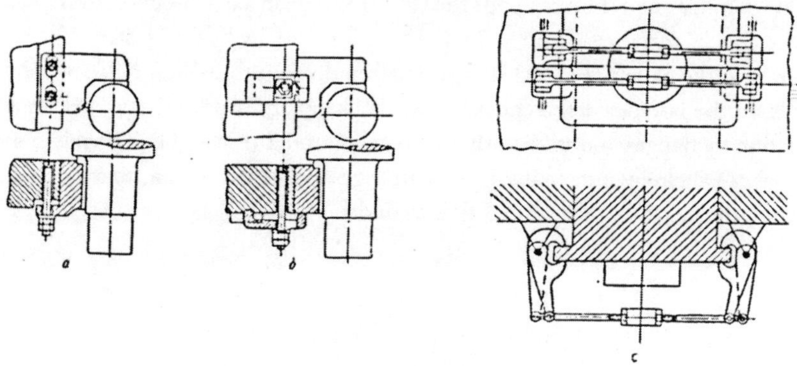

Fig. 4.31 Axial adjustment of rolls rarely moved

a- by flanges on the chocks; b- by strips; c- by a double lever system

When the roll is lifted and lowered frequently the arrangements shown in Fig. 4.32 are used.

Axial adjustment by moving the lining box in the chock by bolts 1, Fig. 4.32 [a] is widely employed when the vertical movements of the roll are considerable as, for example, in the blooming mills and in plate mills. When the roll counterbalancing is disengaged (during roll change) the box is held in the chock by keys, 2. On the inner side the chock has collar, 3, for taking axial stresses.

The axial adjustment of rolls by a single lever system (Fig. 4.32[b]) is simple and convenient. In it the chock, 1, is moved by the adjustment of a tie-rod, 2.

Mill rolling flat products do not require axial adjustment of the roll, Only axial fixing is needed.

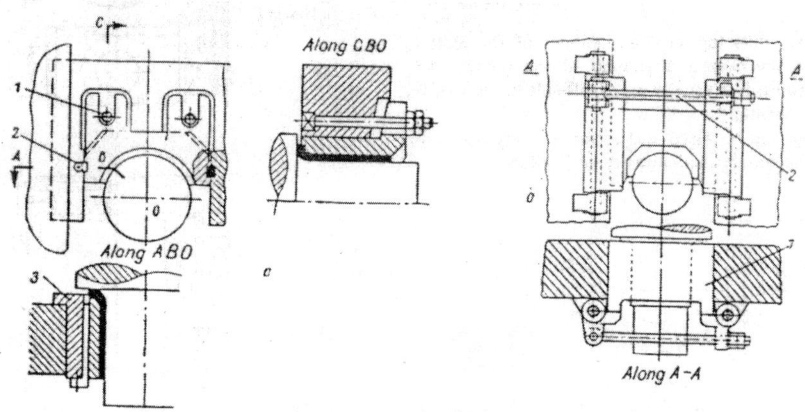

Fig. 4.32 Axial adjustment of rolls moved frequently

a- by moving the lining box; b- by a single lever system;

With roller or oil–film bearings the roll is fixed in an axial direction through one of the chocks (on the side opposite to the drive) usually by fixing strips fitting in grooves in the chock, (Fig. 4.33) or, as is often done in four-high mills, by hinged pawls. In the latter case the chocks of the back-up and work rolls are held in one direction by flanges and in the other by pawls 1 and 2 (Fig. 4.34). The position of the pawls is held by index pins 3. This arrangement is convenient to service since rotation of the pawls takes only a few seconds.

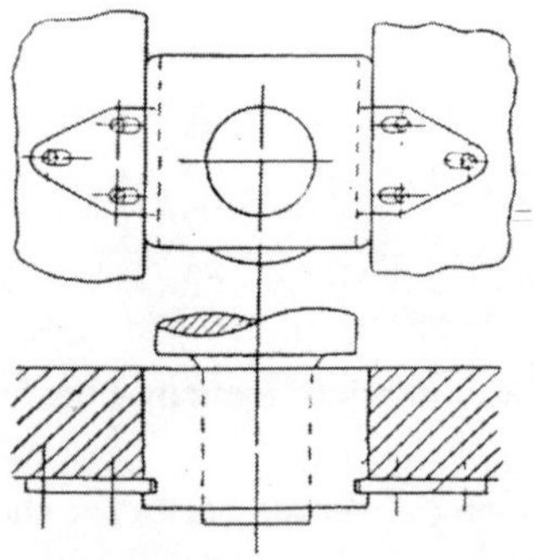

Fig. 4.33 Axial fixing of plain roll mounted in roller or oil–film bearings by triangular fixing strips

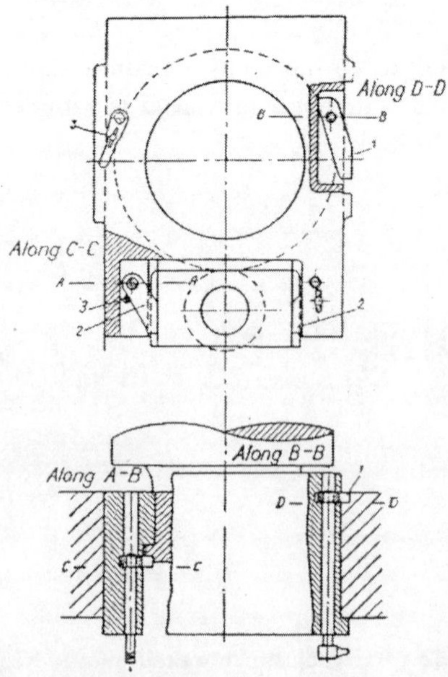

Fig. 4.34 Diagram of axial fixing of rolls in four-high mill by hinged pawl

4.1.6 Roll change device

Rolls in a rolling mill need to be changed and replaced by a new set of rolls, kept ready as spares, when they become worn out or when a different product is to be rolled. The roll neck bearings also get changed during roll changing. As the mill remain inoperative during roll change, different design of roll changing device have been attempted to reduce roll change time. Some of the standard roll change devices and their use is described below:

A. ***Stands with open housing***

In these stands with removable housing caps, rolls are changed by overhead crane. In open train stands, removable cap housings are usually used to achieve quick roll changing which need to be done often to produce different products in section mills.

B. ***Closed top housing***

In these stands, rolls are changed by withdrawing and inserting roll-chock assembly in an axial direction form and to the housing. Roll changing is always done from the non-driven end of the mill stand. The window gap width in the non-drive end is kept marginally higher than that of the drive-end window. This ensures easy roll changing, i.e., during withdrawal or entry, the drive end chock passes easily through the non-drive side window gap.

The three methods which have been developed over time, and presently in use for roll changing are:

(i) By a *roll change coupling* hung from overhead crane. The cylindrical coupling is provided with axial machined holes on both sides which fit with the non-driven end of the rolls. During a roll change, one new roll is hung in one end of the roll change coupling and then the other end of the coupling is introduced to the end of the roll to be changed. The old roll is then taken out by axial movement of the crane, then coupling is rotated by 180° and new roll is introduced by opposite direction movement of the crane. This arrangement is shown in Fig. 4.35 (a).

(ii) By a *bracket roll change rig* hung from overhead crane. In this design, bottom end of the C-bracket is provided with a machined hole which fits non-drive roll end. Figure 4.35(b) illustrates this method.

(iii) By *floor mounted roll change rig:* This method is adopted in all modern mill stand design. This enables set of roll changing in one go. It can remove and replace set of 2 rolls in a 2-Hi mill, all 4 rolls in a 4-Hi mill or the two work rolls of a 4-Hi mill in one operation. There are quite a few design versions of this method, suitable for the design details of the mill stand. This method of roll changing is particularly suitable with Hydraulic mills where roll gap adjustment is done by use of hydraulic cylinders in place of screws. In a modern hydraulic 4-High mill, work roll changes can be completed within 10 minutes, while entire set of 4 rolls may be changed between 15–20 minutes.

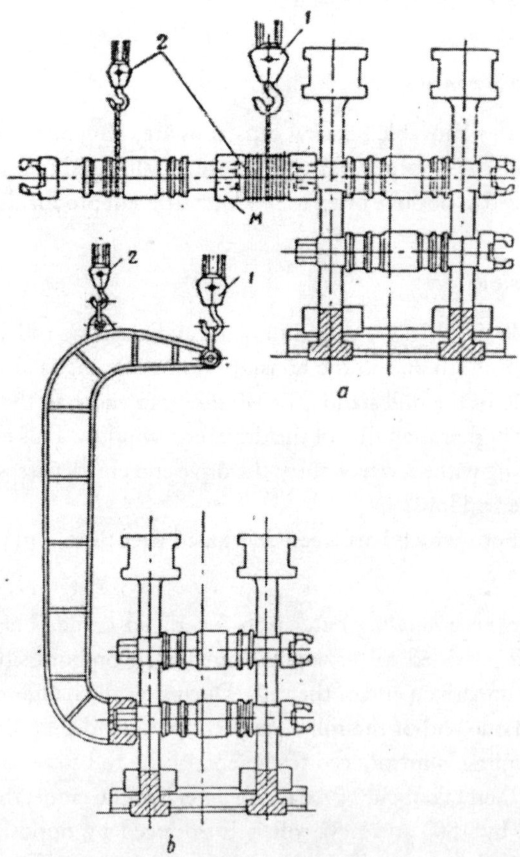

Fig. 4.35 Diagram of roll change with crane

a- with a coupling (M); b- by a bracket; 1- main hook crane; 2- auxiliary hook of crane.

In 2-Hi mills (including blooming or slabbing mills) with conventional screw-down, the roll changing is done by rack and pinion or chain drawing mechanism. Figure 4.36 shows such a roll changing arrangement utilizing rack and pinion mechanism.

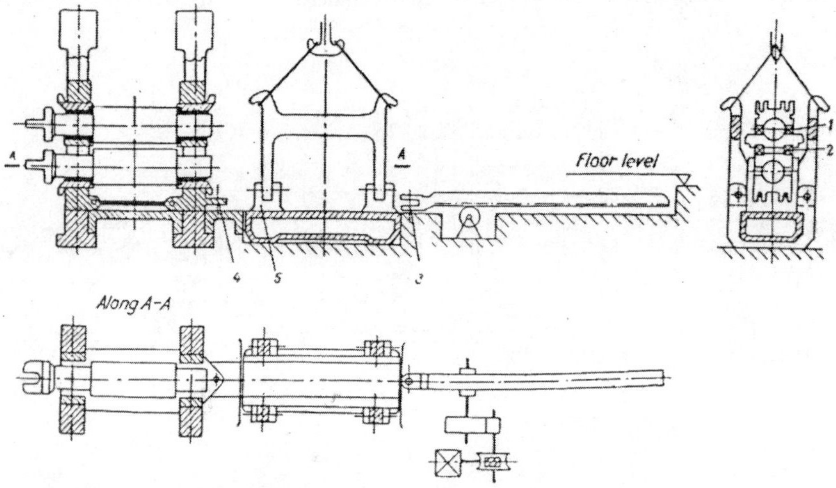

Fig. 4.36 Diagram of rack and pinion mechanism for changing roll with roll box

Before roll change, the top roll chock assembly is raised from the bottom one, and suitable packing pieces are inserted between top and bottom chocks, and the top chock assembly is lowered on the packers. The axial fixing devices are removed. Hydraulic connection, if any, to the chocks are disconnected. The rack mechanism is then driven-in and connected with the bottom chock through lugs 3 and 4. By reversing the rack, the total set of roll chock assembly is brought-out of the mill stand over a metallic slideway or a roll box 5. The roll box along with roll chock assembly are removed by overhead crane, and replaced by a new set of roll-chock assembly, which is introduced into the stand by the same rack.

In a 4-Hi mill with hydraulic roll balance, similar rack mechanism may be used. During work roll change, which is more frequent, the entire set of rolls with chocks are raised up and the top roll chock assembly is locked in position in the housing by suitable keep plates. The roll change rig in form of a low height trolley, running on wheel guided by rails, part of which is fixed inside the mill housing, and its extension in front of the mill is grouted at floor, is pushed in the mill stand. Packers are placed between two set of work chocks and also between bottom work roll and back-up roll chocks, and then the whole set of two work rolls and the bottom back-up roll along with the chocks are lowered on trolley and drawn out. Then the two work roll chocks assembly are removed by crane and replaced by new set of work roll chock assembly. This is called *3 roll change.* When all the rolls are to be replaced, set of all 4 rolls are taken out. A long stroke hydraulic cylinder may be used conveniently when a wheeled trolley type rig is used. Figure 4.37. shows a sliding (not with wheel) trolley mechanism for changing rolls in a 4-High mill, with rack operation. The entire set of chocks are raised by two or four hydraulic jacks.

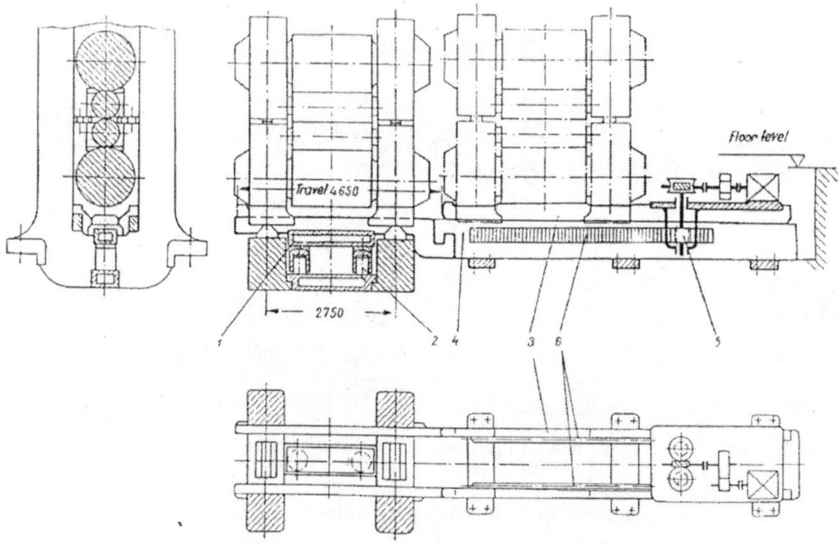

Fig. 4.37 Diagram of sliding trolley mechanism for changing back-up rolls in a four-high mill

In a modern hydraulic mill, the roll load cylinders provided below the bottom chock serve the propose of raising the entire roll cock assembly.

When there are several identical stands in one bay, as in a continuous mill, the sliding trolley mechanism may be made transportable to serve each stand.

4.1.7 Sendzimir cluster mills

Sendzimir cluster mills are used for cold rolling of strips of different metals and alloys including very high strength materials down to thin or foil gauges. Constructional features of the multi-roll Sendzimir cold rolling mills are so unique and different from those of the conventional 2 Hi/4 Hi mills discussed in the previous sections, that they rightfully demand a separate mention.

The fundamental difference is the way in which the rolling force is transmitted from the work rolls, via the intermediate rolls, to the back-up assemblies and finally to the rigid single piece housing. The single piece housing permits the support of all the rolls including the small sized work rolls throughout their length, thereby roll bending is almost eliminated and mill stretch is negligible; hence extremely close gauge tolerance can be maintained throughout the full width of the material being rolled.

In contrast, the rigidity of the conventional mills is governed by the size of the work rolls or backup rolls (for 4-Hi mills) which are supported at their necks in two separate housings. Under rolling pressure these rolls acts as prismatic beams supported at two ends and thus undergo bending causing variation across the width of the material.

Sendzimir cold rolling mills have been built as 6 roll, 12 roll and 20 roll mills. The 20-Hi mills also known as 1-2-3-4 type mills being most popular, and the constructional features of all types being somewhat similar, following discussion will be limited to type 1-2-3-4 mills.

The arrangement of the rolls in a 20-Hi Sendzimir mill is shown in Fig. 4.38[3].

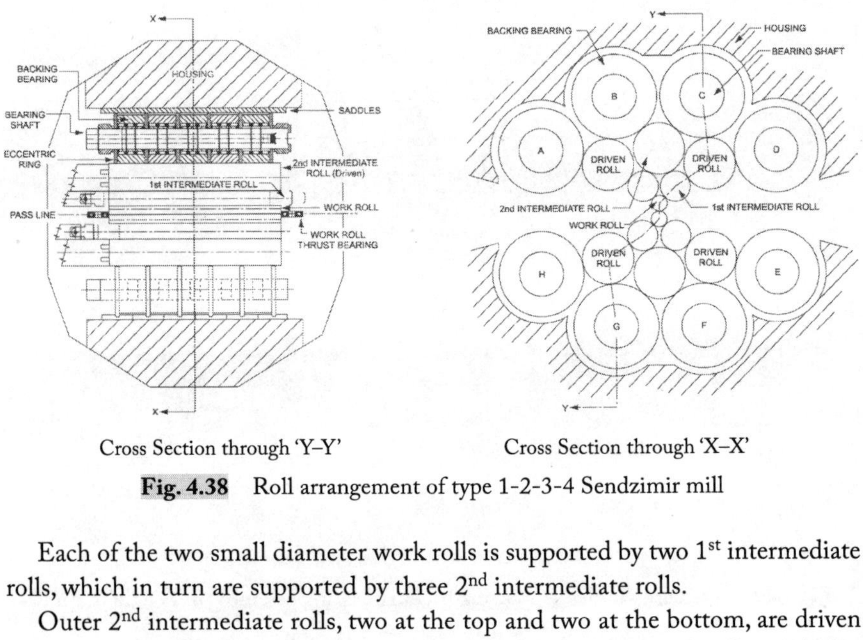

Cross Section through 'Y–Y' Cross Section through 'X–X'

Fig. 4.38 Roll arrangement of type 1-2-3-4 Sendzimir mill

Each of the two small diameter work rolls is supported by two 1st intermediate rolls, which in turn are supported by three 2nd intermediate rolls.

Outer 2nd intermediate rolls, two at the top and two at the bottom, are driven by a four output pinion stand. The work rolls along with all other rolls are driven by these drive rolls through friction contact.

The six number 2nd intermediate rolls are supported by eight number backing bearing assemblies. Each backing bearing assemblies consists of a stationary shaft with a number of roller bearings (specially designed by bearing manufacturers for Sendzimir mills to suit the design requirements) being equally spaced from each other by the components called saddles. These saddles act as spacers between the inner races of the bearings. The shaft is eccentrically mounted on the saddles.

The foot of the saddles are cylindrical in shape which sit on the cylinder shaped faces bored in the housing. The rolling load is thus transmitted from work rolls through the 1st and 2nd intermediate rolls and then to the number of roller bearings distributed along the eight stationary shafts. Finally the rolling load is well distributed throughout the inner bore of the housing through the saddle footings. The phantom view of the Sendzimir mill depicted in Fig. 1.16 shows the rolls inside the single piece box type housing.

Figure 4.39 shows the cross section view through one backing bearing assembly showing the stationary shaft mounted eccentrically on the saddles and the series of backing bearings supporting 2nd intermediate roll.

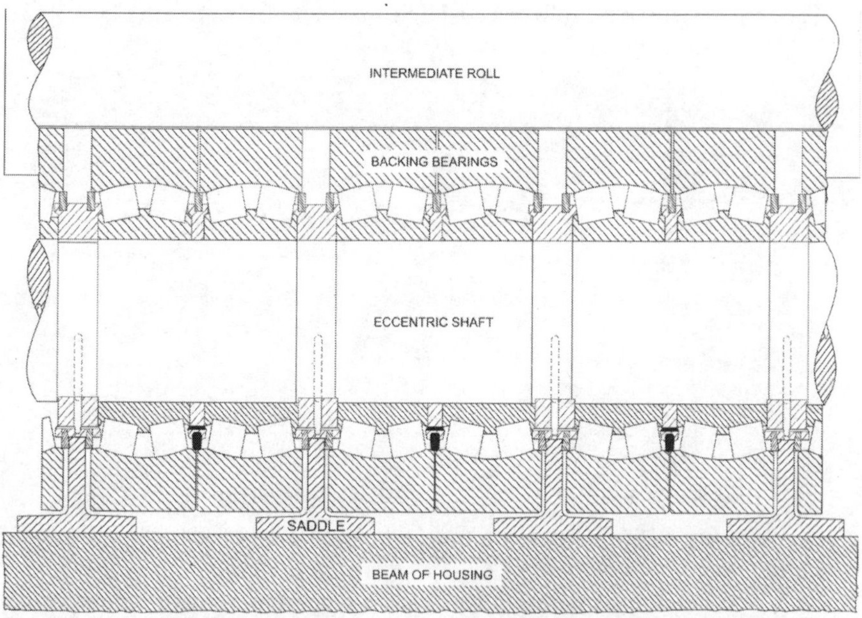

Fig. 4.39 Pressure saddles supporting backing bearing assembly in Sendizimir mill

Sendzimir mills have been designed in a number of models, each model using specific size of roll. Each model is again built in a few specific lengths, dictated by number of saddles used per backing roll, to accommodate different widths of strips. Table 3.2 indicates the basic specifications of the different models of 20-Hi Sendzimir cold mills.

The housing of a Sendzimir cold mill consists of a monoblock steel casting roughly of octagonal cross section. The casting is bored at the center, throughout the length of the casting, creating a hole with a distinctive shape consisting of four circular arcs at top and similar four circular arcs at the bottom side to accommodate the entire set of backing rolls and the saddles. Thick walls of the housing makes it very rigid under rolling forces.

The top work roll and top intermediate rolls are held in position by springs fixed to the housing from both sides. Because of small size of work rolls, it is possible to change them manually within a minute.

Screw-down arrangement: Adjustment of the gap between the working rolls is achieved through rotation of the top and bottom middle pair of the backing shafts B,C, F, and G in Fig. 4.38. By rotating these eccentric shafts, the position of the backing bearings, with respect to the housing, can be changed to closely control the distance between the work rolls. These shafts are provided with pinions at both ends which are engaged by racks whose movement are controlled by servo operated hydraulic cylinders.

Lateral roll adjustment: Lateral adjustment of the 1st intermediate rolls provides a means for rolling strips of various widths with a minimum of set up time between strips of different widths, thicknesses and hardnesses. The 1st intermediate rolls are provided with minute taper at the opposite ends as shown in Fig. 4.40. This feature adds greatly to the flexibility in shape control of the strips.

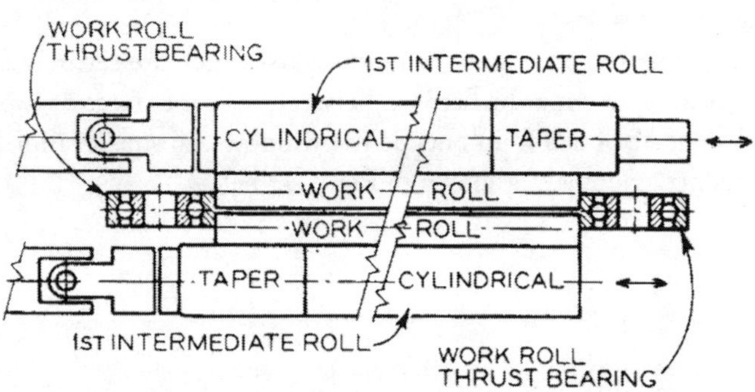

Fig. 4.40 Lateral adjustment of 1st intermediate rolls

4.1.8 Guides and repeaters

Guides are devices fitted on the mill stand at the entry and exit side for directing the roll stock during its entry to the mill bite and delivery from the rolls.

In flat rolling mills, the guides are in the form of vertical flats whereas in section mills, they are in the form of trough similar to the shape of the rolling stock. Fig.

4.41. shows the idea of provision of entry and exist guide with respect to the roll of a rolling stand.

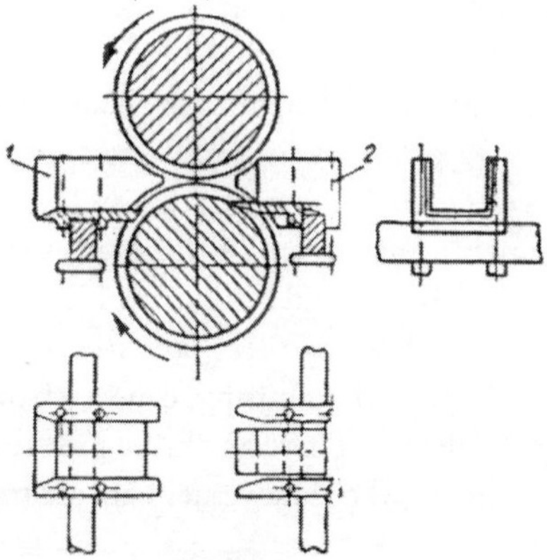

Fig. 4.41 Guides

1- entry; 2- delivery

In large reversing mills such as blooming, slabbing and plate mills, guides are often not used. Instead the rolling stock are directed by the manipulators fixed on roller tables on both sides.

Guides are mounted on cramp (or guide) bars, which are securely fastened to the mill windows. These may be fitted in between the inner faces of two windows, or may be fastened by bolts from outside the windows. In smaller mill the second method of fixing is popular, as shown in Fig. 4.42 below.

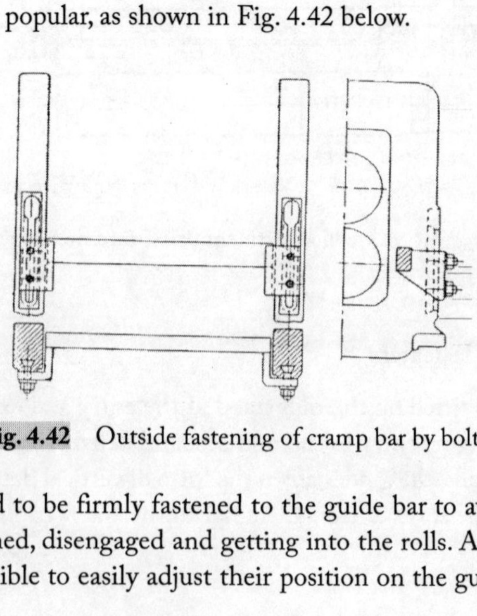

Fig. 4.42 Outside fastening of cramp bar by bolts

The guides need to be firmly fastened to the guide bar to avoid the chance of their getting loosened, disengaged and getting into the rolls. At the same time, it should also be possible to easily adjust their position on the guide bar in relation to the roll grooves. Figure 4.43 shows a very common method of fixing a guide on the cramp (guide) bar.

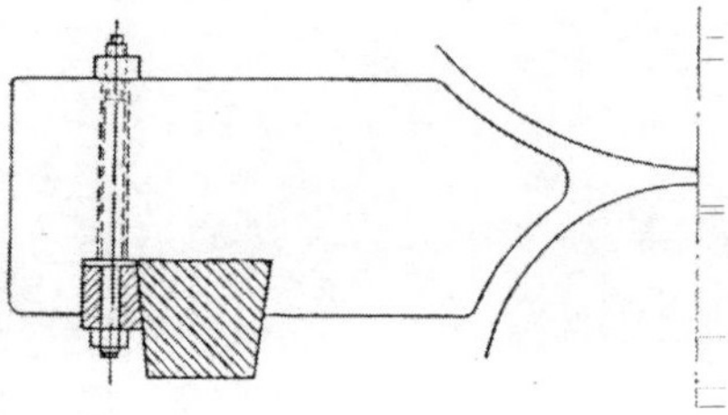

Fig. 4.43 Fastening of side guide on cramp bar

A. ***Non-twist guides***

For flat and often diamond passes, entry guides are made in the form of horizontal bars called guide plates. With rectangular passes and in a number of other cases, when there is no need for great accuracy in the lateral positioning of the rolling stock, entry guides are made channel shaped with their bottom usually in the shape of the section of input stock.

For other sections like an oval, where accurate direction of the piece and non-twisting is essential, the entry guides are box shaped (Fig. 4.44) in which friction side guide blocks are fitted. The working surface of these bocks is having same shape and size of the cross section of the input stock. The entering mouth is slightly flared for easy entry of the stock into the guide box. Relative position of the side guides within box is adjusted by screws.

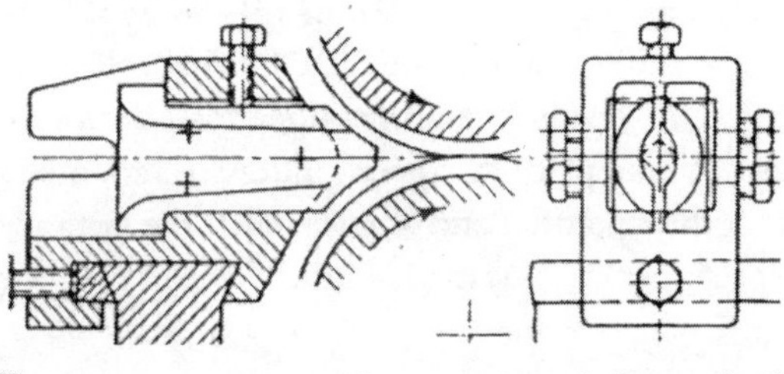

Fig. 4.44 Non-twist entry guide consisting of a box and side guides

The friction side guides are made from chilled cast iron or from cast steel. The friction surface are often built up with hard alloys or chromized.

In order to reduce side guide wear and to avoid marking of the product, *roller guides* are widely used, particularly for high speed bar and rod mills. Figure 4.45 shows a roller guide for ovals, it consists of two pairs of roller bearing mounted rollers, gap between which can be individually adjusted. The face of the rollers is machined to suit the stock size and shape. Rollers are made from hardened alloy steel (or even tungsten carbide for very high speed mill). The rollers are kept cool by spraying of water.

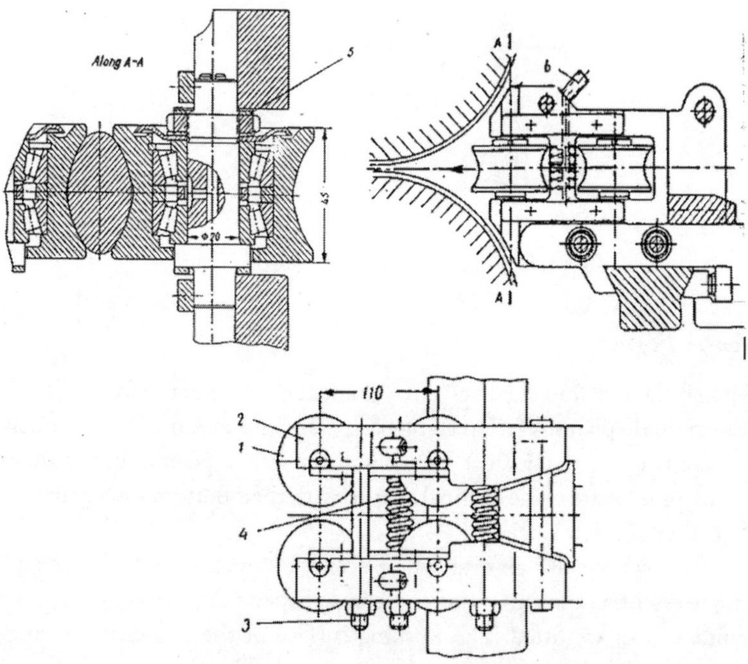

Fig. 4.45 Roller entry guide for ovals

Delivery guides in general consist of a stripper guide and side guides. Stripper guides are generally employed with bottom roll, to avoid the rolled product from collaring (i.e., wrapping around the roll). During rolling sections, generally overdraught is employed (top roll slightly larger than bottom roll). So that top roll collaring does not take place. Stripper guides are mounted with a clearance with side guide, and their pointed end almost touch the bottom roll surface. The working surface of stripper guide is made in the shape of outgoing rolled product.

B. ***Twist guides***

These guides are fixed at the exit end of a mill stand, which automatically twist the stock, usually through 90° or 45° between the stand and the next stand when rolling sections in continuous mills and in looping mills. The friction twist guides are with an internal spiral surface, fastened in a box. The disadvantage of friction twist guides is the rapid wear of the twisting surface. For this reason, *roller twist guides* have been designed. Figure 4.46 shows a roller twist guide.

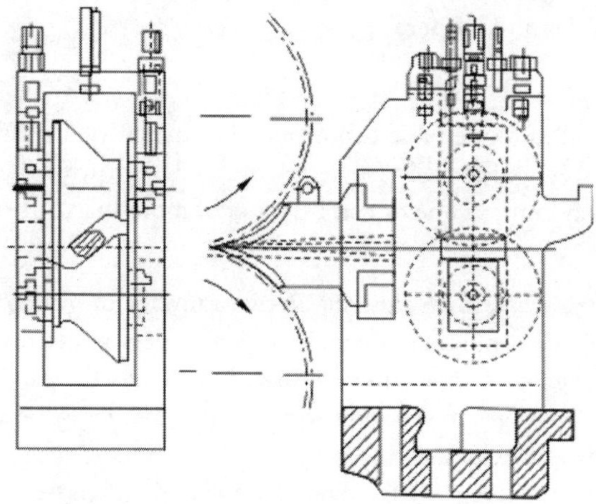

Fig. 4.46 Roller twist guide fixed at mill exit

C. *Repeaters*

Repeaters turn the front end of the rolled piece (usually through 180°) as it comes out of one pass and direct it into the next. They are used in section mills principally for light sections and generally where the stands are in train when rolling is done with looping. In these mills repeaters are generally used in conjunction with twist guides for directing simple shapes: squares, diamonds, and large ovals. With oval widths less than 20 mm, repeaters are not used.

There are various arrangements of repeaters but in all cases it is necessary to allow for growth in the size of the loops since the speed at which the metal emerges from one pass is kept marginally greater that than at which it enters the next. The most common repeaters are open U-shaped troughs connecting the entry and delivery guides of the passes (Fig. 4.47). When the loop increases after the entry of the piece into the next pass it forces itself out of the open trough onto the looping floor.

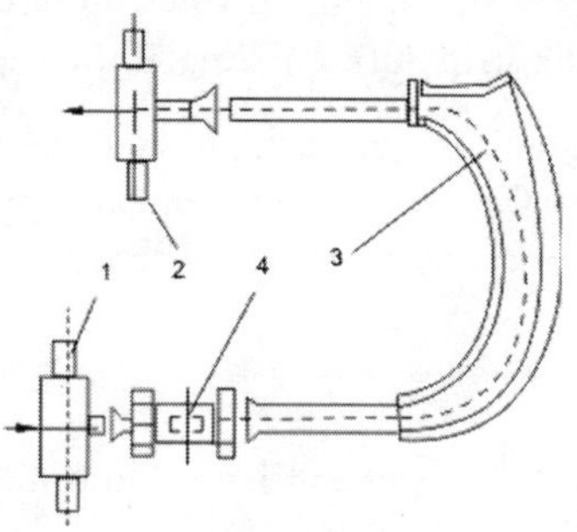

Fig. 4.47 Repeater with open looping trough
1 and 2- 1st and 2nd stand, 3- repeater; 4-roller guide

4.1.9 Guide tables and loopers

A. ***Guide Table***

In a cold strip mill, a guide table is positioned at the entry end of the mill stand. The fabricated steel guide table frame is suitably bolted with the mill housings. A guide table may contain one or combination of more than one of the following mechanisms.

(a) *Roller strip guide* that consists of two number of guide frames each consisting of a number of small vertical rollers made from hardened steel. The guide frames can be adjusted to change the gap between two sets of roller edges to suit strip width. Adjustment may be manual or by a motorized mechanism.
(b) *Pinch roll* consisting one fixed bottom roll and one moveable top driven roll. When the leading end of the strip is placed (manually or by other means) between the pinch roll, the top roll is lowered on the bottom and drive is started. The strip moves forward and enters the roll bite. The top roll is raised when strip is drawn by the mill rolls.
(c) *Feeding table:* When introduction of the strip into roll bite is difficult, a feeding table device may be used. This consists of two flat metallic pieces between which the strip edge is pinched, and then the pieces with the strip held in it, is given a forward motion, which pushes the strip end into the mill bite. The plates are then separated and moved back to their original position.
(d) *Shearing unit* this is a small shear fitted on the guide table at the outermost position. This is used to shear the uneven front end of the strip, so that a square strip end is made to enter the rolling stand.
(e) *Strip wiper* a double sided strip wiper consisting of felt type of wiping material fixed in frame may be used for wiping dirt from the surface of the strip entering the stand. The strip pressure also helps to create a small back tension in the strip during rolling, if the wiper is kept engaged.

Arrangement is provided through rack pinion or a hydraulic cylinder to move the entire guide table away from the housing, so that any part of it does not obstruct the roll chock assembly movement during roll changing.

B. ***Loopers***

Loopers are used between stands of continuous hot mills such as strip, skelp and light section mills.

In continuous hot rolling, it is not desirable to have any tension in the stock as it may get stretched and affect section accuracy in the succeeding stand or it may get snapped if the tension is too high. Tension in stock is avoided by adjusting the speeds of different stands so that the stock forms loops between stands.

The size of the loop is kept within limits by the operator by fine adjustment of rolling speed of the stand. The speeds of the stands may be controlled automatically by sensing the size of the loop by sensors like photocells.

In continuous strip mills, loopers are usually rollers which rise after entry of the strip in next stand. Pivoted roller movement may be by a motor or a pneumatic cylinder.

In skelp or section mills, a looper is often in the form of two hinged plates, as shown in Fig. 4.48 After the stock has entered the next stand, the plates are lowered by a pneumatic cylinder, so that the stock may form a loop under its own weight.

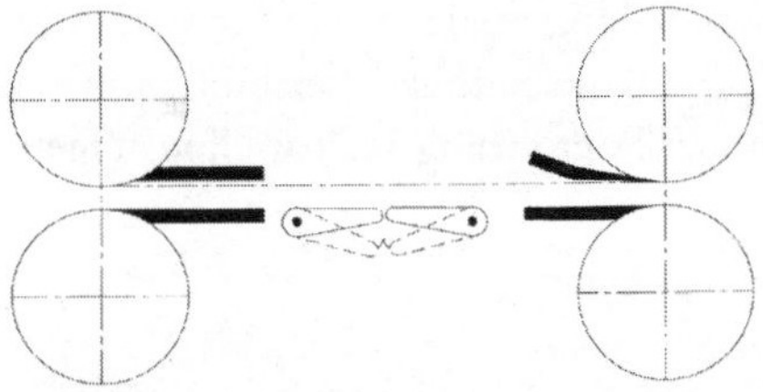

Fig. 4.48 Scheme of a looper in skelp mill

4.2 Mill Drive

The torque and power required by the rolls for rolling is provided by the mill motor via gear box, pinion stand, and universal spindles and couplings. A flywheel may be incorporated for certain mode of rolling. All these components together is called mill drive. These items along with the electric motor and motor control devices will be discussed in this chapter.

4.2.1 Mill couplings

In a mill drive couplings are used between motor and gear box, gear box and pinion stand. Outputs of the pinion stand are connected to the drive end of the stand rolls by spindle and couplings assembly. A flywheel, if used, is put between motor and gear box through flywheel shaft and coupling.

The mill couplings are generally of three types:

(i) *Rigid couplings* (usually flanged type): These are rarely used for very small rolling mills.

(ii) *Flexible coupling* with spiral springs (*Bibby* coupling), or with pins (*BAMAG* coupling). These couplings protect the drive from shock loads at the same time provide some amount of flexibility against non-alignment between motor and gear box shaft. In recent times these have been replaced by *gear couplings* with their greater flexibility.

(iii) *Gear couplings* are the most used couplings now. These are also used as the two couplings for the universal spindles.

(iv) *Safety couplings* are sometimes used between flywheel and the gear box.

A. ***Gear couplings***

These are most widely used couplings as main motor couplings. These are very compact, easy to manufacture with high precision with low friction losses, and offers great deal of flexibility. Figure 4.49 shows half sectional view of a gear coupling. The teeth space of the sleeves engages with the teeth space of the rims. The teeth of the sleeves are crowned along the length of the teeth and each of the teeth are barreled on its faces. This allows complete flexibility of two halves of the coupling throughout 360^0 rotation. The crowning and barreling of teeth is shown in Fig. 4.50.

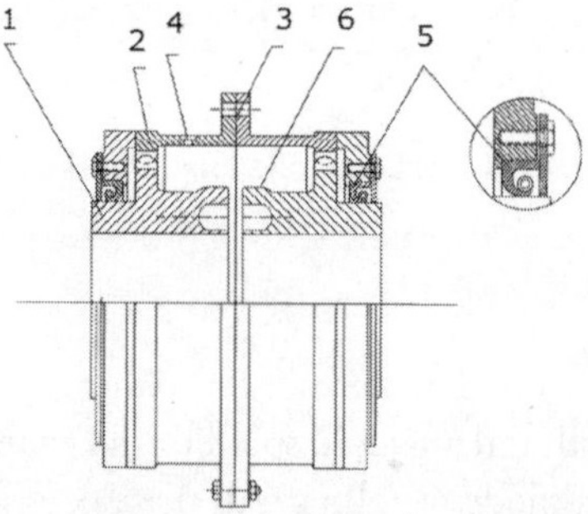

Fig. 4.49 Geared coupling

1- sleeves; 2-rims; 3-joining flanges; 4-plug for lubricant; 5-seal; 6-hole for removing sleeves

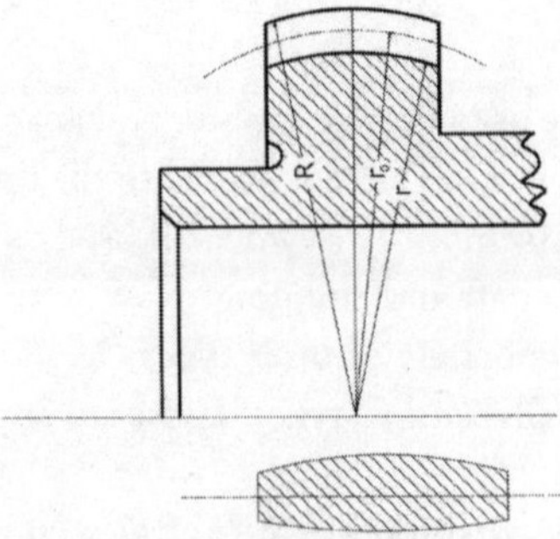

Fig. 4.50 Improved shape of teeth in a geared coupling with barreling (elliptical cross-section)

The bore diameters of the sleeves are made to suit the shaft diameters on both ends. A light force fit is generally employed for fitting the sleeves onto shafts with keys. The couplings are filled with high viscosity oil, which needs change on regular basis (say every 6 months). The oil is retained by oil seals. The rims are fastened together by fitted bolts.

B. ***Bibby type flexible couplings***

Figure 4.51 shows this coupling which is both flexible and largely compensating. Because of its excellent qualities, it has been widely used in rolling mills. However, they are rather difficult to produce and have recently been largely replaced by gear couplings.

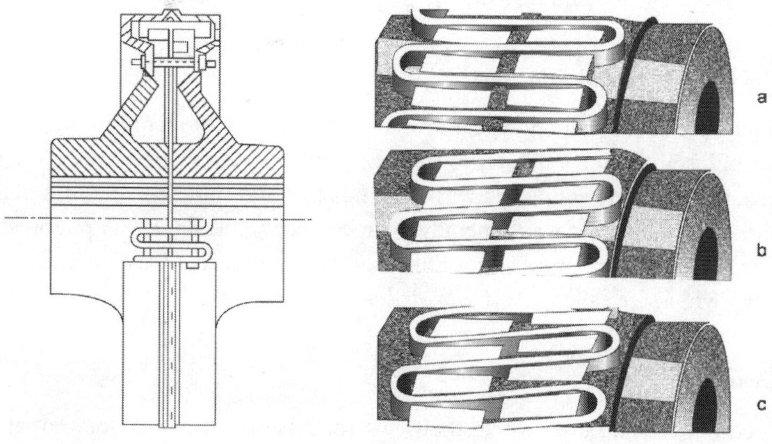

Fig. 4.51 Flexible coupling with spiral springs (Bibby) showing deformation of springs under loads

a- nominal; b- normal overload; c- under very heavy and impact loads

C. ***Selection of couplings***

Couplings are selected based on its torque transmission capacity at different degree of misalignment (i.e., angle between the two connected shafts), and by the maximum size of the shaft it can accommodate.

Manufacturers of couplings give all these data along with overall dimensions of their couplings in catalogues.

The maximum torque to be transmitted is based on either according to maximum theoretical torque of the shaft, or according to the motor torque rating with allowance for overloading. In first it is sum total of torque required per roll. In latter case, the torque of the coupling is taken as below:

(i) Non-reversing mill without flywheel:

$$T_c = (1.5\text{–}2)\, T_m$$

(ii) same as (i) with reversing:

$$T_c = (1.5\text{–}2)\, T_m$$

(iii) Coupling between motor and flywheel:

$$T_c = (1.5\text{–}2)\, T_m$$

(iv) Coupling between flywheel and mill gear box or mill, as may be the case.

$$T_c = (4\text{–}6)\, T_m$$

where, T_c = coupling torque, T_m = nominal motor torque.

4.2.2 Gear box and pinion stand

The purpose of the reduction gear box is to reduce the motor rpm to suit roll rpm of the mill stand. *Pinion stand* is to split the gear box output speed into more than one shafts which are connected with roll ends through spindle–couplings to drive all the work rolls together. Generally no speed change is done in a pinion stand. In smaller mills, sometimes a gear reduction and pinion stand is combined in one housing which are called *combined gear box cum pinion stand.*

A. ***Gear Box:***

Reduction gears are used where their cost (or losses in friction) does not exceed the difference in the cost between slow-speed and high-speed motors. On non-reversing mills, therefore, reduction gears are practicable where the speed of the rolls is less than 200 to 250 r.p.m.

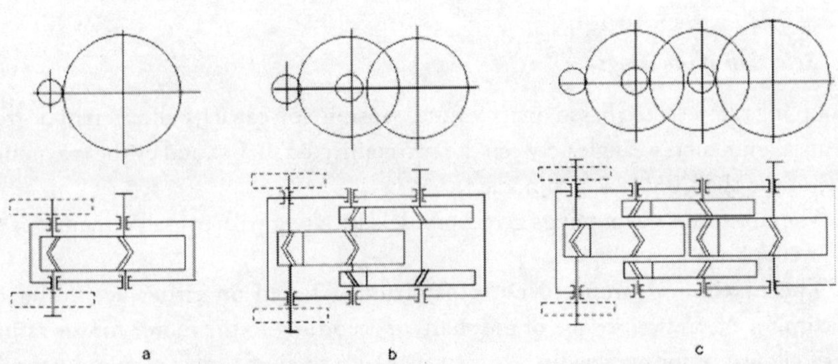

Fig. 4.52 Diagram of 1, 2 and 3 stage reduction gear

At higher speeds a slow-speed motor with a pinion stand without reduction is satisfactory. In large mills which have to be reversed often (blooming, slabbing, plate mills, breakdown stands), reduction gears are not used as the slow-speed motor alone is more easily reversible.

Single, two, and occasionally three-stage reduction gears are used (Fig. 4.52) according to the range of r.p.m. of the rolls: 200–250; 40–50 and 10–15. At high power (500–3000 h.p.) the gear ratio of each stage is not usually above 6. If flywheels are used (shown by dots) they are disposed on each end of the shaft of the small pinion) to distribute the load among the bearings and the shaft itself.

Group reduction gears, where there is only a small distance between the axes of the driven rolls, are made with cylindrical gears. In Fig. 4.53(a), the reduction gear transmit rotations to two lines of the mill with counter-rotation of the rolls.

In Fig. 4.53(b) a multi-line reduction gear is being used to transmit rotations to some of the lines of a continuous train of stands, for example, a light section mill. To maintain the same rotation of the rolls and allow for the small distance between the lines, the shafts of the wheels are arranged in two planes and connected in series with the upper and lower pinions of the pinion stand. The meshing gear ratio is distributed according to the rate of rolling.

Where the distance between the lines is large, group reduction gears are used with bevel wheels (Fig. 4.53(c)).

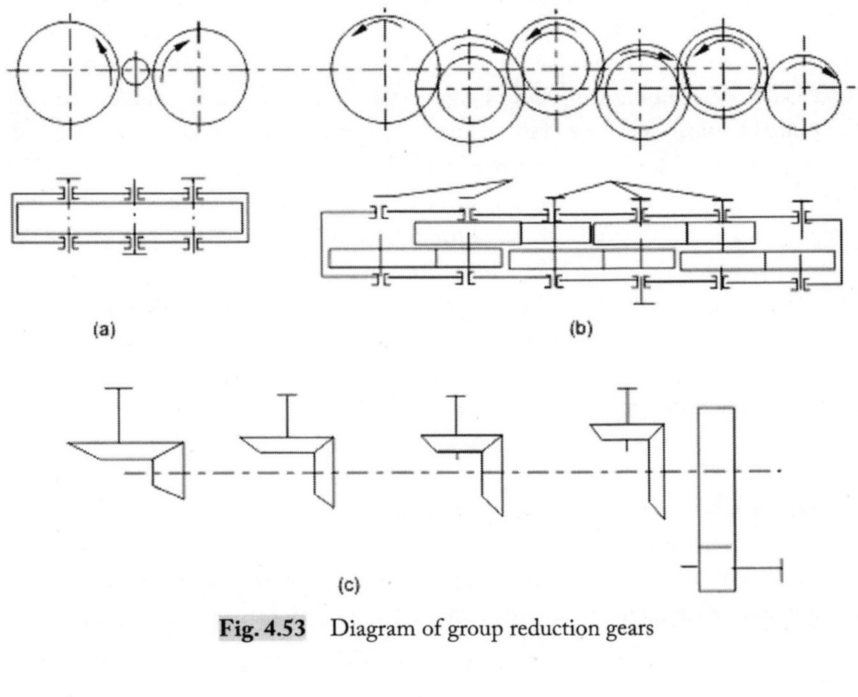

(a) (b) (c)

Fig. 4.53 Diagram of group reduction gears

There is an increasing tendency to avoid the use of group reduction gears, particularly where there are a large number of lines, by driving each line separately from its own motor. Huge multi-line reduction gears are being eliminated altogether or else replaced by less complicated, frequently two-line, reduction gears. The resulting increase in the cost of electrical equipment is compensated by the use of simpler mechanical equipment, a slight increase in productivity and simplification in the alignment of the roll in continuous mills.

The gear wheels of large reduction gears on the main drive are always double helical as these operate more smoothly and there is no axial thrust on the bearings. In two-and three-stage gears, one is usually composed of a pair of helical gears (see Fig. 4.52). In this type of reduction gear only one shaft, usually the slow speed one, may be fixed in the axial direction.

The design of the gear wheels depends mainly on the diameters d and d_v of the gears and shafts. This dependence is roughly as follows:

(1) Where $d < 2d_v$ the pinion and the shaft are integral;
(2) Where $d = (1.8–3)\ d_v$ the pinion is made from solid cast or forged blank;
(3) Where $d > 3d_v$ the gears can be either cast or fabricated.

The gears are usually fitted to the shaft by a press fit; to avoid angular displacement prismatic keys are used.

Various materials may be used for gear wheels. Small gear wheels (below 3 d_v) are usually made of forged steel: at medium loads, of carbon steel with UTS = 60–70 kg/mm^2 and elongation 16–18 percent, and at high loads, of special steel with UTS = 80–90 kg/mm^2 and elongation 14 percent. It is recommended that the pinion teeth should be surface-hardened (flame or induction) to 400–500 HB.

The housing of large reduction gears (divided horizontally) is produced in the form of a rigid rectangular casing of frame or sometimes of welded construction. The bottom is cast integral with the housing. For very large reduction gears this is sometimes welded because of the difficulties in casting. Great care must be taken to see that the joint between the bottom and the housing is completely sealed as once the gear has been erected, the bottom is not accessible from outside. For this reason solid casting with the housing is preferable. Where a detachable bottom is to be used the fastening bolt should be made accessible from inside the reduction gear when the wheels have been dismounted.

Another of the peculiarities of large reduction gear is the necessity of making separate caps for each bearing. The upper casing of the gear is lightly welded and fastened by bolts and flanged on the caps of the bearings.

It is a good thing to use roller bearings (with tapered or angular contact and sometimes cylindrical rollers) for the reduction gear shafts, providing they can satisfy the requirement for r.p.m. and load. The usual thing is to have one bearing to each journal with a double row of rollers (spherical or double row tapered roller bearing).

Spray lubrication is the most convenient form, with a general circulating system for both gearing and bearings.

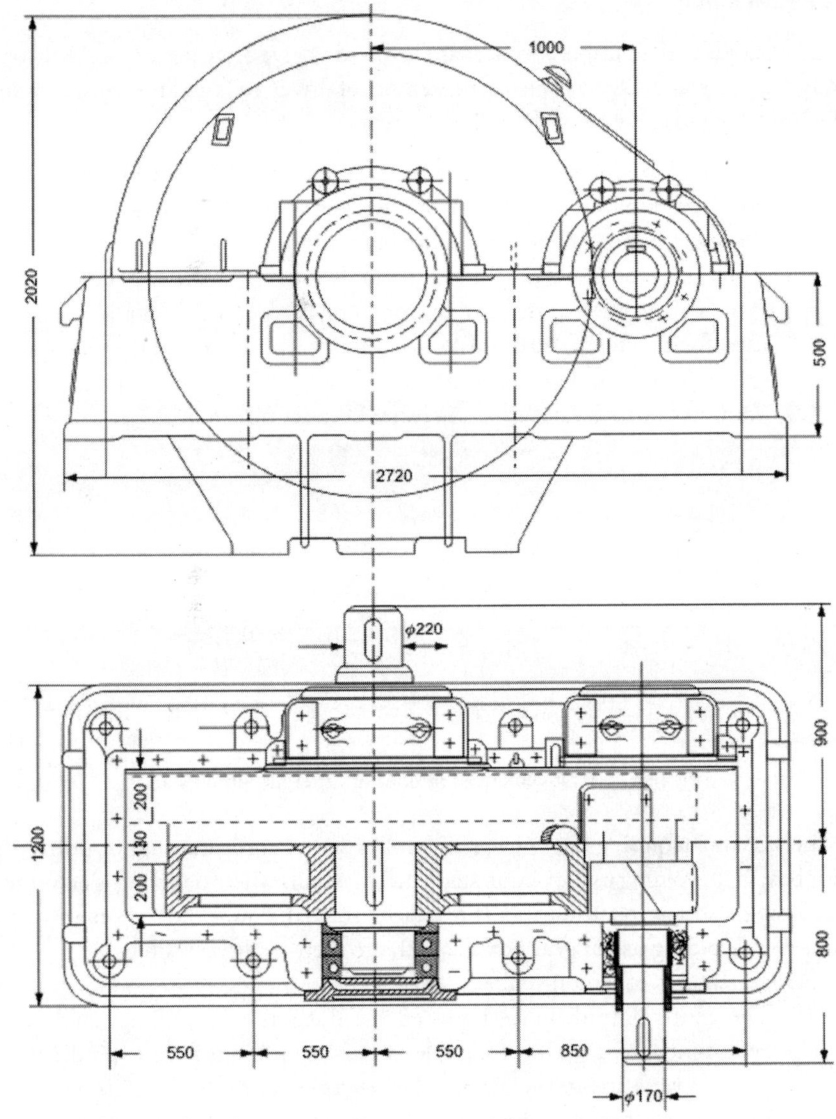

Fig. 4.54 Dimensioned drawing of a reduction gear box with gear ratio i = 6.35, maximum torque on slow-speed shaft 8.73 tm

Figure 4.54 is a typical mill gear box dimensioned assembly drawing showing disposition of input and output gears with shaft and bearing arrangement, including cast steel casings of the gear box.

B. ***Pinion stand:***

Pinion stands with multiple output shafts are used to give drive to each of the driven rolls of a mill stand. Depending on the number of driven rolls and their disposition, the pinion stands may be of the following types:

(a) Two-high pinion stand to drive the two driven rolls of a 2-Hi or 4-Hi mill. In view of the arrangement of drive, and convenience of dismantling, the lower pinion should be the driving pinion.
(b) Three-high pinion stand, in which middle one is the driving pinion.
(c) Pinion stand (two or three-high) for a universal mill. The top pinion is used for driving vertical rolls. However, the vertical rolls are generally driven from an independent motor.
(d) Four output pinion stand for a cluster (Sendizmir) mill.

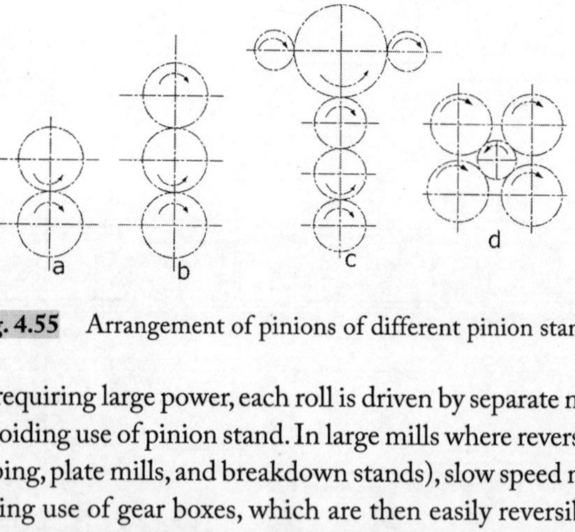

Fig. 4.55 Arrangement of pinions of different pinion stands

In large mill requiring large power, each roll is driven by separate motor/s through gear box, thus avoiding use of pinion stand. In large mills where reversing is necessary (blooming, slabbing, plate mills, and breakdown stands), slow speed motors are often used even avoiding use of gear boxes, which are then easily reversible.

Design features of pinion stands:

(i) The center-line distance between the outputs of a pinion stand is determined from the most suitable working conditions for the spindle at possible variation in the distance between the mill rolls. In mills where the distance between the rolls only changes when they are reground, the pinion stand output center-line distance d_o is calculated from the equation.

$$d_o = \frac{D_n + D_o}{2}$$

where, D_n and D_o are the new and old roll diameters, respectively.

(ii) The pinion stand housing are made from cast steel or fabricated from steel plates. The most common design of the housing is that they are divided horizontally at center lines of the pinions. The pinion shafts are mounted preferably on roller (tapered or spherical roller) bearings. The pinions are generally forced oil lubricated by a pump, with bottom part of the housing acting as oil reservoir.

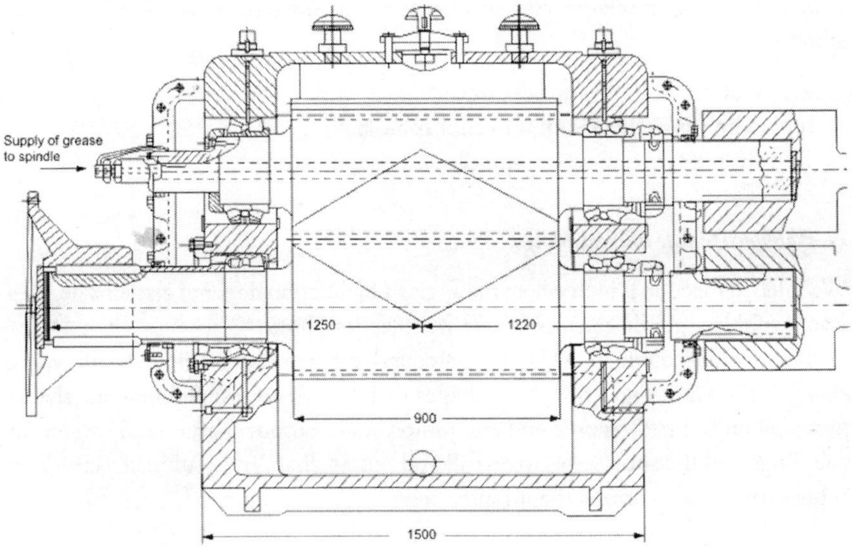

Fig. 4.56 Sectional elevation view of a 2-Hi pinion stand

4.2.3 Spindles

Transmission of power from the pinion stand to the mill stand rolls is made through the *spindle and couplings*. Each roll end is connected with one coupling which is fitted at one end of the spindle while the coupling on the other end is fitted on the pinion stand shaft.

As the rolls need to be moved at different roll gap setting as well as after each roll grinding, it is clear that the angular position of the spindle gets changed and power has to be transmitted even at such misaligned position between roll centers and pinion stand centers. Therefore, it is essential that the mill spindles should be universal, as shown in Fig. 4.57.

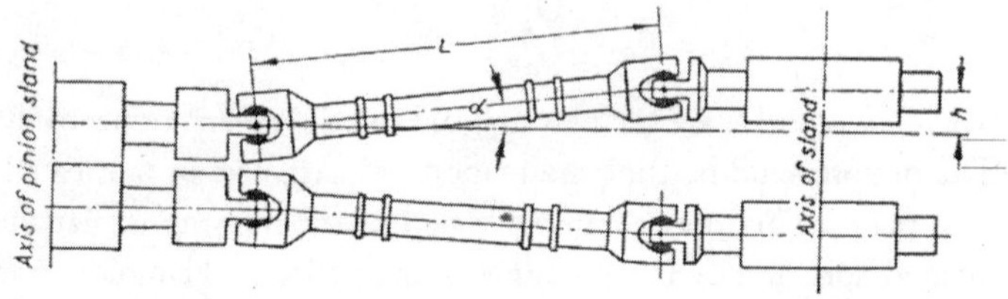

Fig. 4.57 Diagram of arrangement of universal spindles

Different design of universal spindle and couplings are used in rolling mills depending on their relative cost versus precision and ease of use. The most used spindles are:

(i) Wobbler spindle and couplings
(ii) Palm (or slipper) joint universal spindle
(iii) Geared typed universal spindle and couplings

A. ***Wobbler spindle and couplings:***

Wobbler connection is used where the angle of inclination does not exceed 2 degrees. Each wobbler spindle consists of two wobbler couplings.

The wobbler couplings are hollow cylinders having its bore machined with typical shape as shown in Fig. 4.58. These shapes match with corresponding male shapes provided on roll end, spindle end and pinion stand output shaft end. Between the couplings and the spindle, between roll and pinion shaft ends, sufficient clearance is provided which permits the misalignment.

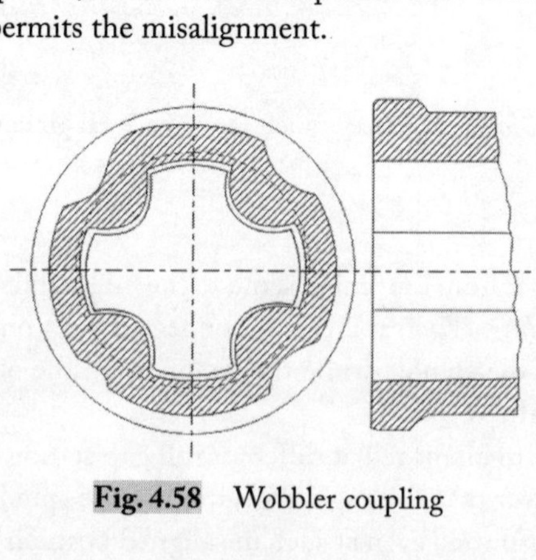

Fig. 4.58 Wobbler coupling

With wobbler spindle and couplings, roll change become easy, and were popular particularly for mill in train. The present trend is to replace them with universal spindle couplings.

In order that couplings should not laterally move on spindle during working, wooden distance pieces are fastened by chips or wires. A typical wobbler connection for a 3-High mill along with spindle balancing arrangement is shown in Fig. 4.59.

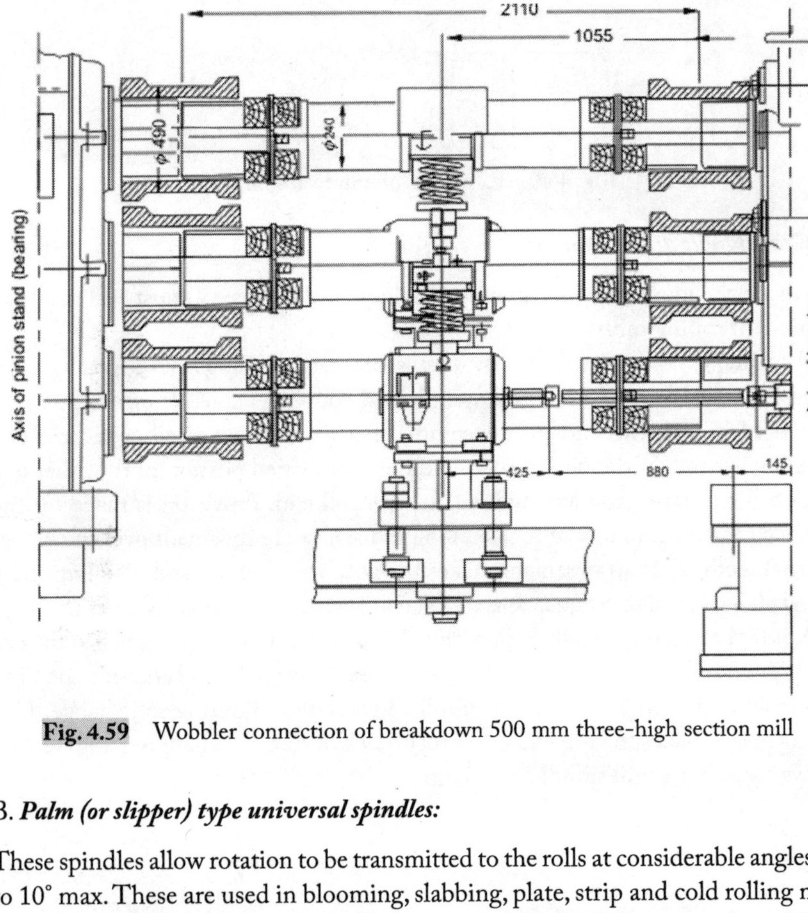

Fig. 4.59 Wobbler connection of breakdown 500 mm three-high section mill

B. ***Palm (or slipper) type universal spindles:***

These spindles allow rotation to be transmitted to the rolls at considerable angles up to 10° max. These are used in blooming, slabbing, plate, strip and cold rolling mill.

Two cylindrical ends of the spindle are made of large diameters and machined from the end to form a spherical (more than hemispherical) groove. These are called palms.

Then a wide rectangular groove is cut at the spindle end along the axis of the spindle. Into these flat grooves the flattened (also called spade) ends of the roll and pinion shaft enters. In the gap between the spherical gap and flat roll / pinion shaft ends, textolite pads are introduced along with 2 nos. flat slippers pad in between and connected to the flat ends of rolls/pinion shafts, as shown in Fig. 4.60:

The elongated slot in the spade allows axial movement between the spindle and the roll / pinion shaft, which is required for different angular position of the spindles.

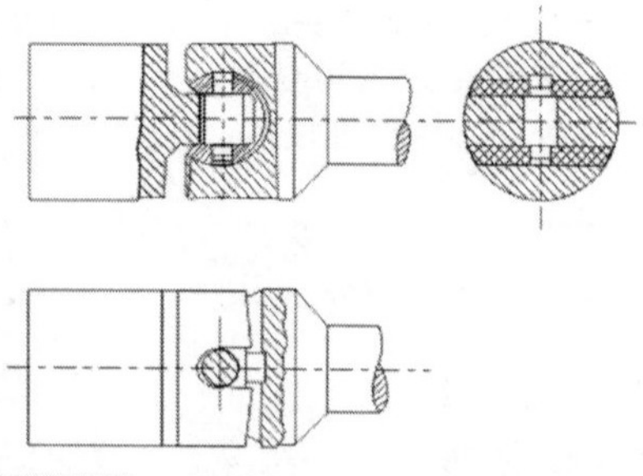

Fig. 4.60 Palm joint of universal spindle

C. ***Gear spindle and couplings***

These are the most used mill spindle couplings and are used almost exclusively in all modern rolling mills.

Advantages and constructional details of gear couplings have already been discussed in section 4.2.1. A gear spindle and couplings consists of two *half gear couplings* fitted on two ends of the spindle shaft. The non-geared portion of one coupling is fixed on the pinion shaft and the non-geared portion of the other one is splined or shaped to have running fit with roll end. Provision is made for the external gear teeth in one of the coupling to have axial movement over the longer internal teeth. This arrangement makes the gear type spindle and couplings fully universal having high torque capacity even at high angle of misalignment.

Misalignment is generally kept within 3–4 degrees. The coupling half at the roll end is provided with internal spring arrangement to keep the roll end side coupling to remain in line with spindle, when roll end is removed during roll change. This arrangement makes the roll end entry very quick during roll changing. Figure 4.61 shows a gear type mill spindle coupling.

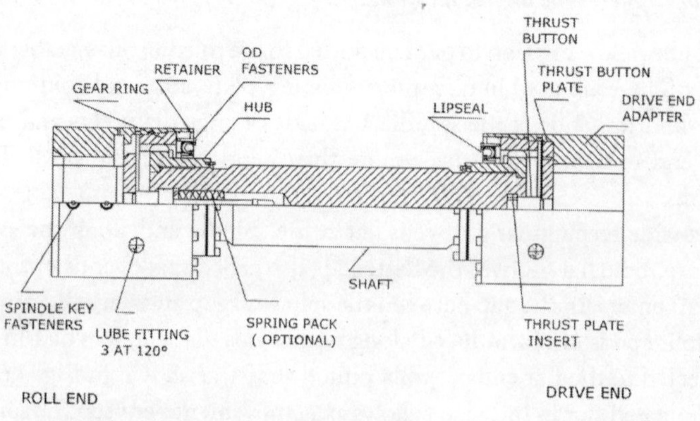

Fig. 4.61 Mill spindle with two half gear couplings at two ends

D. ***Spindle balancing***

For mills with roll diameter greater than 450 mm, the spindles are generally supported, at about the middle of the length, on bearings to avoid load of the spindles to act on wobbler joints or universal joints. All wobbler spindles and also universal spindles with small movements (less than 50–100 mm) are usually balanced by springs (Fig. 4.59). When springs are arranged from one side, as shown in Fig. 4.62, it is easier to remove the spindle (as in mills in train). For large movement of the universal spindles, hydraulic or mechanical counterbalancing is employed.

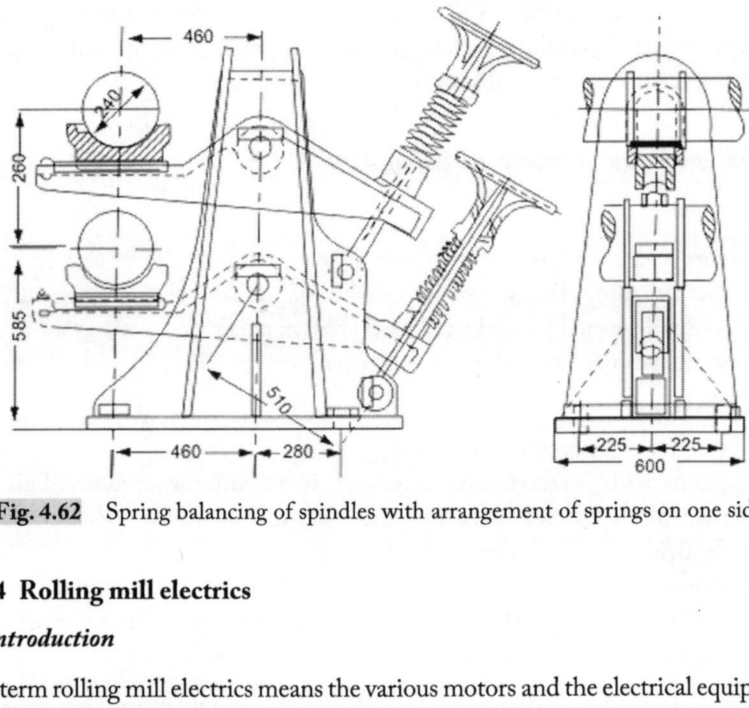

Fig. 4.62 Spring balancing of spindles with arrangement of springs on one side

4.2.4 Rolling mill electrics

A. ***Introduction***

The term rolling mill electrics means the various motors and the electrical equipment for controlling different operating features of a mill like starting, stopping, output torque, speed, etc. Torque capacity of the main mill motor is of prime importance as it must deliver sufficient torque to the mill rolls to make the rolling possible.

While top motor speed determines the maximum rolling speed of a mill stand, arrangement of speed control devices makes it possible to roll at variable speed with desirable torque–speed characteristics.

When rolling long products or strip in continuous mill, the speed and tension control between stands and between stand and coiling / decoiling equipment becomes absolutely critical for producing good quality rolled product. In foil rolling, coordinated control between back tension and rolling speed is vital for foil thickness uniformity.

From the point of view of electrical drives, rolling mills with their drives may be grouped as follows:

(a) Fixed speed non-reversing mill with AC drives.
(b) Fixed speed reversing mill with AC or DC drives.
(c) Variable speed mill with DC drives.
(d) Variable speed single stand strip mill with coiler / decoiler – DC Drive
(e) Long product multi-stand mill- DC or AC with VVVF drive
(f) Tandem strip mill- DC or AC with VVVF drive

In this section the selection of different types of motors and their speed and torque control methods for different types of hot and cold rolling mill stands and equipment have been briefly discussed.

B. *Types of DC and AC motors for rolling mills*

(i) *DC motors*

DC motors have armature and field coils. The armature is a rotor section having Lap or Wave winding , the current being fed through commutators and brush. The Field winding is turned around fixed field poles in a pair.

DC motors can be divided into three types.

- Series field, i.e., the field is in series with the armature. This is used where a high staring torque is required.
- Shunt field, i.e., the field is in parallel circuit with the armature. This is used where the field is independent of armature and is normally used for field reversal for direction reversing.
- Compound field, i.e., a part of the field is in series and the other part is in parallel circuit with the armature. This is the most common practice for precision controls.

Torque–speed characteristics of these three types of DC motors are shown in following figures:

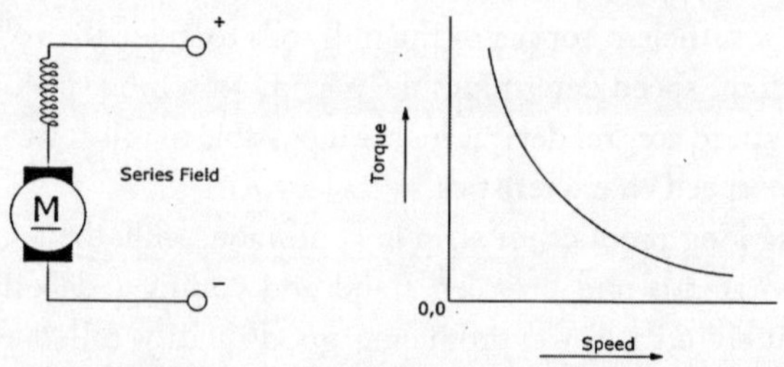

Fig. 4.63 DC motor connection with series field required for high starting torque

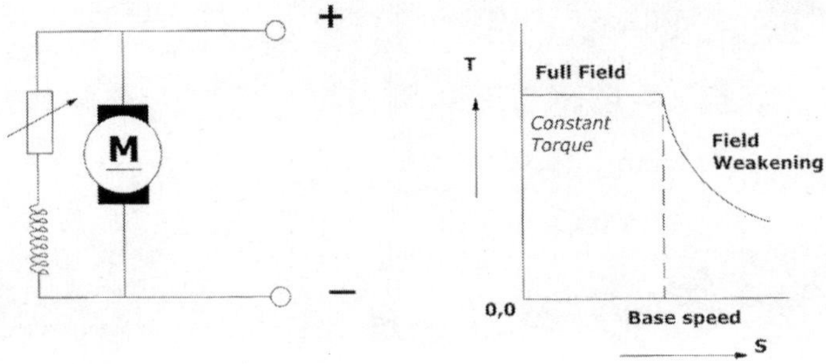

Fig. 4.64 DC motor connection with shunt field required for normal starting torque but speed increase with field weakening beyond base speed

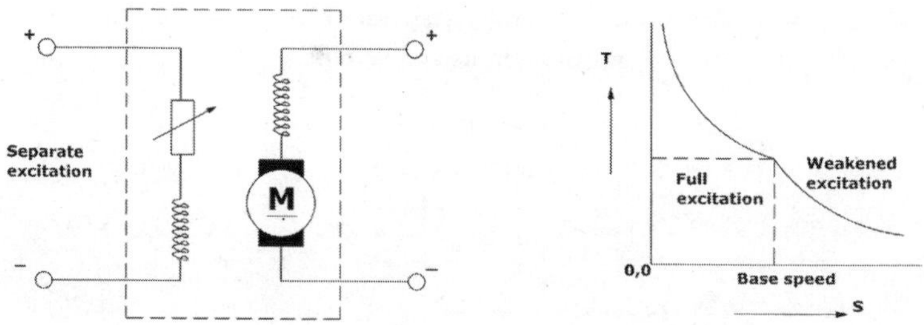

Fig. 4.65 DC motor connection with compound field. Best for high starting torque but good speed control

(ii) *AC motors:*

AC Induction motors used for rolling mills are all of 3-phase design. Here the stator receives the AC 3-phase current and is fixed to the yoke/ frame of the motor. The rotor is a short circuited winding through which the slip frequency current flows to generate a torque. The rotor can be either a squirrel cage design or a slip ring design through which the slip frequency current is fed to an external rotor resistance bank.

AC slip ring motor with motor resistance control offer controlled starting torque and current and are used as main mill drive for roughing and intermediate stands for medium section, bar and rod mill.

The other type of AC motors with large capacities is AC 3-Phase synchronous motors. Here the stator receives the AC 3-phase current and is fixed to the yoke/ frame of the motor. The rotor winding is fed from a separate DC source and is used to control the torque generation. The separate DC excitation system is normally static DC converters and is used to control torque requirement. Synchronous motors are used for main mill drive for heavy plate mills and section mills, offering no fluctuation in mill speed.

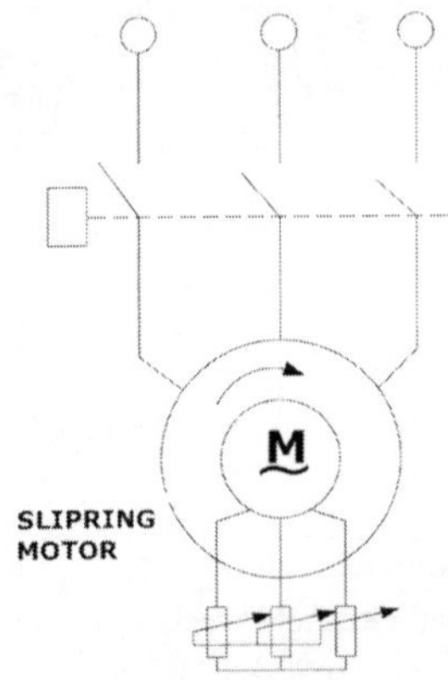

Fig. 4.66 AC slip ring motor with rotor resistance control, suitable for controlled starting torque and current

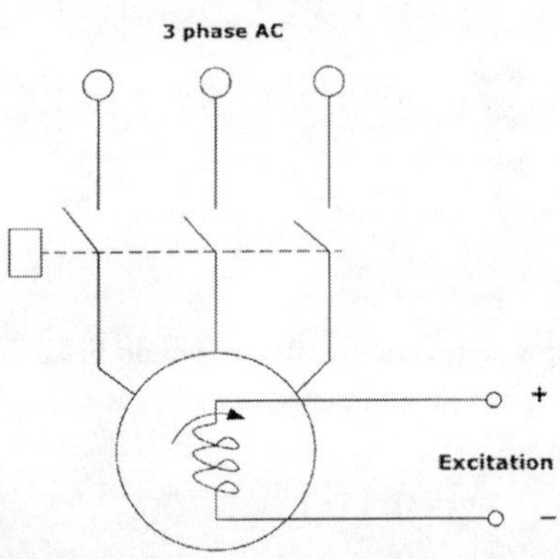

Fig. 4.67 3-phase synchronous motor with separate excitation

For non-reversing rolling mills' main drives, either slip ring motors or synchronous motors are normally used.

C. Motor control schemes

C.1 Thyristor devices for converting AC to DC supply:

A thyristor device is used to convert an AC supply to DC supply like a semiconductor diode. Unlike transistor device using a bias, a conduction gate is used for the thyristor device. The gate pulse can start from 0° to 180° conduction, which is controlled. Hence a thyristor is a semiconductor with control to variable output voltage.

Thyristor device is mainly used for power conversion rather than a control device. The conversion from AC to DC is explained in Fig. 4.68.

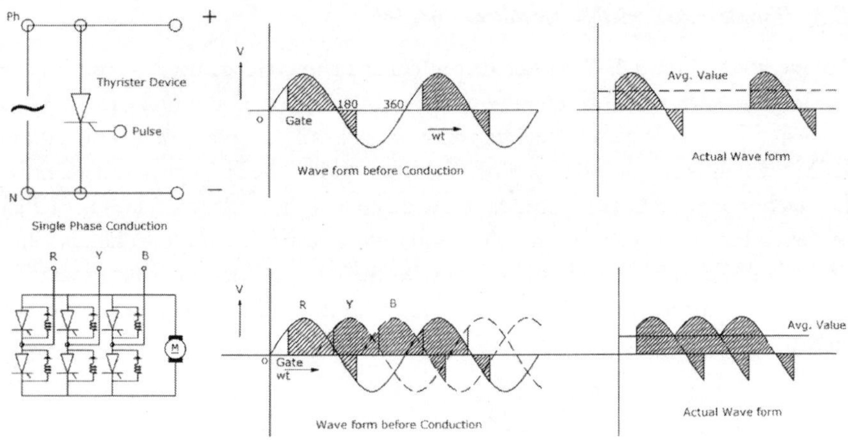

Fig. 4.68 Three functions of a tyristor converter

C.2 VVVF drive for speed control of AC motors:

An AC VVVF drive has two components. First, the 3-phase AC supply is converted to DC supply through 3-phase 6-pulse thyristor converter. The output voltage is known as DC Link Voltage, which is again inverted to 3-phase AC supply with a variable frequency. The output of the Inverter is fed to a 3-phase AC Induction or synchronous motor, to vary the speed by varying the input frequency.

The term VVVF stands for Variable Voltage Variable Frequency where V/f is kept constant up to the line frequency with the torque remaining constant, and above the line frequency the torque reduces with the increase in frequency.

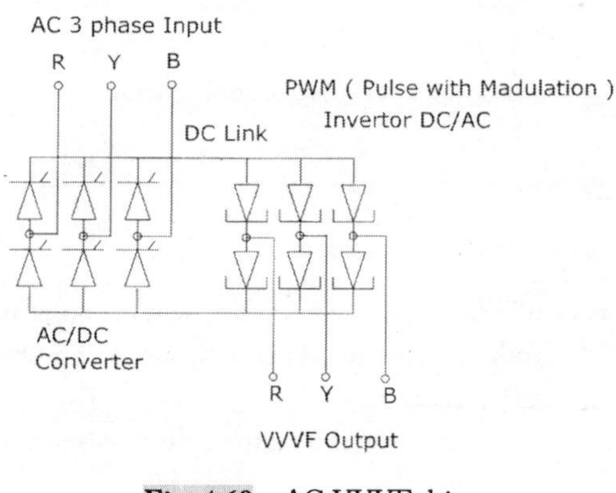

Fig. 4.69 AC VVVF drive

The variable voltage/variable frequency from the inverter is obtained through transistor switching of IGBTs.

C.3 Torque control method for coiler / uncoiler:

Torque control for a DC Motor driven colier or uncolier is used to control the tension and speed for the center wind coiler or uncoiler. The actual strip speed is measured by a tachometer (devices which measures DC Voltage in proportion to RPM) which is mounted on a pinch roll over which the strip pass and a similar DC tachometer mounted on the center wind coiler or uncoiler measures the RPM of the coiler or uncoiler. These two feedbacks are used to compute the diameter of the coil which in turn is used, along with the required tension, to compute the torque set point for coiler or uncoiler current limit. Figure 4.70 shows the torque control scheme.

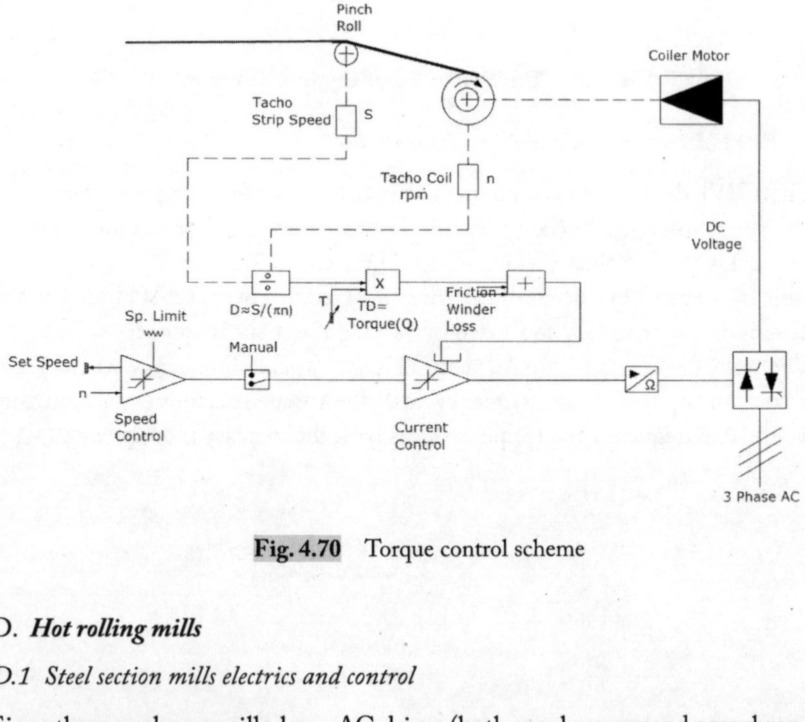

Fig. 4.70 Torque control scheme

D. *Hot rolling mills*

D.1 Steel section mills electrics and control

Since these are heavy mills, large AC drives (both synchronous and asynchronous) are used for various stands. Normally AC synchronous motors are used for the various passes for the different sections. The starter for these large synchronous motors uses HT circuit breakers for the stator side and static excitation for the rotor side.

Other than the main mill drives a large number of auxiliary drives with AC and DC motors are used. These are used for various auxiliary equipment like roller table motors, cooling bed, transfer mechanisms, and so forth.

D.2 Steel plate mills electrics and control:

Like section mills, these mills are also of heavy duty and have many stands between roughing and finishing stands. The mills are of continuous type and also have large AC drives (both synchronous and asynchronous) as the main drives. Like section mills, the starter for these large synchronous motors uses HT circuit breakers for the stator side and static excitation for the rotor side. For the asynchronous AC motors very large static VVVF (Variable Voltage Variable Frequency) drives are used or DC motors with large thyristor converters are also used.

Other than the main mill drives a large no. of auxiliary drives mainly AC and DC driven motors are used. These are used for roller table motors, cooling bed, transfer mechanisms, etc.

In both cases of section mills and plate mills the large AC drives with VVVF control always use constant torque output up to the rated speed corresponding to the line frequency for higher sections or very high rolling torque requirement, and goes to the super synchronous speed with reduced torque output for the lower rolling torque.

D.3 Steel blooming and slabbing mills electrics and control

Since these mills are of very heavy torque requirement to produce blooms and slabs from ingots, generally each roll of the mill is driven by a large AC drive. These large AC drives are again large AC synchronous motors with controls same as those for section mills.

D.4 Steel hot strip mills electrics and control

Usually these mills are of heavy duty and consist of various stands, including roughers, and mainly take the inputs from a slab continuous caster. All these stands are generally in tandem mode, i.e., all in line shortly one after another and run in continuity and hence the stands are usually driven by large synchronous motors. The starter for these large synchronous motors uses HT circuit breakers for the stator side and static excitation for the rotor side. However, in recent mills, large VVVF drives for the synchronous motors are used which offer flexibility in rolling pass schedule.

The controls for these tandem mills are complex so as to meet the desired strip speeds after each pass and also to maintain the proper tension at the winding end and also between stands. For this a mill computer with at least three hierarchies are used.

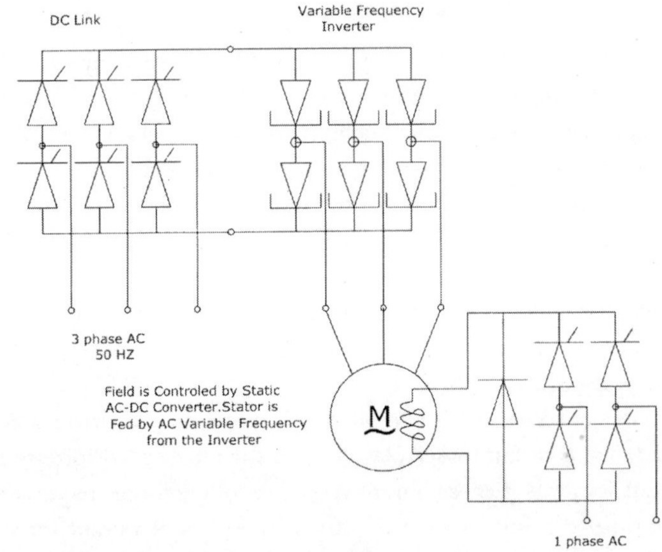

Fig. 4.71 Large synchronous motors with variable frequency converter used for tandem hot strip mills

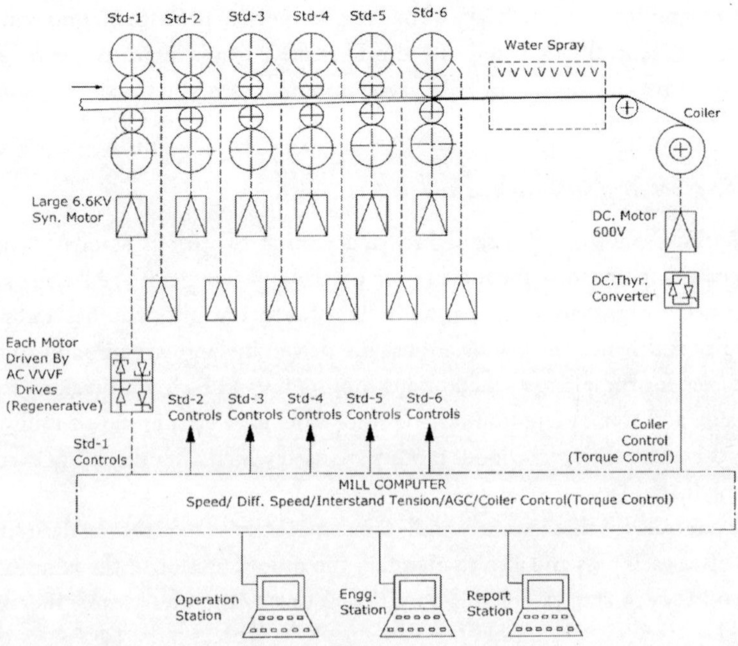

Fig. 4.72 Typical electrical scheme for 6-stand tandem hot strip mill

Hydraulic AGC (Automatic Gauge Control) for each stand is employed to maintain proper gap between working rolls. For this control system sophisticated Roll Load measurement system using load cells under each stand are employed. This is formed into the Mill computer and the whole mill runs in a coordinated manner. The winder control is also a part of the mill computer where *torque control* method is adopted for a continuous strip speed and strip tension. Figure 4.72 shows such a control scheme.

Other than the main mill drives, a large no. of auxiliary drives mainly AC and DC driven motors are used.

D.5 Steel bar and rod mills electrics and control

These mills use a large number of stands consisting of roughing stands, intermediate stands and finishing stands. The roughing stands generally use Asynchronous AC slip ring motors of medium range and the intermediate and finishing stands use DC motors of medium duty.

The starters for the roughing stands AC slip ring asynchronous motors usually comprise of a LT or HT rotor starter such as liquid rotor starter. For the intermediate and finishing stand DC motors, generally a 12 pulse individual *thyristor converter* is used. These drives are also in tandem in nature.

Other than the main mill drives a large no. of auxiliary drives mainly AC and DC driven motors are used. These are used for roller table, cooling bed, transfer mechanisms and shears, and so forth.

D.6 Steel wire rod mills electrics and control

These mills generally have the similar configuration as that for bar and rod mills, but these could be of many strands (3 or more) and shall be with number of finishing blocks as per the number of strands. The output of these mills goes to coiling stands and has a very high speed at the finishing. These mills are of continuous nature and mainly use DC motors for the various stands. For the stand DC motors, generally a 12-pulse individual thyristor converter is used. These drives are also in tandem in nature.

Special controls are required for the coilers for these mills which are usually DC motors. Apart from them, a large no. of auxiliary drives mainly AC and DC driven motors are used.

For the bar and rod mills as well as for wire rod mills, as in the strip mills, a mill computer with at least three hierarchies is used. This mill computer gives out speed set points for the various mill drives as well as control many auxiliaries like flying shears, cooling bed and the coilers.

While speed control method is adopted for the mill drives where voltage is proportional to the strip speed, a current control is used for coiling where the current is proportional to the tension.

E. Cold rolling mills (steel)

E1 *Cold steel strip mills electrics and control*

These mills generally are used to produce finished cold steel strips and have a main reversing stand and two coiling machines and use various passes between the two coilers. By the rolling process, work hardening takes place and hence an annealing process needs to be undertaken after some passes. Generally these mills are very high speed mills and many sophisticated controls are adopted for the gauge finish of the steel strips.

The main stand and the coiling machines are driven by individual DC motors further powered by individual 12-pulse thyristor converters. The main stand drive uses a speed control method while the coiling machines use *torque control* method to maintain the back and front tensions of the strip.

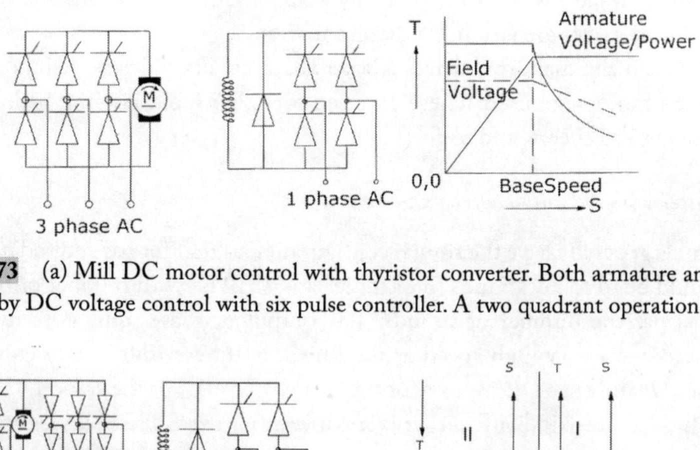

Fig. 4.73 (a) Mill DC motor control with thyristor converter. Both armature and field by DC voltage control with six pulse controller. A two quadrant operation

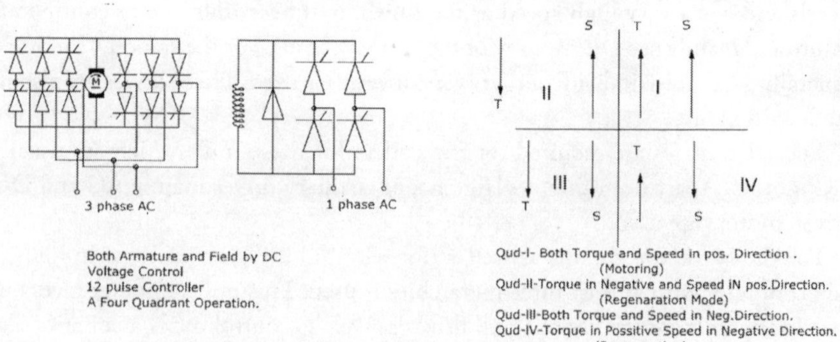

Fig. 4.73 (b) Thyristor Converter with 12-pulse controller

Apart from mill motors, a large no. of auxiliary drives mainly AC driven motors are used. These are used for hydraulic pumps , gauge control, shape control , etc.

The controls for these reversible cold strip mills are complex so as to meet the desired strip speeds after each pass and also to maintain the proper tension at the unwinding and rewinding end. For this a mill computer is used to get various process inputs and give out set points to the main three drives.

Hydraulic AGC (Automatic Gauge Control) is employed to maintain proper gap between working rolls for the desired strip thickness. For AGC system sophisticated Roll Load measurement system using load cell under the stand is employed. This is formed into the mill computer and the whole mill runs in a coordinated manner. The winder control is also a part of the mill computer where torque control method is adopted for a continuous strip speed and strip tension. While speed control method is adopted for the mill drive where voltage is proportional to the strip speed, a current control superimposed with torque control is used for winders where the current is proportional to the tension. The drive and control scheme for a reversing cold strip mill is shown in Fig. 4.74.

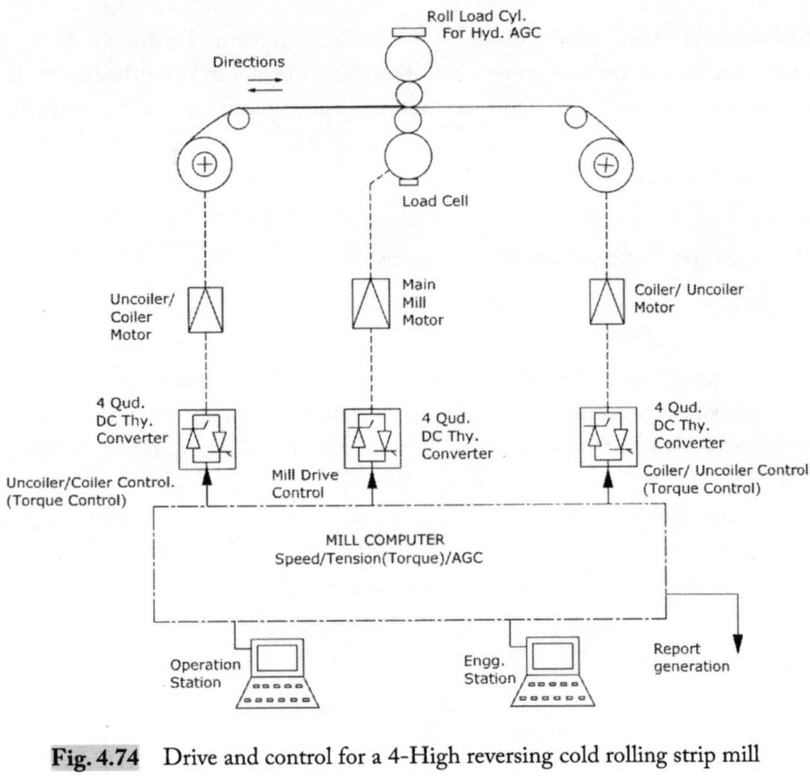

Fig. 4.74 Drive and control for a 4-High reversing cold rolling strip mill

E.2 Cold rolling mills for aluminum

The aluminum cold rolling strip mills are of three categories, Roughing, foil roughing and foil mills. All these mills are non-reversing. The roughing mills have similar controls like the steel cold strip mill while the foil rougher and the foil mills have different controls. The roughing mill has only a mill drive driven by a DC motor, the winders have also DC motors for simple winding.

The DC drives are powered by suitable thyristor converters. While the main drive is run on speed control, the winder drives are run through current control.

The foil rougher uses the same configuration of the roughing mill, so as to use speed control for the mill drive, the winder is driven through current control superimposed by torque control. However, the AGC for this kind of mill is of different type, using Hydraulic AGC for the gap mode and Speed/ tension control for the Dark mode.

The Gap mode is when there is a gap between the two ends of the work rolls, where gap adjustment is possible. Under Dark Mode the two ends of the work rolls touch each other and hence no further gap adjustment is possible.

In a foil rougher mill, GAP mode will be possible up to certain strip thickness and Hydraulic AGC can be used. But under DARK mode Hydraulic AGC will become ineffective and the gauge control can be implemented by either controlling the strip speed or controlling the back tension or a combination of the both. Refer to section 4.3.6 for detail discussion on hydraulic AGC for GAP mode and DARK mode gauge control.

4.3 Auxiliary Equipment and Systems

In a rolling mill, apart from the rolling stands, a host of other equipment are used which are called auxiliary machines or equipment. These machines perform different operations, to make the mill work as a high productive plant. Some of these auxiliary operations are: feeding stock into rolls and accept it after it comes out; manipulating, cutting, straightening, coiling, conveying, bundling, cleaning or degreasing, etc. of the intermediate and finished rolled material.

In modern mills all these operations are performed by mechanized and often automated equipment which have their independent drive and controls.

Some of the important auxiliary equipment are briefly discussed below:

4.3.1 Shears and saws

Several types of shears are used for cutting different types of rolled stock:

A. ***Shears with parallel blades***

In these shears the cutting edges of the shear blades are parallel to each other (Fig. 4.75). These are generally used for cross cutting of heavy sections of hot metal like bloom, slabs, billets, and similar square and rectangular shapes, which are not in motion. These are also used for cold cutting with shaped blades which have the same shape of the metal being cut, e.g. angles, channels, etc.

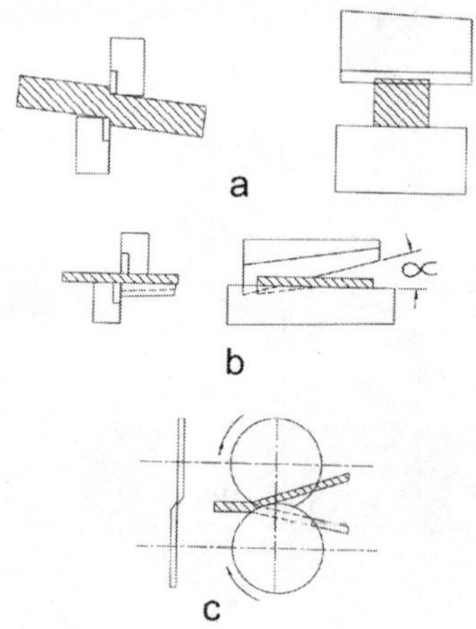

Fig. 4.75 Arrangement of shear blades

a- with parallel blades; b- with an inclined blade; c- rotary shears

There are three main types of these shears:

(i) *Down-cutting shears* in which the top blade moves and the bottom blade remains stationary.
(ii) *Up-cutting shears* in which both blades move but cutting action is done by the up movement of bottom blade.
(iii) *Shear with rotary cutting motion*:

Each of these shears can further be classified based on their general structure like: *closed-type* shears in which the main housing is on both sides of the blades (usually these are big shears), and *open-sided* shears, in which the housing is on one side of the blade (usually smaller shears).

Shears with parallel blades are operated with either electric or hydraulic drive.

Electric shears are convenient to service and are usually of higher speed. Small to medium sized shears are operated by crank mechanism driven from induction motors fitted with flywheel. The motor run continuously and the blades are operated intermittently by means of a clutch.

Hydraulic shears are simpler in design, more powerful and at the same time guaranteed against overloading and breakdown by limiting the maximum oil pressure.

Down cutting shears, when cutting material thicker than 30 mm, the table placed behind the shear lowers itself during the cutting action of the top bladed holders. Figure 4.76 shows the half sectional view of a crank operated down-cutting shear.

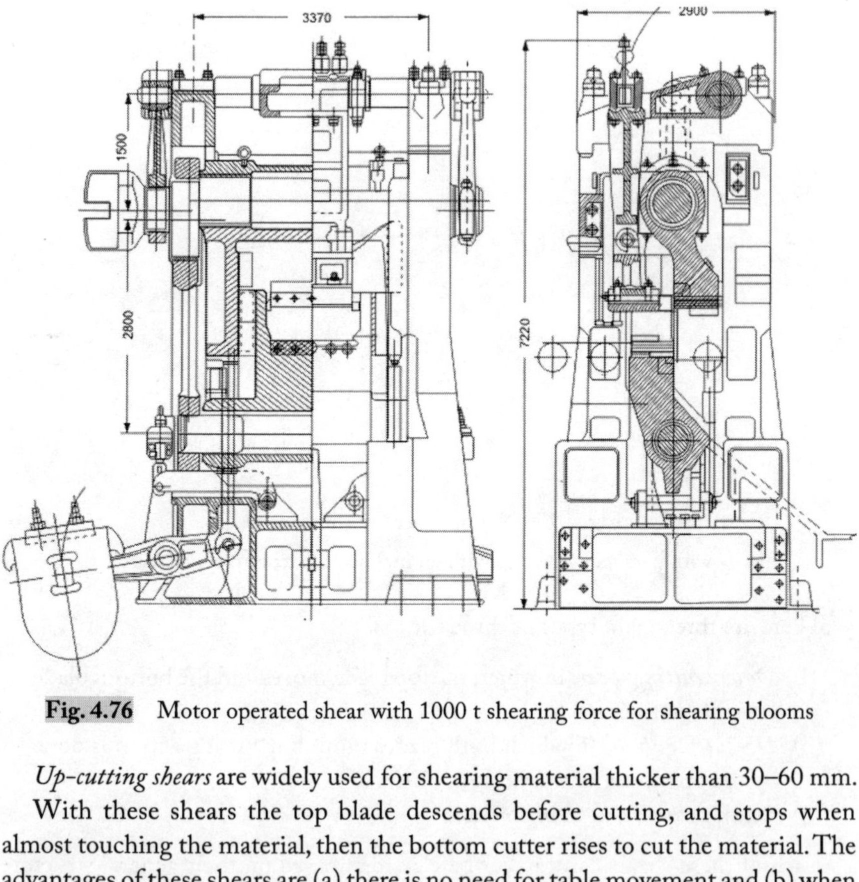

Fig. 4.76 Motor operated shear with 1000 t shearing force for shearing blooms

Up-cutting shears are widely used for shearing material thicker than 30–60 mm.

With these shears the top blade descends before cutting, and stops when almost touching the material, then the bottom cutter rises to cut the material. The advantages of these shears are (a) there is no need for table movement and (b) when cutting long length the part of the pieces on top blade side does not bend. These shears are generally closed type.

Blades for parallel blade shears are usually made symmetrical in shape so that all four edges can be used. These are made from tool steel, properly heat treated and ground. Blades are bolted to holders that move in guides of the shear body, and also protect the blades from overheating form the heat of the piece being sheared. A positive gap of about 0.5 mm is kept between two blades.

B. ***Shears with inclined blade***

These shears are used mainly for cutting sheets, skelp, and sheet bars, etc.

The angle between the blades is chosen depending on the thickness of metal to be cut. It is not more than 10°–12° to avoid the metal being pushed out of shears. For sheet shears, angle is kept within 1–6 degrees. Shears with an inclined blade may be of close or open sided construction with either moveable top or bottom blade.

Shears with moveable bottom blades are usually closed type, used for on-line cross-cutting long lengths of sheet or strip. The bottom blade is generally kept horizontal while top blade has the inclination.

Shears with moveable top blade are made with inclined top blade and generally used for off line shearing of sheets and trimmed edges.

C. ***Rotary shears or slitters***

Rotary shears are used for lengthwise continuous slitting or edge trimming of long strips. Shears used for edge trimming are generally equipped with side scrap cutters for easy disposal of the trimmed portion.

The rotary shear consists of two circular blades with a small overlap and rotated in opposite directions. The comparatively thin strip is fed through the two blades and is slitted at the junction of two blades.

The diameter 'D' of the blade can be approximated from the formula:

$$D = \frac{h + s}{1 - \cos\alpha}$$

where,

h = strip thickness,

s = blade overlap that is kept within 1–3 mm.

α = permissible angle of bite= 10–15 degrees.

In practice 'D' is kept within a limit of D = (40–125) h; when 'h' is smaller, greater value of the coefficient is taken.

The side clearance between two blades is kept within (0.1–0.2) h; and thickness of each blade is kept within (0.06–0.1) D.

Normal design of these shears involves two driven shafts one on top and one on bottom. The circular blades with machined bores and key ways are inserted into the two shafts with sliding keys on it. Requisite number of top and bottom blades is slided axially into the shafts and held in position by putting spacers in between. In certain design of slitter the top shaft is fitted on pivoted arm which allows the engagement of the shear after the strip is threaded through the shear.

In certain design a pinch roll unit may be fitted in front or back of the shear to push the strip into the slitter or pull the shitted material through the rotary shear. Fig. 4.77 shows a rotary 2 shaft multi-knife shear with front pinch roll.

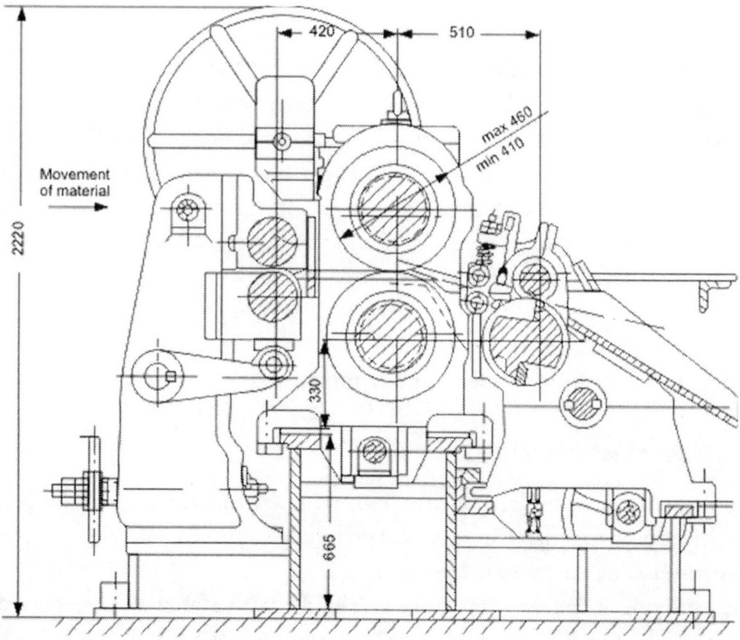

Fig. 4.77 Dimensioned drawing of a Multi-knife rotary shear with side scrap cutter for cold shearing of maximum 6 mm strips / sheets up to 1500 mm wide

D. *Flying shears*

Flying shears are used in rolling mills for cross-cutting the rolled stock in moving condition. These are used on line or off line.

Different designs of flying shear have been developed and used over time. However, the most successful and widely used flying shears are the *two-drum shears* also simply called *drum shears.* These are often mounted behind a continuous mill for hot cutting of strip up to 12 mm thick and for light sections into smaller pieces of predetermined length.

The two-drum shears consist of, as the name implies, two metallic drums of suitable construction, on the periphery of which, one straight blade on each drum is fixed along the length of the drum axis. The drums are connected by gear on one side, such that both can rotate simultaneously in opposite direction. The blades are so adjusted that during rotation with drums, they come together very close when both of the blades are in same vertical plane one above other (see Fig. 4.78). Rolling stock when pushed by the rolls of preceding mill stand or by a feed roller, is moving through the drums of the shear and if the drums are rotating, the moving pair of blades also moves and gets into the moving stock both from top and bottom and shears the stock.

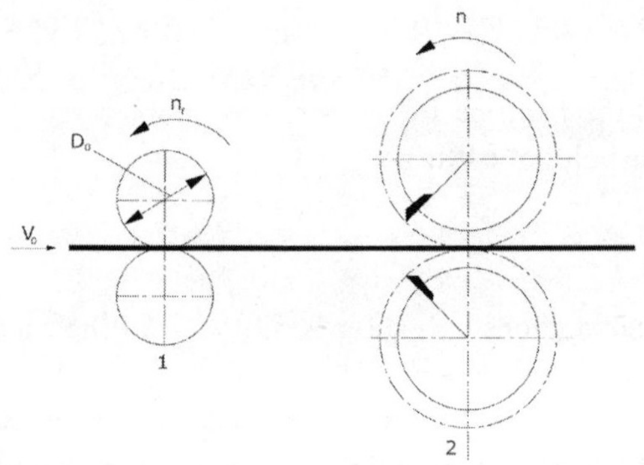

Fig. 4.78 Drum shear with feed rollers 1 and shear 2

It is important that the speed of the drum should be so adjusted that the surface speed of the blade tips is marginally higher than speed of the stock.

If the speed 'v_o' of the stock is constant, the length of cut pieces depends on the time gap, 't', between successive cuts by the relation:

$$L = v_o\, t.$$

To cut a definite length of the material, the shear is stopped after each cut, and again started before next cut is to be performed. Quick starting of heavy shears with large inertia often does not succeed in attaining the speed of rolled metal and then stopping in their initial position after the cut. In this case, instead of the usual shear movement (Fig. 4.79a), a shear movement is employed (Fig. 4.79b) which allows longer time for running up and braking.

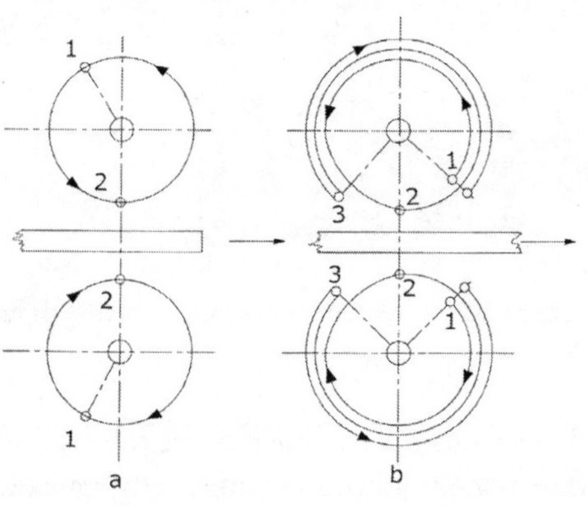

Fig. 4.79 Diagram of the movement of knives in two-drum flying shears operating from stationary position

a- simple; b- complex

With continuously running drums, the cut lengths can be changed by using different diameters of the drums or different speed of the two drums. In both these cases, complicated mechanisms are necessary for momentary speed equalization, of the blades at the point of cutting.

E. *Hot saws*

Cutting complicated sections by a shear is difficult and the cut is not very clean, even using shaped knives (blades). For clean cut of medium to large shapes including tubes, rails, beams, rounds, etc. *hot saws* are used. Apart from usual stationary saws, flying saws are used in continuous mills, where saws move with the same speed as the product while it is being sawed.

The normal hot saw consists of a circular rotary saw with sharp cutting teeth at the periphery of the saw disc, used for hot cutting.

Friction saws with small blunted teeth on a disc are generally used for cold cutting of metals. In small size friction saws, thin grinding discs are also used.

Different designs of hot saws have been built suiting specific requirements. The most popular of them is a *sliding frame saw*, as shown in Fig. 4.80.

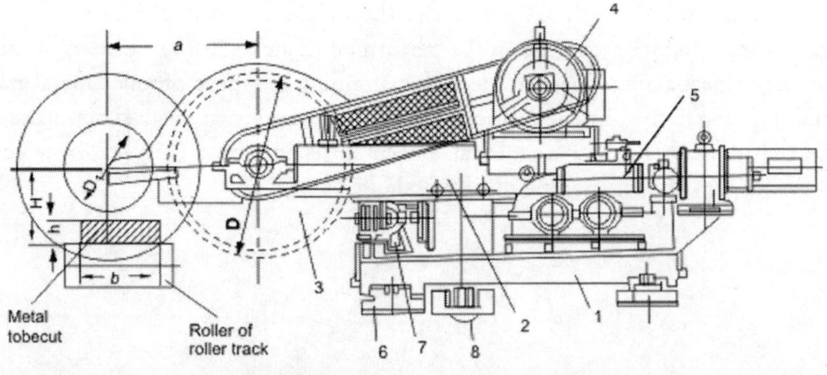

Fig. 4.80 Sliding frame circular saw

1- housing; 2- blade frame (sliding carriage); 3- blade; 4- blade motor; 5- blade feed mechanism; 6- guide rails of saw movement along roller table; 7- motor; 8- rack and pinion

The cutting blade 3 is mounted on a frame 2 which is moved towards the roller table by a feed mechanism 5 on slideways or rollers. For blade life it is very important that the direction of saw movement exactly coincides with the plane of the blade. In this respect slideways are less satisfactory as they wear. For this reason it is advisable to make saws with a roller movement using side rollers as well to prevent side movement. The saw assembly is made to traverse along the moving stock on guide rails, 6, by a separate motor, 7 through a rack and pinion, 8.

The diameter of the circular saw is determined by the size of sections to be cut. Saw diameter in the range of 1000–2000 mm is quite common. The thickness 'S' of the saw can be estimated from the relationship

$$S = (0.18\text{–}0.2)\sqrt{D}\text{, where } D = \text{saw diameter in mm.}$$

Saws are made from alloy steel with ultimate bending stress of 80–95 kg/mm^2 and 12–13% elongation. Among various shapes, tooth profile as shown in Fig. 4.81, called dog teeth, are quite popular. The width of a tooth are made larger towards its point, to improve cutting. Water spray is used for cooling of saw blade and to wash away metal chips from the saw teeth.

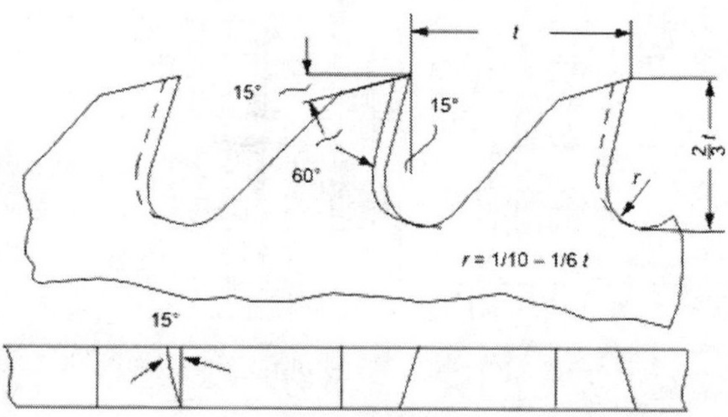

Fig. 4.81 The most common type of saw tooth for hot cutting of steel

4.3.2 Straightening and leveling machines

Final product on delivery from the mill or after cooling gets little bent or distorted. Straightening machines are used offline for making the products straight.

The straightening machines are classified into four types based on their principle of operation and design features:

(i) Straightening press
(ii) Roller straightening machines
(iii) Rotation straightening machines, and
(iv) Stretch straightening machine, also called stretcher leveler

A. ***Straightening press***

Use of these are limited to the straightening of beams, rails and tubes of diameter more than 200–300 mm. Straightening of the metal takes place by pressing the bent portions by the movable top block tool while supporting the metal on two bottom supports. Straitening presses are generally of vertical open structure and crank operated. The tools and supports are changed to suit the shape being straightened.

Fig. 4.82 shows a typical straightening press with 500 T pressing capacity. These presses can also be of horizontal design.

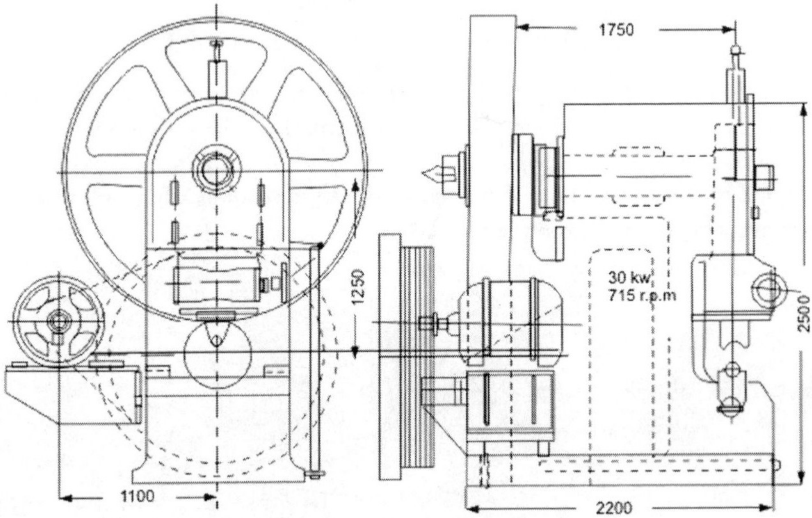

Fig. 4.82 500 T straightening press

B. ***Roller straightening machine***

These machines consists of a number of parallel rollers arranged in two levels, between which sections and flat products like plates, sheets and strips are passed. The gap between the two layers are so adjusted that the passing metal is subjected to successive bending, usually more sharply between the first rollers and less sharply in subsequently ones. The principle is exhibited in Fig. 4.83.

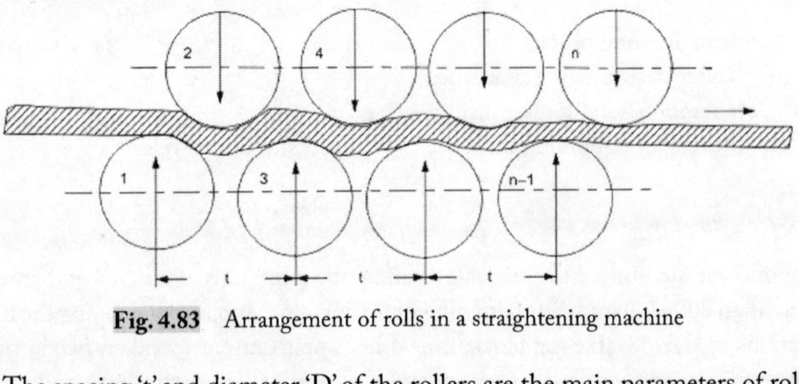

Fig. 4.83 Arrangement of rolls in a straightening machine

The spacing 't' and diameter 'D' of the rollers are the main parameters of roller straightening machine. In practice the relation between them is: D = (0.9–0.95)t

The range of roller spacing, 't', can be theoretically shown to range between:

$$12h < t < 240h,$$

where
'h' is the metal thickness. $t \cong 60h$ is considered to be an optimum spacing.

Leveling machines are a special group of roller straightening machines where considerably smaller spacing and diameter of rolls are used for straightening thin range of sheets and strips. For wide materials, the small bending rolls are generally backed-up by larger rolls. In machines with very small bending roll diameters (25–40 mm), the backup rolls are made with large diameters, and are arranged in chequerboard fashion (Fig. 4.84). The number of bending rolls at top row is usually one more than in the bottom. The leveling machines are classified based on the method of adjusting the top bending rolls.

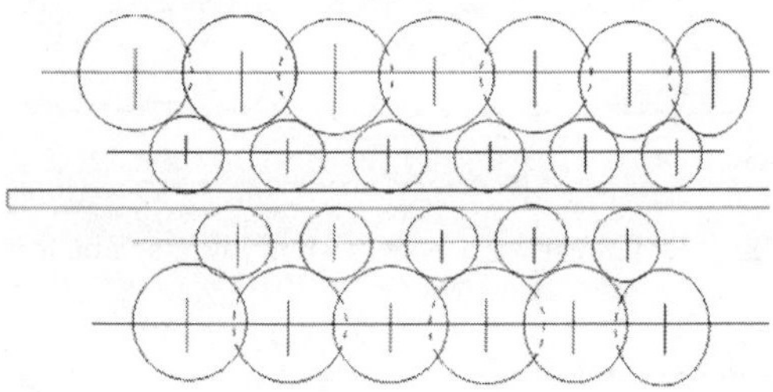

Fig. 4.84 Chequerboard arrangement of backup rolls in relation to bending roll

Figures 4.85 and 4.86 show the side view and front view, respectively, of a 8-roll heavy section roller straightening machine.

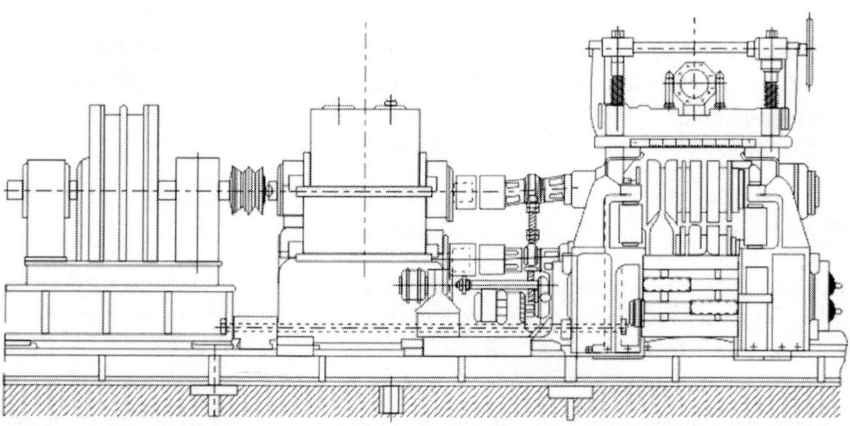

Fig. 4.85 8-roll heavy section straightening machine side view

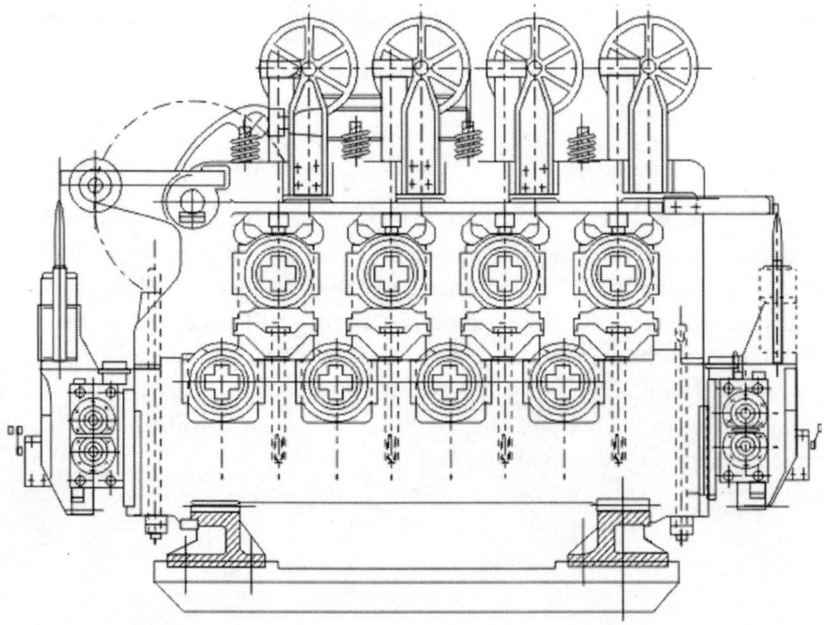

Fig. 4.86 8-roll heavy section straightening machine front view

C. *Rotation straightening machine*

These machines are mainly used for straightening tubes and round rods with higher degree of straightness than would be obtained in straightening press machines with parallel rolls.

In these machines the circular section material is helically advanced through obliquely arranged rolls, some of which are rotated at speed which determines the linear speed of the product. Because of the multiple bendings of the metal between the rolls, the metal continuously changes direction. This ensures the axial symmetry of straightening.

The principle of these machines is schematically shown in Fig. 4.87.

D. *Stretch straightening machine (or stretcher leveler)*

In these machines small cross section products, particularly of non-ferrous materials (like aluminum sections), of definite lengths are straightened by stretching the product held between one fixed and another moveable jaws. The movable jaw head is placed on a sliding base structure and pulled by hydraulic cylinder or by screw and nut mechanism. The straightening takes place by stretching the piece nearly to its yield point.

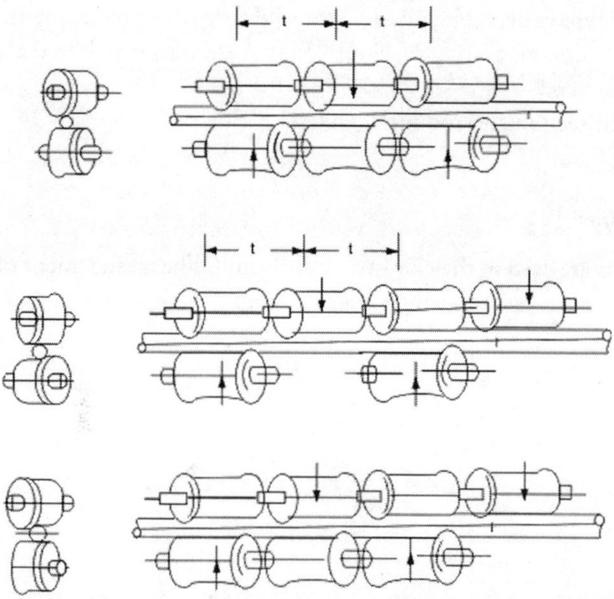

Fig. 4.87 Diagram of rotation straightening machines with rolls set obliquely (driven rolls have rotary arrow; rolls taking part in the straightening process have vertical force arrows.)

It should be noted that the principle of leveling by stretching is also used to a certain extent in cold rolling when tension is applied to the rolled strip by the ***coiler,*** which eliminates strip distortion to a great extent.

4.3.3 Coilers and decoilers

Machines for winding rolled strip into a coil are called *coilers* or *winders*. These are placed in line with a strip rolling mill (hot or cold). In a reversing cold strip mill, two such coilers are placed on both sides of the mill, while one is acting as a coiler the other feeds the strip into the mill, and is termed as *decoiler* or *unwinder*. They reverse their role as the mill reverses.

When a fresh coil of strip is brought to the cold rolling mill, it is accommodated in a machine called *pay-off* unit from which the strip is fed to the mill for the very first pass of a fresh coil.

The light sections like flats and rods from a hot mill can also be made into loose coil shape by using a class of coilers called *rod coilers.*

The three main types of coilers are:

(i) Drum type tension coilers onto which rolled metal strip is coiled under tension

(ii) Roll type coilers for coiling hot or cold strip or skelp coming from a mill or continuous casting machine (for non-ferrous metals and alloys) into a loose coil

(iii) Light section and rod reels (coilers)

A. ***Drum coiler***

These coilers are used in the cold strip rolling mill. The arrangement of coilers and pay-off unit in a reversing 4-high cold strip mill is shown in Fig. 4.88.

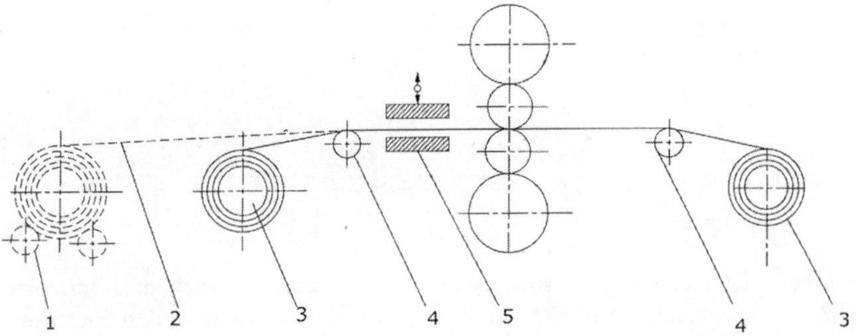

Fig. 4.88 Arrangement of tension coiling drums in a reversing cold rolling mill

1- pay-off; 2- position of strip during first pas; 3- drum coilers; 4- idle supporting rolls; 5- brake blocks (or tension bridle)

During rolling, the strip is coiled and decoiled under constant tensions which helps to keep the rolling force reduced and at the same time the strip is stretch leveled and the coil is accurately formed on the drum.

Matching the rolling and coiling speed and maintaining the tension constant while the diameters of both coils on coiler and decoiler are changing is maintained by proper selection of D.C. motor and its control. The decoiler motor goes to regeneration mode returning energy back to electric circuit.

The most important part of a coiler / decoiler is its drum. After completion of rolling schedule, the finished coil has to be taken out from the drum. This is achieved by collapsing the drum diameter. Therefore, the essential feature of such drum coilers is the facility for *expanding and collapsing* of the coiler drum. At the beginning of coiling, the drum is expanded, and when coil has to be taken out it is collapsed. Another important feature is arrangement for gripping the front end of the strip in the drum before coiling starts. In order to keep necessary strength of these drums under coil weight and strip tension, the drum diameters needs to be large. Expanded drum diameters have been standardized to 16 in. (406 mm) and 20 in. (508 mm). Metallic cylindrical cores, used for coiling of thin/foil gauges, are mounted on these drums and held in position by expanding the drums.

In compact modern design, the coiler drum is overhung from coiler body. The extended drum shaft is mounted on heavy antifriction bearings and act as the output shaft of the coiler gear box. The coiler DC motor is connected to the input shaft which is connected to the drum shaft through double helical gearings of required ratio. A general view of such an overhung coiler with expanding mandrel (drum) is shown in Fig. 4.89(a).

Fig. 4.89 (a) General view of coiler with overhung expanding mandrel

Typical specification and constructional details of a collapsible drum of a drum coiler is given below:

Drum design	: 3 segment, concentric collapse with strip gripping facility.
Drum actuation	: Double acting rotary hydraulic cylinder
Drum diameter	: Expanded 508 mm Collapsed 496 mm
Drum face length	: 530 mm (for upto 425 mm strip width)

The coiler drum is of the concentrically collapsing type and comprises three forged steel segments mounted on a central shaft and operated by three pairs of opposed wedges. The outer wedges are located in radial slots in the drum shaft and the inner wedges are connected by a forged steel spider and thrust rod assembly to a rotary hydraulic cylinder, mounted on the drum shaft at the rear of the coiler. The thrust rod assembly passes through and is supported in the drum shaft central bore.

The drum segments are secured to the coiler shaft by studs fitted with Belleville washer assemblies. These assemblies ensure positive collapse of the segments when the inner wedges are withdrawn by actuation of the rotary cylinder.

To eliminate marking when coiling thin strip the bronze inner wedges and the spheroidal graphite iron outer wedges are designed to completely fill the gaps between the segments with the drum fully expanded [see Fig. 4.89 (b)].

A rubber gripping element is incorporated in one of the outer wedges and operates automatically with the expanding and collapsing of the drum. The grip closes when drum is expanded and opens up when drum is collapsed.

The coiler drum shaft is extended into a totally enclosed fabricated steel base and cover which accommodates the reduction gear drive. The drum and gear shafts are supported on roller bearings which, together with the gears, are forced oil lubricated by a motor driven pump from a base sump.

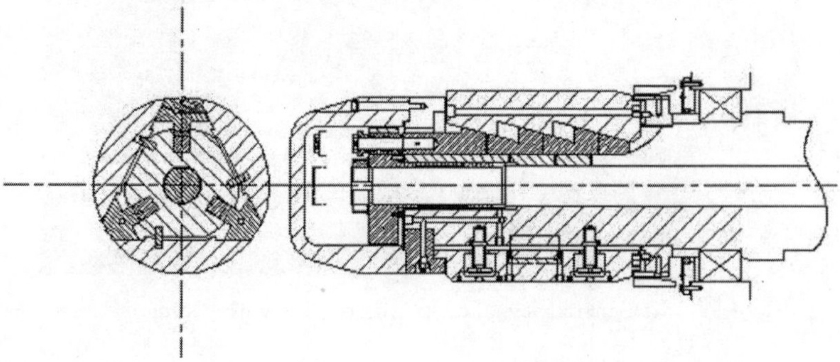

Fig. 4.89 (b) Sectional view of an expanding /collapsing mandrel of a coiler

B. Roll type coilers

These machines are used for hot and cold coiling of strip from 1.5–6 mm or more thick. The incoming strip is formed into a loose coil by bending the strip between rollers, as is done in roll bending machine.

In these coilers no speed regulation is necessary as coil diameter builds-up. Some of the typical designs of such coilers are:

(i) Three roll machine also called upcoiler
(ii) Multi roll
(iii) Roll and drum

Figure 4.90 shows a roll type coiler with three wrapping rolls installed at the end of a continuous coil build-up line for cold coiling strip 1.5–6 mm thick and up to 2300 mm wide with coil weight up to 25 ton. The strip is fed by pinch rolls, between solid wrapping roll 2, 3, 4 coming out from which it forms into a coil on top of rolls 5 and 6. The plate 7 is gradually moved to the right during coiling for gradually increasing the coil diameter. All the rolls are driven. The two bottom rolls are adjusted for height. Both edges of coil are guided by rollers 9 mounted on moveable plates, 10. When coil is finished, 7 is lowered, and the finished coil is ejected by the pneumatically operated pusher, 11.

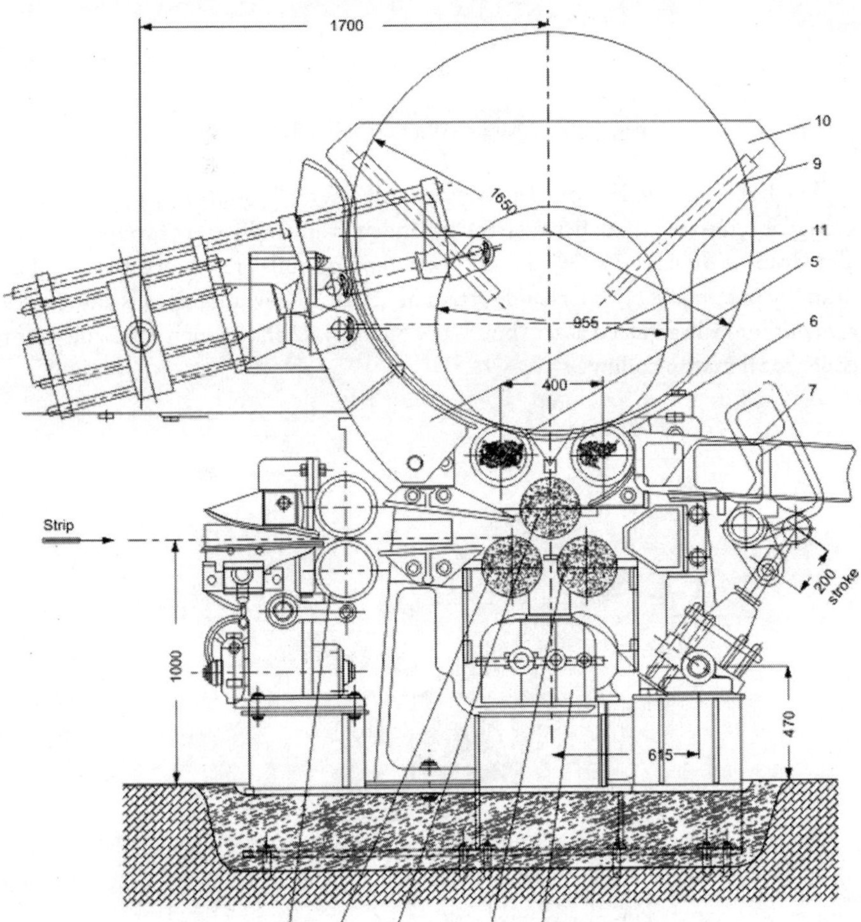

Fig. 4.90 Roll type coiler with three wrapping rolls (upcoiler).

Multi-roll coilers are only used for coiling of hot rolled strips. Schematic view of a multi-roll down coiler is shown in Fig. 4.91.

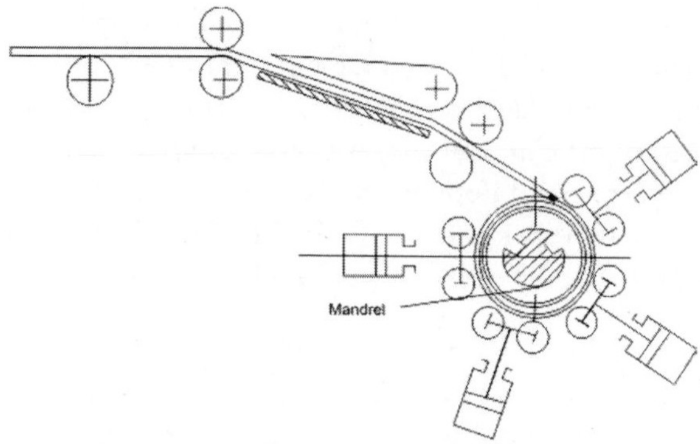

Fig. 4.91 Diagram of multi-roll down coiler

The hot strip is fed by the driven pinch roll assembly, guided by the adjustable strip guide, into the space between the expanded drum and series of wrapping rolls. The drum is driven by a DC motor. The wrapping rolls are pressed to the coiler drum by pneumatic cylinders and system of levers. The wrapping rolls and guides are continuously adjusted as coil builds up. Figure 4.92 shows such a roll and drum coiler for hot strip coiling.

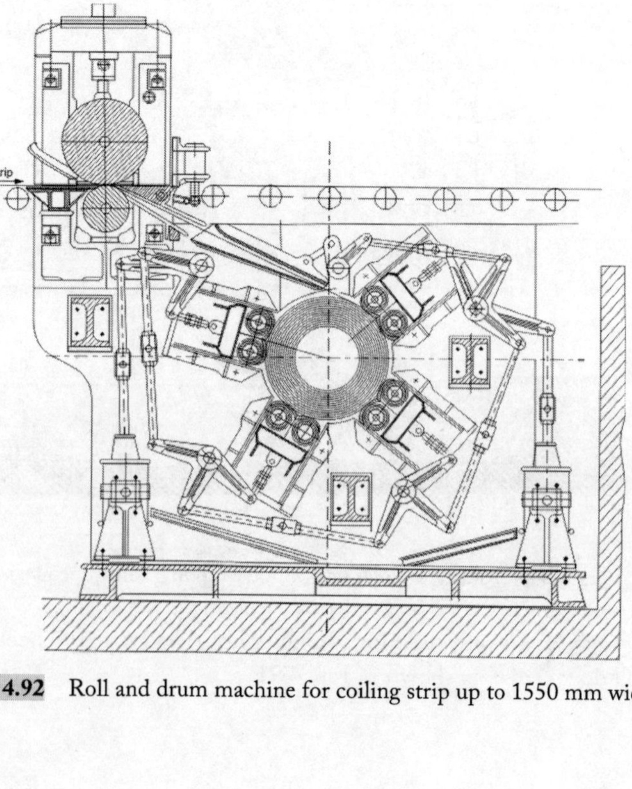

Fig. 4.92 Roll and drum machine for coiling strip up to 1550 mm wide

C. ***Rod reels or rod coilers***

These reels are used for hot coiling of:

(i) wire rod and small steel sections up to 20 mm and
(ii) narrow bands and strip up to 3–10 mm thick.

These rod reels are placed directly at the end of mill stand. Depending on the sectional shape of the metal, there are two types of reels available:

(a) *Laying reels* (also known as Edenborn type reel coilier) which are only used for round metal which can be twisted through 360° with each revolution of the reel. The laying reels may be vertical or positioned at an inclination. Figure 4.93 shows schematic diagram of a vertical laying reel.

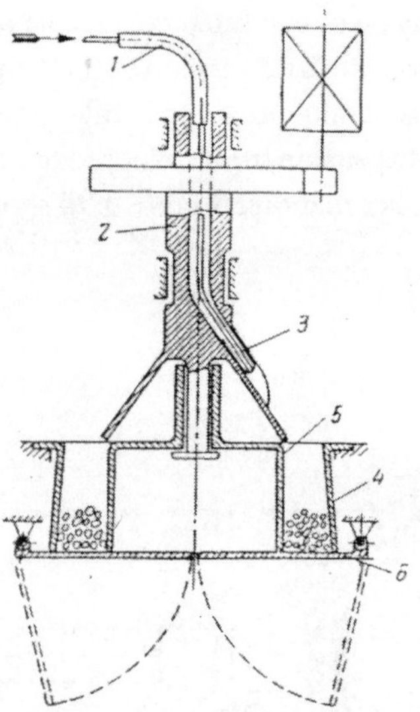

Fig. 4.93 Diagram of laying rod reel, Edenborn type

The main advantage of these reels is that it allows any coiling speed up to 25 /sec or more of the rod. The rod coming from the mill passes through a tube, 1, a hollow rotating spindle, 2, and a spiral tube, 3 on the cone shaped part of the spindle, and lays itself in turns on the annular space between cover, 4 and drum,5. Coils are dropped onto a moving slat type conveyor by opening flap, 6.

When the laying reel is positioned in an angle with the vertical, the bottom part of the reel consisting of the drum, cover and the flap is dispensed with.

The rod comes out of the rotating spiral tube as circular coils, which are allowed to fall on a moving steel slat type conveyor. These inclined laying reels are also known as *laying head.* The diameter of the coil can be controlled by changing rotational speed of the laying head with respect to finishing speed of rod.

(b) *Pouring reels* effect coiling without twisting the metal, and therefore, used both for rod and light non-round section. In a Garret type pouring reel, the metal directed through a tube enters an annular space tangentially, which is also rotating. After coiling is finished the coil is ejected out.

D. ***Drum type pay-off unit***

These are for paying off the coils after they are built up in a roll coiler, or after annealing a coil. They may be of overhung collapsible mandrel design like a drum coiler discussed earlier. To create back tension in the strip during paying off, a water cooled slipping brake is generally provided. Collapsible mandrel is made of four segments and its range of expansion from collapsed to expanded dimension is more than that provided in a coiler mandrel. Figure 4.94 shows such a unit.

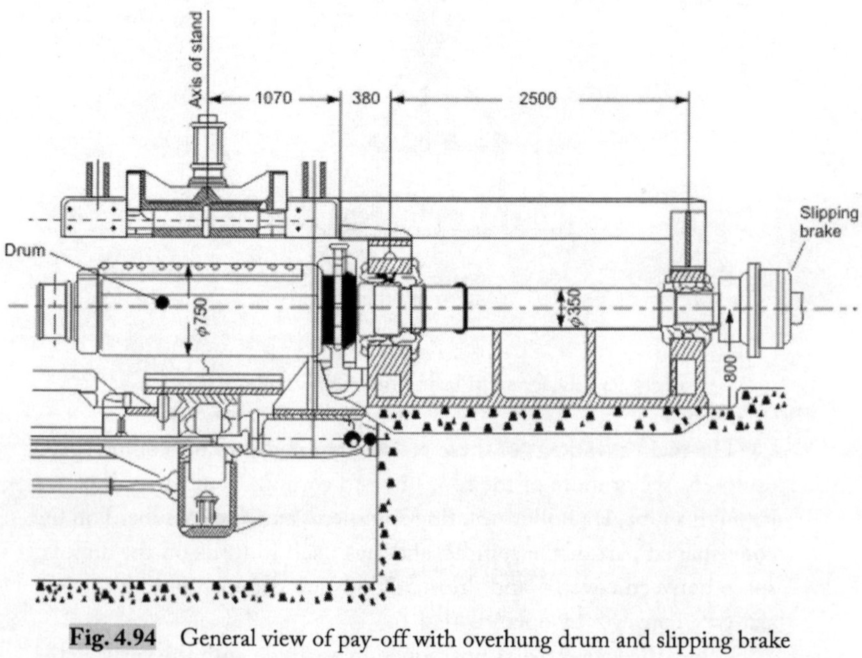

Fig. 4.94 General view of pay-off with overhung drum and slipping brake

In a *double cone type mandrel* design, the conical drum has more range to fit loose coil I.D. Double mandrel payoff unit is shown in Fig. 4.95.

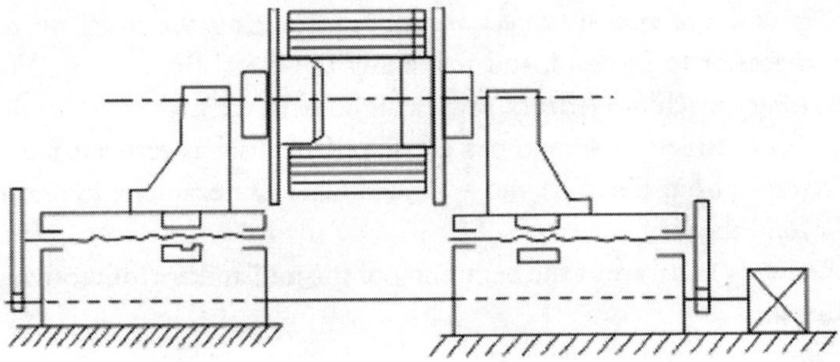

Fig. 4.95 Operation of double cone decoiler

The coil is held on its belley on a coil car, pushed up to right height by the coil car jack (hydraulic) and the two conical mandrels mounted on moveable structures, are actuated from two ends to grip the coil from both ends. Small deviation of bore diameter form coil to coil is accommodated because of conical shape of the two mandrels of the decoiler. A slipping brake is used to one or both of the mandrel shafts to provide tension in the strip during pay-off.

A *decoil box* is used for paying off coils with large sized bores or whose bore diameter varies over an wide range. The variation of coil bore may be caused during taking out a coil from coiler mandrel when some inner layers may collapse or slip out, which have to be cut off the coil. The thick gauge strip coiled on an *upcoiler* have generally large and unequal bore which cannot be accommodated by a collapsible drum type or double cone type pay-off unit.

A decoil box consists of two free rollers on which the coil may be placed by overhead crane or by manually rolling the coil. The two sides and the back is covered by steel fabricated plates, forming the box. After the coil is put into the decoil box, a round bar is passed through the coil bore, and fixed in the side plates. This bar prevents the coil from jumping off the rollers when the strip in pulled by the mill.

4.3.4 Roller table

Roller tables are roller conveyors, which move material by the rotation of their rollers. They are usually the main means of moving metal in rolling shops, and their weight sometimes reaches 40–60% of the total weight of the mill equipment.

Classification of roller tables by their functions gives the following types:

1. *Main tables* are situated on both sides of a stand. The first rollers of these tables in primary mills are often located in the stand housing.
2. *Main extension tables* are continuations of the main tables and take part in the work when the length of the piece is greater than that of the main table.

3. *Approach and run-out tables* are for transporting the metal to reheat furnaces or to the mill, and for taking the metal from the mill to the auxiliary machines (shears, saws, staighteners, etc.).

   One table can sometimes act simultaneously as run-out table (for example, from the mill) and approach table (for example, to the shears beyond the mill).
4. *Receiving tables* are at the beginning of the mill and are for receiving the metal.
5. *Piling tables,* usually made with inclined rollers, are for moving the piece simultaneously along and across the table in order to form packs before or after the shears or to move the metal into pockets or onto cooling beds. These tables are fitted with end stops which can be lowered.
6. *Lifting and depressing tables* not only move the metal in a horizontal direction but also raise or lower it. They are used in three-high mills as *tilting tables* or *lifting tables.*

Design of roller tables: Roller-tables can have: (1) group drive; (2) individual or semi-individual drives, where each roller or each pair of rollers is driven by a separate motor; (3) idle rollers, transporting roller tables with idle rollers being made with slight inclination so the metal moves along them under the action of the force of gravity. For this reason they are called gravity tables.

Group drive is only used in tables working in particularly difficult conditions such as the transporting of massive short pieces whose weight is concentrated on a few rollers. They are also used in main and approach table of beam and plate mills, etc. Roller table are made up from separate sections, each with its motor, up to 6m in length and sometimes more. The sections of main table of breakdown stands are usually equipped with two motors, one of which is spare.

The rollers are generally driven through a line shaft with mitre (bevel gear) wheels. Figure 4.96 shows such a line shaft with bevel gears group drive arrangement for a heavy roller table. When the roller spacing is very small the drive is through intermediate idle mitre wheels.

Individual or semi-individual roller drive, where each roller or each pair of rollers is driven by a separate motor, are widely used for transporting long pieces whose length is spread over a large number of rollers and, in consequence, the advantage of group drive in the use of motor power is practically lost. The higher cost of the electric drive in these tables is fully compensated by their advantages in operation and in manufacture in view of the very great simplification of design. Besides the obvious simplifications in doing away with complicated gearing, it also becomes possible to use separate housings for each roller (or each pair) instead of the integral long and heavy frames. This is particularly advantageous with large roller spacing.

Three types of motor are in use: the usual secured by feet, flange mounted and special roller table motors with hollow shafts.

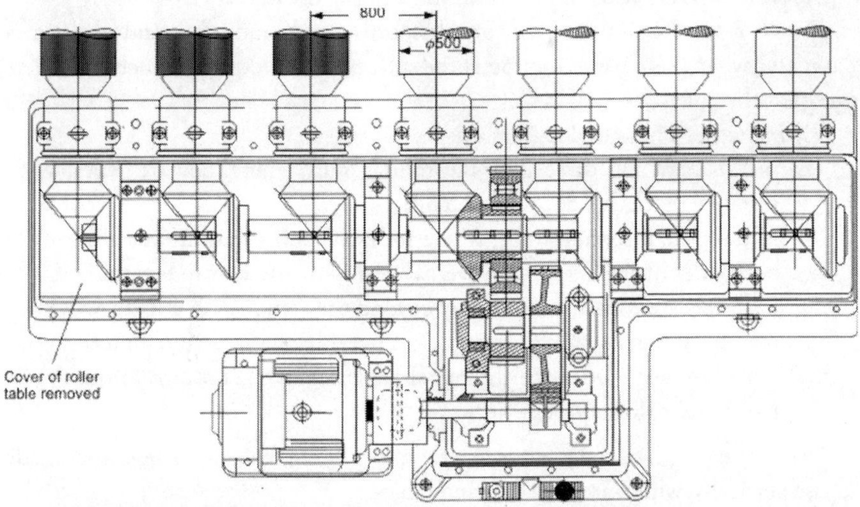

Fig. 4.96 Typical group driven receiving table of 1150 blooming mill

Figure 4.97 shows an individually driven roller through a flange mounted motor and geared coupling.

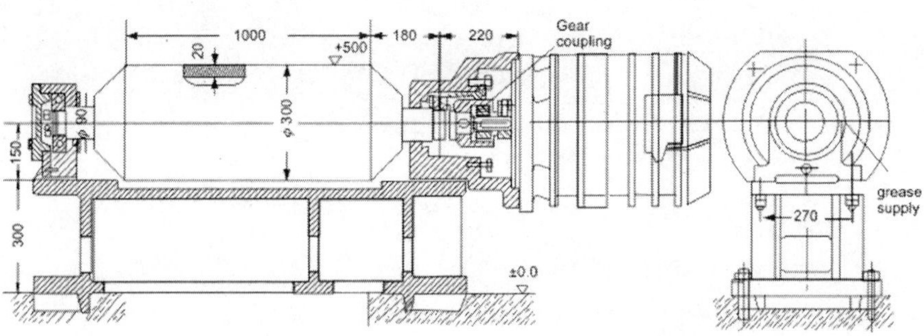

Fig. 4.97 A typical dimensioned drawing of a Roller with separate drive by flanged motor through gear coupling

Design of Rollers and their bearings: The shape of the roller barrel depends mainly on the function of the table.

Cylindrical rollers are the most common; they are used in both section and flat rolling mills.

Stepped rollers are used as the first (stand) rollers of primary mills; their diameter varies in relation to the depth of the passes in the rolling rolls.

Ribbed rollers are used in sheet and plate mills for main tables and for cooling beds in these mill; they are also used in heat treatment furnaces.

Tapered rollers are used in plate mill for rotating the plate.

Double-tapered rollers are used for transporting rods and tubes and sometimes flat metal, when it is desired to protect the surface by not letting it touch the rollers except at the edges.

Tilting tables. These tables are used:

(a) in two-high and three-high flat rolling mills, where they are generally installed on both sides of the stand;

(b) in three-high section mills, where for heavy mills they are installed on both sides of the stand; however when roll diameters are less than 600 mm they are usually placed only beyond the stand (with a lever or step manipulator in front of the stand). In three-high light section mills tilting tables are not used since the metal passes from the bottom to the top pair of rolls through repeaters.

When the length of the piece is less than 5–10 mm, tables are sometimes made to rise vertically, which are called lifting tables.

The mechanism used for tilting the table is usually crank or eccentric operated. The balancing of the lighter tables for two-and three high strip mills is often effected by springs and for heavy ones, it is done by weights (Fig. 4.98).

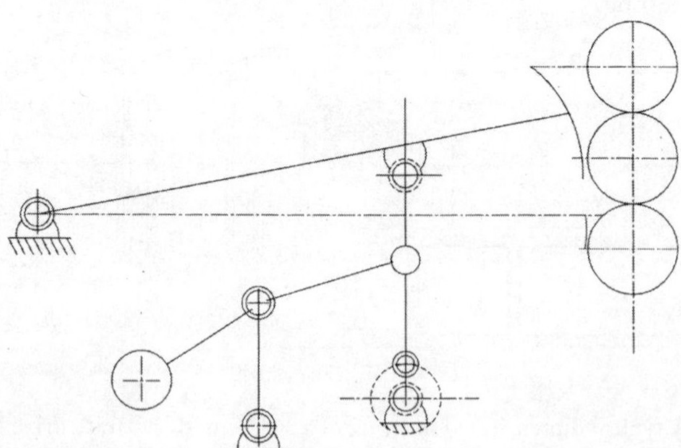

Fig. 4.98 Diagram of light tilting table with balancing arrangements by weight

The length of tilting tables is determined by two factors:

1. With a piece length of less than 10–15m the tilting table should also serve as a main table and its length should be not less than two-thirds the maximum size of the piece.

2. The inclination of the table (in the raised position). For the metal to enter the rolls reliably and for reducing its bending on passing to a horizontal table (which is important for sections), is made not more than 1:10 or below 6°.

In tilting tables all the rollers are invariably driven, having separate or group drives. With a group drive the motor is only mounted on the table frame if it is small. Usually the motor is mounted on the foundations in order to reduce the weight of the moving part of the table. Different schemes of roller drive is shown in Fig. 4.99.

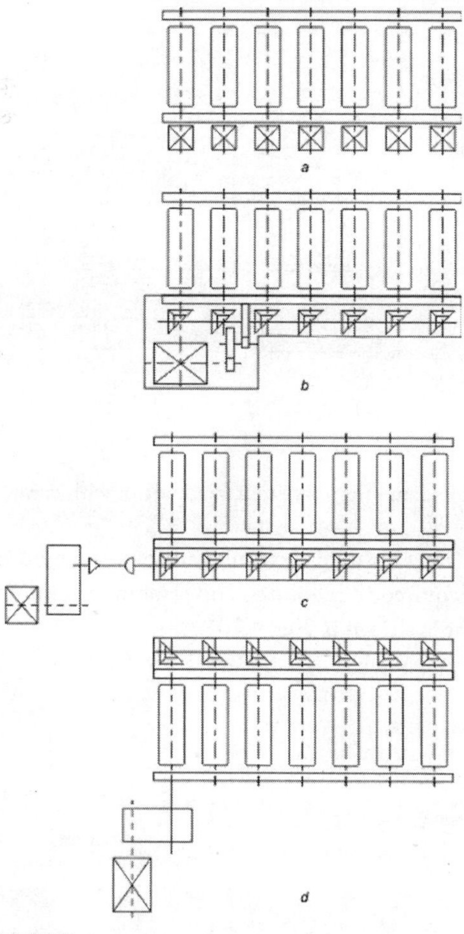

Fig. 4.99 Diagram of roller drives for tilting tables

a- separate; b- group, by a motor on the table frame; c- group drive motor at floor, through an universal spindle; d- the same, through the tilting axis of the table

4.3.5 Miscellaneous equipment

A. ***Manipulators and turning mechanisms***

Manipulators are equipment used over the main roller conveyor for moving the heavy stock laterally so that it is in the right position in relation to the mill rolls. Figure 4.100 shows the diagram of a primary mill manipulator.

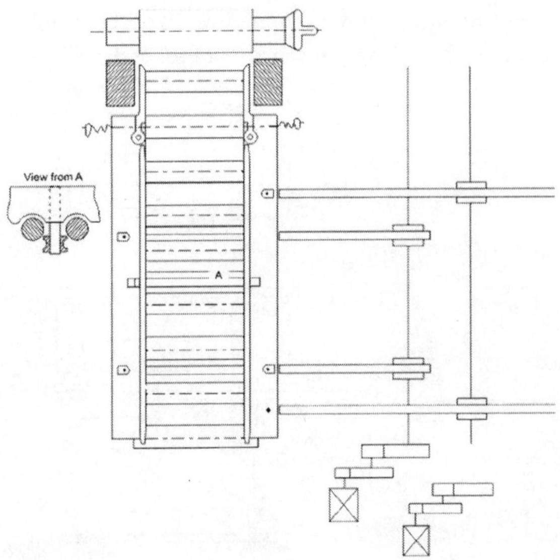

Fig. 4.100 Diagram of primary mill manipulator with drive located on side

Vertical manipulators are used for turning (or tiling) the stock through 90° about its axis. These are required in blooming and slabbing mills. Mechanism of such a turning manipulator is shown in Fig. 4.101.

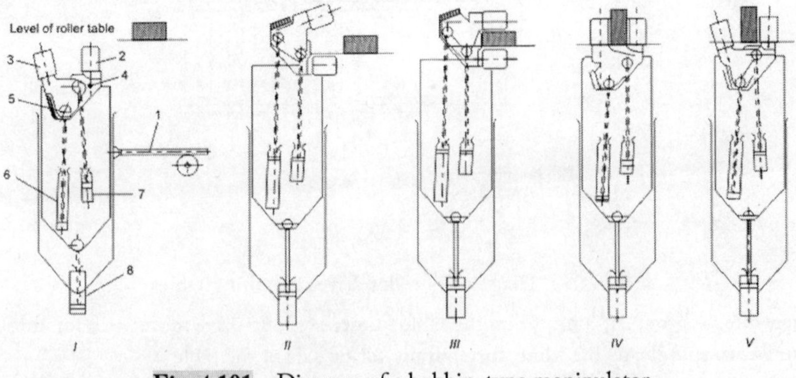

Fig. 4.101 Diagram of a bobbin-type manipulator

Position of manipulator: I- retracted; II-before gripping the piece; III-gripping the piece with simultaneous horizontal movement of manipulator; IV-tilting; V-separation of rollers. 1- travelling rack of manipulator; 2 and 3- tilting and clamping rollers; 4 and 5- pivoting points of tilting and clamping roller levers; 6 and 7- cylinders for operating tilting and clamping rollers; 8- cylinder for raising manipulator frame.

B. Cooling beds

Cooling bed is a structural platform on which cut to length hot sections/ rods are placed and allowed to cool when the metals are transported laterally, i.e., at right angles to its movement on delivery from mill rolls.

Different mechanisms used for these lateral movements are:

(i) cable transfer skids
(ii) pushers
(iii) movable rack
(iv) roller arranged at angle.

4.3.6 Automatic gauge and shape control system

A. ***Hydraulic Automatic Gauge Control (HAGC) system***

Maintaining strip thickness within a close tolerance throughout the length of the strip is a vital requirement of a hot or cold strip mill. It is important to first understand where the gauge error comes from during rolling and how it is tried to be controlled in a conventional screw down mill before discussing the automatic gauge error reduction system being incorporated in the modern high speed strip mills.

For the first pass through a strip mill, the roll gap is set by the screw down mechanism equal to the desired output thickness and the strip is introduced into the mill. As soon as the strip enters in roll bite mill housing gets stretched corresponding to the roll separating force. The gap between the rolls gets increased by this amount of mill housing stretch. The mill stretch is proportional to the stress in mill housing and length of the mill housing. The stress can be reduced by increasing the cross section of the mill housing, and the old mills were all built with large cross section mill housing which made them heavy and costly.

By measuring the thickness of the strip at output, an experienced roller can reduce the mill gap to an extent where the mill stretch will get compensated and the desired output gauge would be achieved. This condition will result in a strip with fairly close gauge tolerance provided the thickness of the input material remains constant, and yield stress value of the material is uniform. Any deviation of thickness in the input gauge, when comes to the mill bite, the corresponding variation in mill stretch will result in similar errors in output gauge. If a thickness measuring indicator is fitted, the operator can try to adjust the screw down to correct the error. As the response time of the operator as well as the screw gauge system is high, a substantial length of strip would have been rolled with the error. In a fast mill, such screw down gauge correction may actually result in greater deterioration of output gauge.

The hydraulic automatic gauge control (HAGC) system developed by mill builders guarantee very accurate gauge control of the rolled strip even at a high rolling speed. HAGC system consists of the following subsystems and procedures for a 4-Hi Rolling Mill:

(i) Two large hydraulic cylinders, called *roll load cylinders,* are fitted between the two housings and back-up chocks, at the top or bottom side, which replace the conventional screw down. The oil comes in or out of the cylinders from the hydraulic oil power pack through a high response electro hydraulic proportionate value (servo valve). LVDT type displacement transducers are fitted to each cylinder which can measure position of the cylinder ram up to one micron accuracy. Hydraulic pressure sensors mounted on the cylinders and/or load cells mounted between the housing and hydraulic cylinder calculate the roll separating force and covert it to electrical signal. One or preferably two accurate thickness measuring gauges are mounted on the exit and entry side of the mill. Scheme of a 4-Hi mill with roll load cylinders at the bottom side of the stand and a load cell mounted at the top of top back-up chock is shown in Fig. 4.102. The basic control scheme is shown in Fig. 4.103.

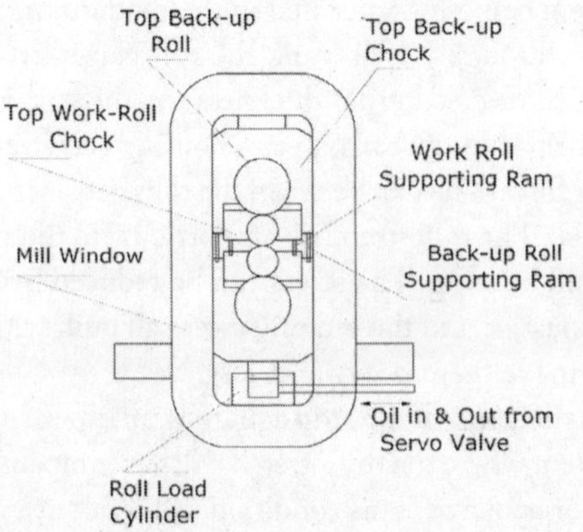

Fig. 4.102 4-Hi stand with roll load cylinder and hydraulic roll balancing arrangement

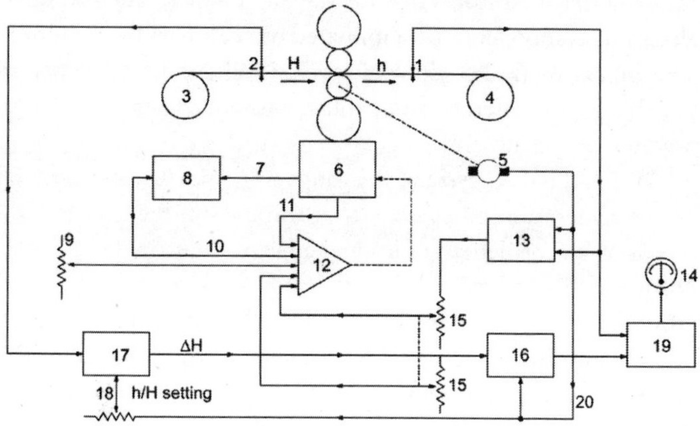

Fig. 4.103 HAG control scheme

(ii) The mill stretch characteristics, i.e., mill stretch versus roll separating force is pre-determined experimentally and the data is fed into the computer controlling the system.

(iii) The exact roll gap is created by the roll load cylinders and read by the load cylinder positioning sensors, after roll gap zeroing is done by extending the cylinders and touching the two work rolls under a small pressure.

(iv) As soon as rolling starts the oil pressure in cylinder increases to balance roll separating force to keep the roll gap constant after compensating for mill stretch. This constant roll *gap control* virtually makes the housing infinitely stiff. The response time is about 300 milliseconds to achieve the correction for mill stretch. After mill stretch compensation, any variation from set gauge is sensed by the exit side thickness gauge which sends signal for correcting the roll gap further so that output gauge remains at set gauge. This scheme is same as operator's manual control of gauge, but very fast due to high response of servo valve and hydraulic cylinders.

(v) Any variation in input gauge is sensed by the entry side thickness gauge and the information is fed forward to the controller, which calculates anticipated variation in mill stretch and takes corrective action when the variation comes to the mill bite. This is called *'gauge Control'* through feed forward.

(vi) The exit side thickness gauge checks the error of the gauge and may take corrective measure to make the error zero. Gauge error feedback ensures the absolute accuracy of the hydraulic cylinder control. Changes in the roll gap zero datum and any initial gauge error due to inaccurate setting of the anticipated load value for the gap control, and also drift of the electronic components are eliminated by such feed back control.

(vii) The *tilt control* facility allows a tapered roll gap to be created through which tapered material can be rolled without causing side buckle (side waviness in rolled strip).

With a HAGC system, the gauge tolerance as close as ± .003 mm can be achieved in thin gauge strips throughout the length of the strip, excepting a short length of leading edge of the strip.

(viii) During rolling of strips below 0.1 mm, dark mode rolling condition prevails where gap control is no more effective when 'speed / tension AGC' control is made active. This mode permits control of thickness gauge by controlling tension, rolling pressure and speed in the following two selectable modes:

(a) tension control spilling over to pressure
(b) tension control spilling over to rolling speed control spilling over to pressure control.

In the first mode, the strip back tension is varied up to about ± 20% of set tension to control strip thickness (thickness reduces when tension is increased and vice-versa). When tension control approach saturation (i.e. ±20% of set value), the control switches over to rolling load by increasing or reducing roll load cylinder oil pressure till desired strip thickness is achieved.

Similarly in the second mode, the control starts with varying back tension which then switches over to speed (higher rolling speed reduces strip thickness and vice versa) which in turn spills over to pressure control.

In both mode of controls, the exit thickness gauge compares actual gauge with set gauge and gives feed back for activating necessary corrective controls.

B. Shape control system

Flatness of a cold rolled strip is very important for its subsequent processing. A strip is considered to be flat when a cut piece of the strip lies flat on a flat surface. A strip with a good thickness profile, rolled under advanced thickness control techniques like HAGC control, may not contain a good shape. This is particularly true for wide strips.

The reasons for bad shape generated during rolling are as follows:

(i) Sheet camber: Variation of sheet / strip thickness along its width is called camber. There may be three types of sheet camber viz. convex, concave and triangular or tapered. These cambers are generally originated during hot strip rolling.

When such a cambered strip is cold rolled through a rectangular roll gap, the portions of strips whose thickness is more, will undergo more reduction and therefore its rolled length will be more than the portion which has less thickness. The additional length is accommodated in form of waves, and the shape of the rolled product deteriorates. A convex strip will result in waves at the middle of the strip, known as full center, a concave strip will result in side buckled strip while a tapered shape will produce waves pronounced at one side gradually decreasing towards the other.

(ii) Roll camber: Even if the input strip is of good rectangular shape, if the roll gap between the working rolls is not perfectly rectangular, i.e., the rolls do not remain perfectly straight cylinders during rolling, then also shape problem in rolled strip will result. This variation in roll profile, called roll camber, may be caused due to (a) bending of rolls under rolling force, (b) thermal cambering caused by heat generated during rolling.

The resultant effect of roll bending and thermal camber is tried to be negated by intentionally giving a permanent camber to the rolls during grinding. However, with variation in input strip width and camber, it is impossible to remove waviness in rolled strips by giving a fixed roll camber. During wide strip rolling the thermal camber in rolls do not also follow a steady pattern.

The modern system of shape control in a 4 Hi cold mill with hydraulic roll load and roll balance system works as follows:

Under tension, thin gauge strip invariably appears to be perfectly flat, though this may not be true which becomes apparent when tension is removed. A perfectly flat rolled strip being coiled under tension shall have equal stress all along its width. Such a strip will remain flat even when the strip tension is removed.

In the shape control system, the tension along the strip width is measured continuously by application of special sensors. One of the most popular sensor is a sensor roller unit which comprises a series of rotor rings, each separately supported on an air bearing around a static arbor. The sensor roller is placed on the path of the rolled strip with a very small wrap angle. Each ring gets displaced according to the tension in the section of the strip touching the rings. Higher the tension higher will be the displacement. These displacements are indicators of tension profile in the rolled strip, which can be measured by suitable transducers and can be viewed in a video monitor.

The shape control is done in the following steps:

(i) Necessary roll bending is caused by operation of hydraulic rams housed in work roll chocks and back-up chocks during rolling to take care of general shape of input strip and effect of roll bending and thermal camber in roll. This is done automatically by a computer by checking the tension profile. This part of the control is called *roll bending* control.

(ii) The tapered strips are rolled through a tapered roll gap by differential gap control from the HAGC. This is known as *tilt control.*

(iii) The third step of control is by selective control of roll coolant spray. If certain sections of the roll are heated more, the roll gaps will reduce in these sections and the tension in the section of strip being rolled under these section will become less. These sections will be sensed by the sensor roller, and the coolant sprays of these zones will spray more coolant in these sections and the localized thermal cambers will get controlled and the shape of the strip will remain under control.

Figure 4.104 shows constructional details of a particular design of a tension measuring sensor roller[4].

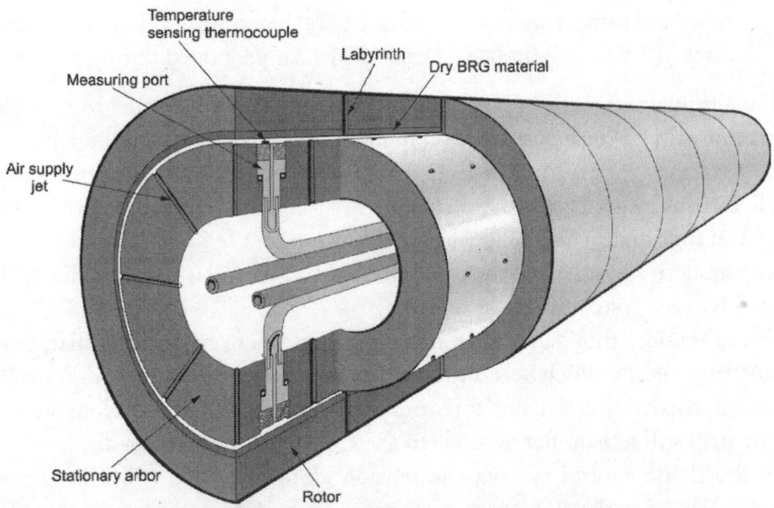

Fig. 4.104 Tension measuring sensor roll

All modern thin gauge strip mills and foil mills are fitted with a shape control system.

C. ***Edge Drop Control (EDC)***

Strip cold rolled under HAGC gives excellent gauge tolerance throughout the length and across the width excepting at the two edges where the thickness drops. The reason for this drop in edge thickness is due to the change in roll flattening profile from the loaded area to the unloaded area at the strip edge. This phenomenon is known as *edge drop.* For stringent quality requirement, the two edges are trimmed which reduces the plant yield.

The modern technology to control edge drop is to reduce the work roll barrel at the strip edges by a tapered ground portion and shifting the taper by axial adjustment of the two work rolls till the edge drop is minimized to a set value. In a tandem cold strip mill, the edge drop control is progressively done in all the stands.

4.3.7 Roll lubrication and cooling system

A. Purpose of Roll Lubrication and Cooling System:

Lubrication between the roll and stock surface reduces the friction between them and as a consequence it:

- Reduces the roll separating force and rolling power.
- Reduces the roll wear.
- Improves the surface of the rolled material.

During rolling of metals, the rolls get heated up. In case of hot rolling the heat from the preheated stock flows to the rolls and also form the heat generated by the deformation of material during reduction.

During cold rolling, the two sources of heat which heats up the rolls are:

- Frictional heat generated at strip and work roll contact and also between work roll to backup roll contact. For a back-up roll driven mill, the later source of heat generation is not negligible.
- Heat of metal deformation.

In both hot and cold rolling, cooling of rolls are necessary for the following reasons:

Hot rolling mil rolls need cooling primarily to maintain its hardness and thus longevity and secondly to maintain the correct shape of roll which is important for hot strip rolling. In cold rolling the primary purpose of roll cooling is to maintain correct shape of the roll and thus ensuring flat thin gauge cold rolled strip and foil.

Water with high specific heat is the ideal cooling medium to take the heat away from rolls and keep its temperature within limits. However, water is an inadequate lubricant for today's high speed mills and different metals. Alternatively, lubricants consisting of oil and other additives are very good for reduction of friction in rolling but are poor in transferring heat.

A balance between the two functions of lubrication and cooling is achieved by mixing suitable lubricants in water. Such a medium is called a ***roll coolant***. The oil in water type roll coolant has additional benefit of inhibiting rusting of ferrous mill parts.

Besides cooling of rolls, control of thermal profile and providing lubrication, coolants may be applied for achieving the following purposes also:

(i) to wash away small particles of metal and scale,
(ii) to control temperature of the product between stands of a continuous hot mill.
(iii) to achieve desired metallurgical properties by product surface cooling (for example in tandem hot strip mill).

B. ***Different roll coolants***

The roll coolants and their methods of application vary depending on whether it is hot or cold rolling and also depending on the product being rolled.

(a) Cooling and lubrication in hot rolling:

Air cooling (i.e., no liquid coolant) is practiced in many cases despite their several limitations like changing temperature of rolls causing problem of thermal crown, and risk of roll damage due to thermal stress. However, it is still used where temperature loss of the product needs to be avoided, as in the case of stainless steel rolling.

Plain water is generally used as coolant in hot steel strip rolling. Water as a cooling medium is effective when the roll temperature is below 100°c. If the roll temperature is higher than 100°c, water coming into contact with the roll immediately forms a layer of steam and prevents effective heat transfer (Leidenfrost effect). However, higher the roll temperature below 100°c, less amount of water needs to be used.

Trials have been made with water coolant mixed with different lubricants in hot steel strip mills, but for reasons like environmental, economical and lack of real product improvement, lubrication has not become a standard practice.

In high speed section mills, particularly when rolling non-symmetrical sections, application of lubricants sprayed directly into the roll gaps with nozzles are practiced. This improves the roll groove surface, and reduces excessive wear in roll collars.

Secondary use of water spray for breaking surface scale, washing out metal particles / scales and for metallurgical treatment of the strip surface in a continuous hot strip mill has already been mentioned.

During hot rolling of aluminum, its alloys and other non-ferrous metals, emulsion based roll coolants are used. The main function of rolling oil is to reduce the sticking of metal to the roll surface and to improve quality of finished product.

(b) Cooling and lubrication in cold rolling:

Modern high speed mills use roll coolants added with different lubricants to suit the varying metals being rolled, different rolling mills (single stand reversing to multi-stand tandem strip mill, foil mill, etc.) and rolling conditions (speed, width and gauges).

There are three basic types of water based roll coolants which are used: emulsions, dispersions and stabilized dispersion.

Emulsions are normally produced by combining a surfactant into the base lubricant which is then mixed with water resulting in a white colored solution with little or no free oil. This is by far the most common type of roll coolant used in cold rolling.

In *dispersions*, the dispersant acts as a protective layer on the surface of oil particles which eliminate coalescence of oil. For the dispersion to stay steady, agitation of the coolant is necessary. With uniformity of particle size, the oil content in the coolant can be reduced and thereby cooling effect is also enhanced as more water goes to the mill. A dispersion rolling solution also acts as a detergent and keeps the roll stand parts in contact with solution clean.

Stabilized dispersion is a hybrid of emulsion and dispersion.

A very important aspect of using lubricants in cold rolling is their effect on surface cleanliness. The major problem arises from adherence of the lubricant on the strip surface and their effect after annealing of the material.

Four categories of rolling lubricants in use are:

(i) Conventional technology where tallow or white grease, diluted in mineral oil is used as lubricant. The nature of animal fat used dictates the strip cleanliness performance.

(ii) Semi-synthetic technology where a part of animal fat is replaced by synthetic esters and natural vegetable oil (Palm oil). The strip cleanliness is slightly improved with a small increase in product cost.

(iii) Highly synthetic technology where animal fat is fully replaced by synthetic esters and oxidation inhibitors are added.

(iv) Advanced synthetic technology is achieved by using all synthetic formulations—synthetic esters and additives, and no natural hydrocarbons are used.

A mixture of mineral oil and kerosene with additives but without water, is used as a roll coolant in cold strip and foil rolling of aluminum alloys. This rolling oil reduces rolling friction and also washes away aluminum dust produced during aluminum rolling. Mineral oil and its fume being highly combustible, special precautions are to be taken for containing the fumes through *fume extraction arrangement* and to avoid any source of heat and fire near the rolling stand.

C. *Roll coolant system*

Roll coolant is supplied to the mill stands through a circulating system. The coolant is pumped by low-pressure high volume pumps to the mill stand though a single, large diameter supply pipeline. After the coolant is sprayed through series of nozzles connected to distribution headers, it is collected in the catchment tray placed below the mill stand and returned (generally by gravity) to the treatment area through a large diameter return line. The returned dirty coolant is filtered, cooled, re-constituted by addition of requisite make-up water, lubricants and other additives and returned to the clean reservoir from which it is pumped again to the mill.

(a) Coolant distribution in the mill stand:

It is at the work roll bite that most heat is generated, so it is necessary to apply large amount of roll coolant along the line of contact between strip and work rolls. This is done by spraying from above and below the strip.

In case of 4-Hi mill, generally the top support roll is provided with coolant spray. The bottom support roll is drenched with coolant falling from the bottom work roll and thus generally not provided with independent spray.

If the mill is reversing type capable of rolling in both directions, the coolant spray bars (distribution headers) are duplicated on both sides of the mill. The coolant is applied only on the ingoing side of the mill depending on the direction of rolling, while the coolant flow in other side is kept closed.

In a tandem cold mill, roll coolant is applied only on the entry side to avoid coolant falling on the outgoing strip before coiling and thereby creating condition of staining during annealing.

In case of hot mill, cooling of rolls is generally done by spraying coolant (usually plain water) from both sides of the mill for following reasons:

(i) Higher temperature involved need more cooling.
(ii) The inhibition to heat transfer due to formation of steam can be largely counterbalanced by spraying coolant form the exit side.
(iii) Unlike cold mills, there is generally no interference by instrumentation at the exit side.

The purpose of spray system is to remove heat from roll surface before heat can flow to the core and heat built up occurs. It should maintain a heat balance across the full width of the roll as well as between top and bottom rolls, to prevent uneven thermal camber in the rolls.

The amount of heat removed from the roll surface is a function of the spray nozzle distance from the roll. Smaller is the distance better is the heat transfer. However, for practical reasons like threading the strip around roll, for space required during roll changing and lastly to allow space for coolant exit path, typically the gap between the nozzle and maximum diameter of working roll is kept in the range of 75 mm.

Two types of nozzles are generally used—flat spray and full cone. For equivalent pressures and flow rates, flat fan type nozzles have shown better cooling efficiency.

Spray applied at close to 90° to the roll surface yield most effective results. Heat transfer also depends on the coolant pressure. However, there is practically no gain above 10 bar (140 psi) pressure. Therefore, coolant pumps with a pressure between 7–10 bar at full flow is chosen.

Spacing of nozzles in the coolant header is crucial to ensure desired distribution of coolant over the entire barrel length of the roll. Nozzles can be placed to have either an even distribution over entire length or a heavier distribution at the middle. If space is available two rows of nozzles are used per roll so that the chance of running a dry spot on the roll, due to clogging of a nozzle is minimized.

Adjustment of thermal crown (increase in roll diameter due to thermal expansion) can be made by manipulation of the coolant flow patterns. In many rool coolant installations, manual flow control valves are provided to control flow in different segments of the spray headers. Reducing flow to the central sprays would cause a heat build-up in the central area of the rolls and consequently a positive crown will be set up. This method of crown control is slow but effective.

When the spray system is integrated with the shape control system, as discussed in section 4.3.6, a selective electro-pneumatic control of spray nozzles can be effected to control localized thermal camber of the work rolls.

Roll wipers are used against the two support rolls of a 4-High mill to prevent carry over of roll coolants on the exit side of the strip to minimize possibility of strip staining in subsequent annealing of coiled strip.

(b) Recirculation of roll coolant:

Mainly *re-circulating type roll coolant system* is used for cold rolling mills. Direct application system with separate lines for coolant and lubricants are also used but rarely. Emulsion roll coolant is used for cold rolling of ferrous alloys while mineral oil (kerosene) based roll coolants are used for cold rolling of aluminum and its alloys.

The general scheme of an emulsion roll coolant system is shown in figure 4.105. The system consists of a rectangular steel tank partitioned inside to form a clean coolant tank-1, and a dirty coolant tank-2. The volume capacity of the total tank is about 15 to 20 minutes of full pumping capacity of the coolant feed pump-3, to the mill stands. The feed pump is centrifugal type with delivery pressure of about 10 bar. Number of pumps depends on the number of mill stands. Standby pumps are always kept in the system so that rolling of coolant flow may continue even when operating pumps to be stopped for maintenance. At the output of each feed pump a plate type heat exchanger -4, is kept to cool the roll coolant to 50–55°c before entering mill. Heat exchange is done between hot coolant and cooling water in the heat exchanger. The cooling water flow rate is controlled automatically to maintain the desired coolant temperature.

The heated water at the exit of the heat exchanger is pumped back to cooling tower to cool before reuse.

Depending on the demand, an auto backwash filter -5, may be used in each feed pump delivery line. The filtration level for a strip mill is kept in the range of 70 to 80 micron level. The pressure and thus the flow rate of the emulsion flow is controlled through pressure control valve. The feed line is provided with a flow meter.

The sprayed coolant is collected in the catchment tray -8, positioned below the mill stand -7, and flows by gravity into the dirty tank. In this dirty tank is mounted a magnetic separator-10, which separates the fine ferrous particles generated in the process of rolling and are carried back with the sprayed coolant. The ferrous debris are collected by the series of magnetic rods and are scraped and collected in sludge bin to be disposed off. The dirty chamber is equipped with a motorized agitator -9, to keep the emulsion in constant motion to avoid separation of oil form water. If agitation is stopped for a considerable period of time, say 15 minutes, the oil starts separating out. It is important to run the agitator continuously even when the mill is not rolling.

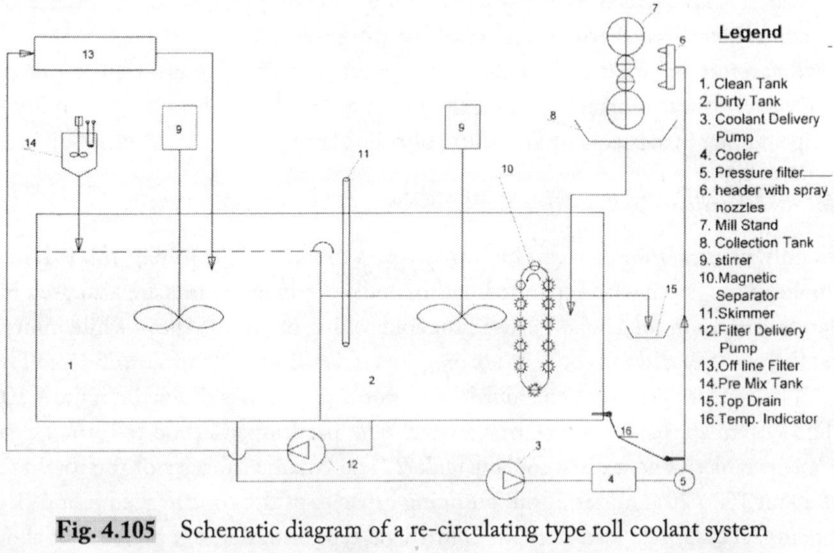

Fig. 4.105 Schematic diagram of a re-circulating type roll coolant system

This tank is also provided with one or more number of belt type oil skimmer -11, The function of the belt skimmer is to separate any tramp oil (hydraulic or lubrication oil) that is leaked from mill stand and carried by the roll coolant. During operation of the belt skimmer, the agitator is stopped to allow the tramp oil to separate and float up to the coolant surface. Then the skimmer is started to remove the floating tramp oil. The tramp removal is done in steps so that the agitator need not be stopped for more than 12–15 minutes.

The dirty oil containing fine metallic particle and dirt are filtered in an offline paper filter -13. These paper filters may be pressure type or vacuum type. The filtered coolant is then pumped to the clean tank.

The clean tank is also equipped with a stirrer for purpose stated earlier. Emulsion heating arrangement is provided in this tank. The heating may be done by electrical heater or steam heater either inside the tank or in a bypass heater. This heater is used for heating the initial quantity of emulsion prepared and brought to this tank. The levels in the two sides of the tank is so maintained that clean coolant form clean tank overflows over the partition into the dirty tank. The tank is provided with connections for level switches, temperature indicators and switches, demineralized (DM) water filling, by pass heating line and draining.

An auxiliary pre-mix tank -14, with stirrer and heater with a dosing pump is kept near the main tank for occasional feeding of emulsified oil and other additives in the tank.

Procedure for starting the system

The tanks are to be cleaned by cloth and then both tanks and connected pipelines are cleaned with water, which is to be drained out. Then the emulsion tank is filled with DM water through the DM water filling line. This line leads to the clean side of the tank. During first filling, both sides of the tank are connected by opening the valve in the pipe connection in the two sides of the tank. Upon completion of DM water filling this valve is closed.

The cold DM water is then heated to the operating temperature by in-tank or by pass heater. The agitator is simultaneously started to ensure a uniform temperature, the desired temperature is 50–55°c.

Base oil and additives are dosed into the system by means of dosing pump. The dosing pump is stopped once pre-set concentration of oil is dosed. Alternatively, the emulsion may be prepared in the pre-mix tank and then emptied into main tank.

In process coolant testing

The emulsion is to be assessed for its efficiency during rolling operation at regular intervals. Coolant samples are taken either from the spray or from feed pipe line. The samples are tested for the following characteristics of the coolant:

(i) total oil content, (ii) pH-valve, (iii) ash content, (iv) total iron content, (v) free and solute iron content, (vi) saponification number, (vii) bacteria, yeasts, and mould fungi presence.

Level of filtration has to be lower as metals are rolled to thinner gauges. Thus, selection of right kind of filtration plant becomes more and more critical. For aluminum foil rolling with mineral oil based roll coolant, specially designed off line filters with coated paper filter or polymeric filter is used to obtain filtration of 1–2 micron level.

Suitable mill enclosures and fume hoods at the mill stand fitted with powerful fume exhaust arrangement is provided to reduce coolant / oil fumes in the vicinity of mill area. Proper fume exhaust system is important to contain fire hazards when mineral oil (kerosene) based roll coolants are used.

Estimation of roll coolant requirements

Quantity of coolant flow into the mill stand depends on the rate at which heat has to be removed to maintain steady temp profile of the rolls. The total heat generated at the mill stand due to metal deformation and roll friction is dissipated in different ways, e.g., in form of strain energy in the rolled product, heat conducted to structural components, heat convected and radiated to the surrounding and balance heat carried away by the roll coolant. The coolant flow rate is therefore proportional to the rolling power.

In a modern mill, the rise in temperature of the roll coolant is kept within 10°c and based on this, the maximum flow rate of coolant in lpm may be roughly estimated to be 1.2–1.4 times the mill motor power in kw.

References

1. "Bearings in Rolling Mills", Publication No. 2768E, SKF, 1971.
2. "Timken Tapered Roller Bearings: Rolling Mill Equipment Supplement", published by the Timken Company, USA, 1972.
3. "Sendzimir Cold Rolling Mills", Catalog of T.Sendzimir Inc., Waterbury, USA.
4. Catalogue on "Vidimon Shapemeter", published by Lowey Roberson Division of Davy-Lowey Ltd., UK.

Suggested Readings

Beynon, Ross E. 1956. *Roll Design and Mill Layout*. Pittsburgh: A. I. S. I.

Ghosh, A., and A. K. Mallik. 2004. *Manufacturing Science*. New Delhi: East-West Press Pvt. Ltd.

Ginzburg, V. B. 1989. *Steel Rolling Technology: Theory and Practice*. New York : Marcel Dekker Inc.

Ginzburg, V. B. 1993. *High Quality Steel Rolling: Theory and Practice*. New York: Marcel Dekker Inc.

Ginzburg, V. B., and R. Ballas. 2000. *Flat Rolling Fundamentals*. New York: Marcel Dekker Inc.

Juneja, B. L. 2007. *Fundamentals of Metal Forming Processes*. New Delhi: New Age International Publishers.

Larke, E. C., and Cook, M. 1963. *The Rolling of Strip, Sheet & Plate*. Second edition. London: Chapman and Hall Ltd.

Lee, Youngseog. 2004. *Rod and Bar Rolling: Theory and Applications*. New York : Marcel Dekker Inc.

Pittner, John, and Simaan, M. A. 2011. *Tandem Cold Metal Rolling Mill Control*. London: Springer.

Roberts, William L. 1978. *Cold Rolling of Steel*. New York: Marcel Dekker Inc.

Roberts, William L. 1983. *Hot Rolling of Steel*. New York: Marcel Dekker Inc.

The United Steel Companies Limited. 1960. *Roll Pass Design*. Sheffield, England

Tselikov, A. 1967. *Stress and Strain in Metal Rolling*. Moscow: MIR Publishers.

Tselikov, A. I., and Smirnov, V. V. 1965. *Rolling Mills*. Trans. M. H. T. Alford. London: Pergamon Press.

Tselikov, Nikitin, and Rokotyan. 1981. *The Theory of Lengthwise Rolling*. Trans. V. Afanasyev. Moscow: MIR Publishers.

Wusatowski, Z. 1969. *Fundamentals of Rolling*. London: Pergamon Press.

Index